Springer Series in Electrophysics
Volume 14

Springer Series in Electrophysics

Volume 1 **Structural Pattern Recognition** By T. Pavlidis

Volume 2 **Noise in Physical Systems** Editor: D. Wolf

Volume 3 **The Boundary-Layer Method in Diffraction Problems**
By V. M. Babič, N. Y. Kirpičnikova

Volume 4 **Cavitation and Inhomogeneities** in Underwater Acoustics
Editor: W. Lauterborn

Volume 5 **Very Large Scale Integration (VLSI)**
Fundamentals and Applications Editor: D. F. Barbe 2nd Edition

Volume 6 **Parametric Electronics** An Introduction
By K.-H. Löcherer, C. D. Brandt

Volume 7 **Insulating Films on Semiconductors** Editors: M. Schulz, G. Pensl

Volume 8 **Fundamentals of Ocean Acoustics**
By L. Brekhovskikh, Yu. Lysanov

Volume 9 **Principles of Plasma Electrodynamics**
By A. F. Alexandrov, L. S. Bogdankevich, A. A. Rukhadze

Volume 10 **Ion Implantation Techniques** Editors: H. Ryssel, H. Glawischnig

Volume 11 **Ion Implantation: Equipment and Techniques**
Editors: H. Ryssel, H. Glawischnig

Volume 12 **VLSI Technology**
Fundamentals and Applications Editor: Y. Tarui

Volume 13 **Physics of Highly Charged Ions**
By R. K. Janev, L. P. Presnyakov, V. P. Shevelko

Volume 14 **Electrophotography and Development Physics**
By L. B. Schein 2nd Edition

Volume 15 **Relativity and Engineering** By J. Van Bladel

Volume 16 **Electromagnetic Induction Phenomena** By D. Schieber

Volume 17 **Concert Hall Acoustics** By Y. Ando

Volume 18 **Planar Circuits** for Microwaves and Lightwaves By T. Okoshi

Volume 19 **Physics of Shock Waves in Gases and Plasmas**
By M. A. Liberman, A. L. Velikovich

Volume 20 **Kinetic Theory of Particles and Photons** By J. Oxenius

Volume 21 **Picosecond Electronics and Optoelectronics**
Editors: G. A. Mourou, D. M. Bloom, C.-H. Lee

This series has been renamed
Springer Series in Electronics and Photonics starting with Volume 22.

Volumes 22–31 are listed at the end of the book.

L. B. Schein

Electrophotography and Development Physics

Second Edition
With 208 Figures

Springer-Verlag Berlin Heidelberg GmbH

Dr. Lawrence B. Schein

IBM Research Division, Almaden Research Center, 650 Harry Road,
San Jose, CA 95120, USA

ISBN 978-3-540-55858-3 **ISBN 978-3-642-77744-8 (eBook)**
DOI 10.1007/978-3-642-77744-8

Library of Congress Cataloging-in-Publication Data. Schein, L. B. (Lawrence B.), 1944– Electrophotography and development physics / L. B. Schein. – 2nd ed. p. cm. – (Springer series in electrophysics ; v. 14) Includes bibliographical references and index. ISBN 3-540-55858-6 (Berlin : alk. paper). – ISBN 0-387-55858-6 (New York : alk. paper) 1. Xerography–Developing and developers. I. Title. II. Series. TR1045.S34 1992 686.4'4–dc20 92-29462

Originally published by Springer-Verlag Berlin Heidelberg New York in 1992

54/3140-5 4 3 2 1 0 – Printed on acid-free paper

Dedicated to my parents

BERNARD AND SYLVIA SCHEIN

Preface to the Second Edition

During the last four years, since this book was first published, the field of electrophotography has experienced some astonishing changes. Most obvious is the emergence of high quality color copying and printing. Successful implementation of color electrophotography required the solution of many technical problems, some of which necessitated the invention of new development systems. In addition, major advances in our understanding of the technology have occurred. Background development was identified in the first edition as one of the major unsolved problems in electrophotography; significant progress has been made (Sect. 12-6.6). The effective dielectric constant problem has been solved, with experiments and theory in agreement (Sect. 12-6.2.2), and a major advance in our understanding of insulator, i.e. toner, charging has occurred with the identification of an experiment that can distinguish between the low and high density limits of the surface state theory (Sect. 12-4.3).

In order to bring these new results to the attention of the electrophotographic community, it was decided to update the 1988 version of this book. To make it as easy as possible for the reader to identify and learn the new results, this second edition is organized as follows. Chapter 11 is devoted to color electrophotography. Chapter 12 updates Chaps. 1-10. Each section in Chap. 12 is identical in subject to the earlier section indicated by the number following 12- and assumes knowledge of the earlier section.

The author would like to acknowledge the IBM Corporation for encouraging and supporting the writing of this second edition and the many people who have assisted in bringing out this book, including Sheila Hill and Pam Hale of the Publication Department at the IBM Almaden Research Center, Deborah Hollis of Springer-Verlag, who has patiently edited all twelve chapters, Lynn Ritter of Dataquest, who kindly supplied the market information, and Robert Durbeck, Peter Castle, Barry Schechtman, and Connie Schein, who critically reviewed Chaps. 11 and 12 before publication.

San Jose, California
March, 1992

L. B. Schein

Preface to the First Edition

Electrophotography (also called xerography), the technology inside the familiar copier, has become increasingly important to modern society. Since the first automatic electrophotographic copiers were introduced in 1959, they have become indispensable to the modern office and now constitute a multibillion dollar industry involving many of the world's largest corporations. By the 1990s, it is expected that electrophotography will be one of the most prevalent printer technologies. This will occur because of the growing need for printers that are quiet, that can produce multiple fonts, and that can print graphics and images. Electrophotographic printers satisfy these requirements and have demonstrated economic and technical viability over an enormous speed range, from 6 to 220 pages per minute, with output quality that approaches offset printing.

Organizations contemplating designing a new electrophotographic copier or printer need to deal with two sets of issues. First, for each of the six process steps in electrophotography there are several different technologies that must be evaluated and chosen. For example, there are three development technologies (two component, monocomponent and liquid); cleaning can be done with a blade or brush; and the photoconductor can be inorganic or organic, either of which can be configured in the form of a belt or a drum. Second, once a technology for each step is chosen, it must be optimized and integrated with the other process steps. This optimization and integration is facilitated by a firm scientific understanding of the technologies being considered. Unfortunately, certain key technologies in electrophotography are not well understood, even after years of industrial practice.

Perhaps the most crucial technologies which are not well understood are those used in the development step, because this step most directly determines the quality of the images. It is in this step that the "blackness" of the lines and solid areas, the cleanliness of the nonimaged areas, the uniformity of solid areas, and the ratio of the "blackness" of lines to solid areas are determined. Those who used Xerox copiers during the 1960s will remember that they would only reproduce the edges of solid areas (Fig. 3.1), a copy quality defect attributable to characteristics of the open cascade development system (Chap. 5). The generally perceived high copy quality of the Eastman Kodak line of copiers introduced in 1976 resulted directly from the introduction of a new development system, conductive magnetic brush development (Chap. 7).

There are several reasons why aspects of the development system are not well understood. First are the scientific reasons. It is known that the proper toner charge is important to good development (toner is the black plastic powder that ends up on the paper); yet our understanding of toner charging, and more generally insulator charging, can be characterized as pre-scientific, with most knowledge being empirical. While our knowledge of the physics of solid area and character development is becoming relatively firm, our understanding of the causes of background development (the black spots on copies due to toner developing onto the white or nonimaged areas) is lacking because not enough attention has been given to this problem.

Second, there have been several recent inventions that are opening up the possibility of further improvements in the development step and which challenge electrophotographic scientists to understand and improve. In 1980, Canon introduced the first monocomponent development system based on magnetic, insulating toner (Sect. 9.5) which was an important factor making it possible for Canon to manufacture and sell the first electrophotographic copier for under $1000. Only a few years ago, in 1985, Ricoh and Toshiba both announced new monocomponent development systems based on nonmagnetic, insulating toner (Sect. 9.6). Also, new tools for measuring toner charge distributions are becoming available which will help characterize and design new toner systems (Sect. 4.4.4). And there has been a virtual explosion in the patent literature on charge control agents, which are toner additives that assist in controlling the toner charge (Sect. 4.4.3).

The primary purpose of this book is to discuss critically the physics of all known electrophotographic development technologies and their associated toner charging mechanisms. To assist the reader who may be new to electrophotography, a tutorial is presented in which the technical history and market of electrophotography are examined (Chap. 1), followed by a discussion of the physics within each of the six electrophotographic process steps (Chap. 2). In selecting the literature to review for this book the following choices were made. I have attempted to reference completely the scientific literature on development physics up to October 1987. In the two chapters on toner charging (Chaps. 4 and 8) the physics of static electricity is thoroughly discussed and related to our understanding of toner charging. Patents are only referenced if they contain important physics unavailable elsewhere. Materials and engineering problems are not discussed because they are beyond the scope of a single book. In the tutorial on electrophotography (Chap. 2), papers and review articles have been selected that will allow the interested reader to find the appropriate literature.

The author would like to acknowledge his management, Don Burland and Bob Durbeck, and the IBM Corporation for allowing and encouraging the writing of this book. The stimulation of my many colleagues at the IBM Almaden Research Center and in Joe Woods's Technology Laboratory at the IBM Boulder facility are also appreciated. Others at IBM whom I would like to acknowledge include Hans Coufal, who initially suggested that a new book on electrophotography would be useful, Lorraine Rodriquez's Publications Department staff, especially Linda Perez, and Vilia Ma's library staff, especially Beverly Clarke.

I have been working on many aspects of electrophotography since September 1983 at IBM and 1970–1975 at the Xerox Corporation. I would like also to acknowledge Mike Shahin at Xerox who suggested I look into the "new" magnetic brush development system, and Mike's and Mark Tabak's continual support during the years it took to sort out the physics of the insulative magnetic brush development system.

Writing a book is a strain on any family. I appreciate the support and understanding of my wife Connie and children, Daniel and Benjamin, during those many weekends and evenings when I was unavailable to them.

The author would also like to acknowledge the people who have critically reviewed chapters in this book prior to publication, including Bob Durbeck, Don Burland, Hans Coufal, Bruce Terris, Peter Castle, Joe Abbott, K Jenkins, Gene Bishop, Campbell Scott, Vlad Novotny, Connie Schein, Lynn Ritter, John Bickmore and Bill Greason.

Finally, I would like to acknowledge my many colleagues over the years, who are too numerous to mention; their comments and criticisms have helped shape my thinking and the field. This book is in fact the sum total of the thinking of all of the people who have shared with me the fun and excitement of working on electrophotographic development physics.

San Jose, California
October, 1987

L. B. Schein

Contents

List of Symbols
(used more than once)

Symbol	Unit	Description
A_c	cm^2	surface area of carrier
A_t	cm^2	surface area of toner
C_t	%	toner concentration: ratio of the mass of all toner particles on carrier to carrier mass
C_{AB}	C/V	capacitance between bodies A and B
C_0	C/V	capacitance between two bodies at 10Å separation
d_s	cm	photoreceptor thickness
d_t	cm	thickness of deposited toner layer
D		optical reflection density
e	1.6×10^{-19} C	charge on electron
E	eV	energy of a trap below conduction or above valence band
E_{air}	V/cm	electric field in air gap
E_{av}	V/cm	average electric field in air gap
E_D	V/cm	average electric field in developer
E_F	eV	Fermi level in metal
E_ℓ	V/cm	electric field in liquid (Chap. 10)
E_n	eV	neutral level of insulator
E_p	V/cm	electric field in photoreceptor
E_{th}	V/cm	threshold electric field
f		distribution function of either Q/r_2 or r
$\widetilde{F}$		fraction of toner removed from carrier bead
F	g/cm s	developer flow rate
F_{es}	dyne	electrostatic adhesion force
F_M	dyne	magnetic adhesion force
F_p	dyne	adhesion force of toner to photoreceptor

F_t	dyne	adhesion force of toner to carrier
h	cm	height above roller at which toner is field stripped (Chap. 6)
H	G	magnetic field
H_t	cm	bead drop height (Chap. 5)
k	0.86×10^{-4} eV/K	Boltzmann's constant
K		developer dielectric constant
K_c		carrier coating dielectric constant
K_E		effective dielectric constant
K_i		dielectric constant of material i
K_{max}		maximum enhancement of the electric field over the air gap value
K_p		polymer dielectric constant
K_s		photoreceptor dielectric constant
K_t		toner dielectric constant
ℓ	cm	carrier coating thickness
L	cm	gap between photoreceptor and electrode
M_c	g	carrier mass
M_t	g	toner mass
M/A	g/cm^2	developed toner mass per unit area
n		number of toner particles developed from a carrier
$n(E)$	eV^{-1} cm^{-2}	number of insulator states per unit energy per unit area
n_i	cm^{-2}	number of ions per cm^2 on insulator surface
n_0		total number of toner particles on a carrier
N		number of rollers
N_c	eV^{-1} cm^{-2}	surface states per unit energy per unit area on carrier
N_ℓ	cm^{-3}	volume density of toner in liquid developer (Chap. 10)
N_t	eV^{-1} cm^{-2}	surface states per unit energy per unit area on toner
p		carrier surface packing
p_t		toner volume packing density
p_v		carrier volume packing
Q	C	charge exchanged during contact electrification
Q_c	C	carrier charge

Q_t	C	toner charge
Q/M	C/g	toner charge-to-mass ratio
r	cm	toner radius
R	cm	carrier radius
R_p		reflectance of paper; D=$-\log_{10}$ R_p
R_T		reflectance of toner
t	s	development time
T	K	absolute temperature
v_t	cm/s	velocity of toner in air stream
v_p	cm/s	photoreceptor velocity
v_r	cm/s	roller velocity
V	V	applied voltage
V_{bias}	V	bias potential on roller
V_c	V	contact potential difference
V_m	V	Mylar voltage
V_r	V	photoreceptor residual potential
V_t	V	toner voltage
V_{th}	V	voltage threshold
V_w	V	width of development curve
U_i	eV	energy of ion on a surface
W	cm	nip width
z	cm	distance between two solids
δ	cm	distance from carrier surface to photoreceptor surface (Chap. 7)
δ_{air}	cm	air gap in which toner develops
δ_r	cm	distance from toner to roller surface (Chap. 9)
ε_0	8.85×10^{-14} F/cm	permittivity of free space
ζ	V	zeta potential
η	poise	viscosity of air
θ		angle used in conductive magnetic brush development theory
Λ	cm	dielectric distance from electrode to carrier charge
μ		permeability of toner
μ_t	cm^2/V s	mobility of toner (in liquid development)
ν		v_r/v_p
π	3.14	pi
ρ	Ωcm	resistivity of developer
ρ_c	g/cm^3	density of carrier
ρ_p	C/cm^3	volume charge in photoreceptor
ρ_t	g/cm^3	density of toner
ρ_{tv}	C/cm^3	charge per unit volume in toner

ρ_T	cm^{-3}	number of trapped charges per unit volume
σ	C/cm^2	toner conductivity (in liquid)
σ_c	C/cm^2	carrier charge per unit
σ_i	$(\Omega cm)^{-1}$	ionic conductivity (Chap. 10)
σ_p	C/cm^2	charge per unit area on photoreceptor
σ_s	$(\Omega cm)^{-1}$	charge per unit area
σ_t	C/cm^2	toner charge per unit area
σ_{tr}	C/cm^2	charge per unit area on the back of paper during transfer
τ	s	release time of a charge carrier from a trap
τ_ℓ	s	time constant for liquid development
ϕ_i	eV	work function
ϕ_I	eV	insulator work function
ϕ_M	eV	metal work function
	newton	10^5 dynes, CV/m

1. Introduction

Electrophotography is the technology used in virtually all copiers commercially available today and it promises to be the most prevalent printer technology of the 1990s. This book has been written to assist both the newcomer and those already in the field to better understand this important and complicated technology and its most crucial subsystem, development.

Chapters 1 and 2 are tutorials written to assist the readers who may be new to electrophotography. The primary subject of the book, development physics, begins in Chap. 3 where all available development technologies are listed and compared. In the following chapters, the current state of our technical understanding is reviewed critically for each of these, along with their associated charging mechanism. Two component development systems are discussed in Chaps. 4–7; work on monocomponent systems is reviewed in Chaps. 8 and 9; and liquid development systems are described in Chap. 10.

In this chapter electrophotography is introduced with a discussion of its technical history and the current and projected markets. The evolution of the subsystems are traced from Carlson's first concepts in 1937 to present-day embodiments. The market for electrophotography really began with the introduction of the first automatic copier by the Haloid (now Xerox) Corporation in 1959. Since then the copier business has evolved into a multi-billion dollar revenue industry with many of the world's largest corporations participating. In addition, the already large electrophotographic printer business is expected to grow even faster in the coming decade as the demand for computer output devices continues to increase.

The only potential non-impact competitors to electrophotographic printing are two related powder marking technologies, magnetography and ionography. In magnetography, magnetic forces replace the electrostatic forces used in electrophotography. In ionography, the latent image is created by placing ions on a dielectric surface, eliminating the need for a photoreceptor. These two technologies and other variants of electrophotography also will be described in this chapter.

Technical details of the physics of electrophotography are reserved for Chap. 2. However, a basic knowledge of the process steps of electrophotography will make this chapter more readable. In Fig. 1.1 the six steps of the electrophotographic process are indicated schematically:

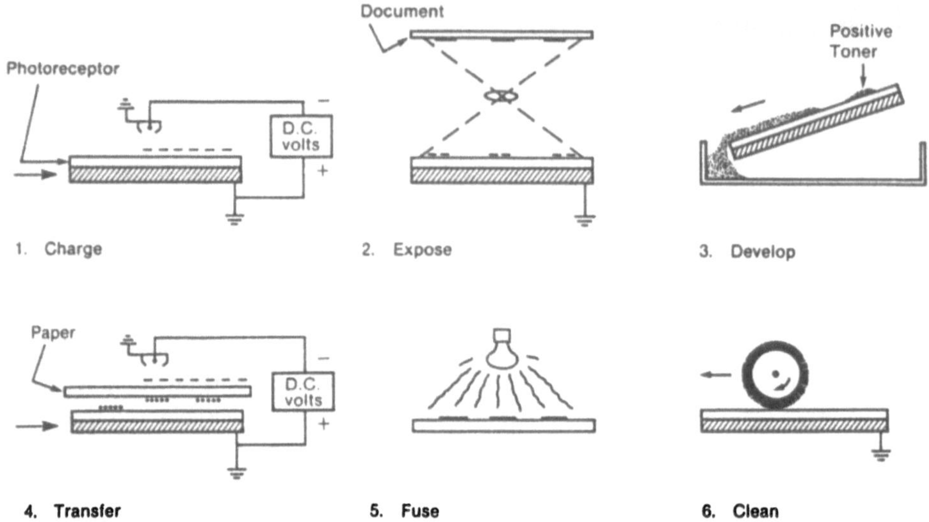

Fig. 1.1. Schematic diagram of the six steps of the electrophotographic process: charge, expose, develop, transfer, fuse and clean

Charge. A corona discharge caused by air breakdown uniformly charges the surface of the photoreceptor, which, in the absence of light, is an insulator.

Expose. Light, reflected from the image (in a copier) or produced by a laser (in a printer), discharges the normally insulating photoreceptor producing a latent image—a charge pattern on the photoreceptor that mirrors the information to be transformed into the real image.

Develop. Electrostatically charged and pigmented polymer particles called toner, ≈ 10 μm in diameter, are brought into the vicinity of the latent image. By virtue of the electric field created by the charges on the photoreceptor, the toner adheres to the latent image, transforming it into a real image.

Transfer. The developed toner on the photoreceptor is transferred to paper by corona charging the back of the paper with a charge opposite to that of the toner particles.

Fuse. The image is permanently fixed to the paper by melting the toner into the paper surface.

Clean. The photoreceptor is discharged and cleaned of any excess toner using coronas, lamps, brushes and/or scraper blades.

1.1 Technical History

Electrophotography [1.1–4] was clearly the invention of one man, *Chester Carlson* [1.5]. He conceived the need for a simple, inexpensive device that would allow office employees to copy any type of document. His background, a B.S. degree in physics and work in the patent offices of Bell Laboratories and P. R. Mallory Company, gave him extensive knowledge of patents related to copying processes.

During the 1930s, when Carlson was searching for a simple copying device, essentially the only copying method available was the Photostat process based on silver halide photography. Turn-around times could be several days, the "copy machine" was only available at a few service centers or county court houses, and the copies produced were reversed (because the customer was given a paper negative) with white letters on a black background. The diazo process (which requires ammonia fumes to develop the blue illuminated diazonium compounds coated on paper) and the earlier blue print process (which produced white lines on blue background by UV exposing iron salts coated on paper) remained engineering copying techniques. Others besides Carlson recognized the need for a better copying process and several alternatives evolved during the 1940s, including Eastman Kodak's Verifax process, a wet process also based on silver halide photography; 3M's Thermofax process, in which a special paper is developed by heat produced by the absorption of light in the printing on the document; and Gevaert's and Agfa's diffusion transfer process, a forerunner of the Polaroid process (without the pod) in which the unexposed silver salts in the positive image on film are caused to diffuse to another sheet of paper where they are reduced with special chemicals forming a positive image.

The two ideas that Carlson brought together in 1937 were: (1) the formation of an electrostatic latent image using photoconductivity to selectively discharge a surface charged insulator, and (2) "development" of this latent image by dusting with powders charged electrostatically. This joining of photoconductivity and electrostatics was a remarkable feat. Electrostatic charging of materials was, and in fact still is, a little understood, highly empirical, mostly ignored aspect of solid state physics. Photoconductivity of insulators was basically an unstudied science at the time of Carlson's invention.

It is clear from *Carlson's* writings [1.2] that he was familiar with prior experiments and patents in which electrostatic images were developed with charged powders. For example, he traced the history of charged powder development from Lichtenberg to Selenyi. In 1777 *Lichtenberg* [1.6] observed starlike patterns on insulators when dust settled onto a cake of resin that had been sparked. In 1936 *Selenyi* [1.7] demonstrated an electrographic recording system in which a charged pattern is written on an insulator (Fig. 1.2) by

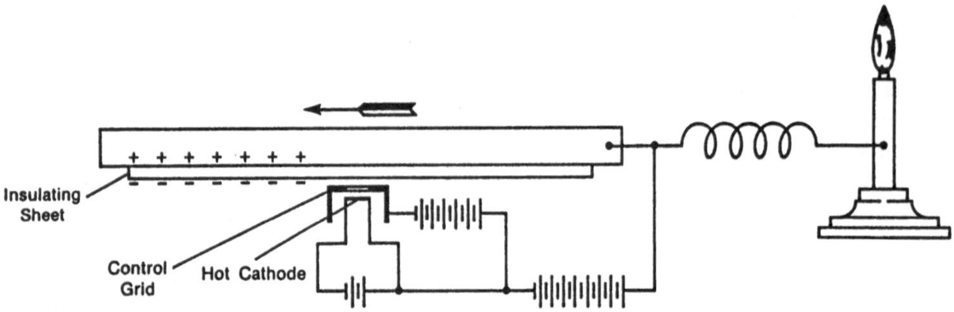

Fig. 1.2. Selenyi's electrographic recording system. A charged pattern is written on an insulator by controlling a cathode current to the insulator surface with a grid. A screw arrangement permits motion of the insulator surface in a spiral path for scanning the full page. By dusting with an insulating powder, the image is developed. The candle is used to erase the latent image [1.7]

controlling a cathode current to the insulator surface with a grid. By dusting with an insulating powder, the image is developed.

It is less obvious how Carlson came upon the idea of using thin photoconductive insulators to form a latent image. Quoting *Carlson* [1.2],

> The difficulties involved with the electrochemical systems led Chester F. Carlson to the conclusion that their relatively high current requirements were incompatible with the small currents available from photoconductive effects. Considering the photoconductor as an energy control element it became apparent to Carlson that the energy controlled by the system could be increased by greatly raising the voltage. This was difficult with electrochemical systems. A brief description of Selenyi's work on the powder development of electrostatic images formed by facsimile scanning appeared in the United States in 1936. Following this lead, Carlson began investigations of electrostatic image formation on photoconductive insulating layers which led him to the invention of electrostatic electrophotography the following year.

By 1937 Carlson had conceived of the process he called electrophotography. It was given a practical form with the help of Otto Kornei in October 1938, filed for patent in April 1939, and issued a patent in 1942 [1.5].

The first photoreceptors were composed of pure sulfur which had been fused and spread onto a metal plate and allowed to harden. Later, plates of sublimed anthracene layers with higher (!) light sensitivity were used. (To give the reader an idea of the sensitivities involved, "clean" anthracene used in experiments today has quantum efficiencies of $\simeq 10^{-4}$ as compared with unity for most modern photoreceptors; Carlson's anthracene was at best 10^4 times less sensitive than currently used photoreceptors.) The photoreceptor was charged "by rubbing it vigorously with a soft material such as a cotton or silk handkerchief." An alternative method was to place a transparent conductive plate parallel to the photoreceptor. When a voltage was applied between the back of the photoreceptor and the transparent conductive plate in the presence of illumination, the top surface of the photoreceptor became charged if the light and then the voltage were removed. Exposure of the photoreceptor to create the latent image was done, for example, by securing the plate to the back of a camera, where the image of the original was focused on the photoreceptor. Development to create a real image was accomplished by sprinkling a fine dust or powder from a can having a cloth or fine wire screen closing its mouth. Pulverized resins were preferred (because of fusing requirements) but gum copal, gum sandarac, ordinary rosin, sealing wax, dyed lycopodium powder, talcum powder, carbon dust, etc., were also used. The dusted plate was then subjected to a "...gentle draft of air by blowing the breath on it or directing air from the nozzle of a suitable blower against the dusted surface to blow off all loose powder not held on the surface by electrostatic attraction." Transfer of the powder to paper was accomplished by carefully laying the paper on the photoreceptor carrying the dusted image and firmly pressing against the surface by a block carrying a felt or sponge rubber pad. To improve the transfer, an adhesive such as plain water, wax or other soft or sticky sub-

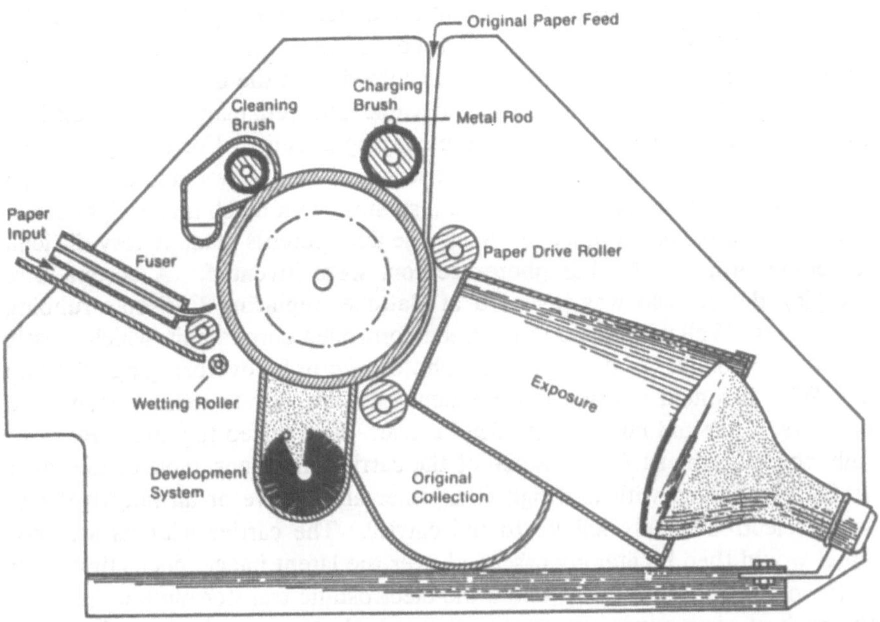

Fig. 1.3. The first automatic copy machine, invented by Chester Carlson [1.8]. In operation, the photoreceptor, coated onto a drum, is rotated first past a charging brush made of a plush-covered roller with a metal rod to drain off any electrical charge. The original is fed into a vertical slot and is driven in contact with the photoreceptor in front of the light source. It is then separated from the drum and deposited in the collection space. The development system consists of floating powder particles created by agitation of a brush; those opposite in sign to the latent image are attracted to it. A blank sheet of paper is fed in from the 9 o'clock position against a wetting roller (to promote transfer) and then against the photoreceptor. The fuser is two hot plates. Brush cleaning is suggested

stances could be applied to the paper. The preferred method of fixing was to melt the resin or wax powder into the paper.

Carlson worked alone until 1944, further developing the process and patenting [1.8] the first automatic copying machine (Fig. 1.3). Beginning in 1940 he tried to enlist commercial support for the invention, unsuccessfully approaching twenty well-known companies including RCA, Remington Rand, GE, Eastman Kodak and IBM. Finally in 1944 Carlson entered into a royalty-sharing agreement with Battelle Memorial Institute and joint development of the process began under Roland Schaffert. A short time later John Dessauer, the director of research at a little company in Rochester, New York, called Haloid (which was barely competing with Eastman Kodak and the Photostat Corporation in the photographic paper supply business for photography and Photostat copying), showed Joe Wilson, Haloid's president, an article on Carlson's electrophotography in *Radio News*. Wilson and Dessauer decided to gamble the company on the new copying technology and signed an agreement with Battelle in 1946. Haloid's funds, plus funds from the U.S. Army with interest in military photographic applications, accelerated progress in developing the technology. The process was first publicly announced and demonstrated by Battelle and Haloid at the annual meeting of the Optical Society of America in Detroit in October 1948. At that time the term xerography, meaning "dry writing," was coined from the Greek.

During the ten years after Battelle became involved, many basic inventions were conceived which made automatic copying a reality. For example, Bixby discovered that amorphous selenium layers prepared by vacuum evaporation onto aluminum were photoconductive insulators with much higher light sensitivity than sulfur or anthracene. Such selenium patents made it very difficult for competitors until other photoreceptors were invented. A corona wire charging device also was invented at Battelle, replacing Carlson's rubbing techniques. Walkup invented the screen controlled corona unit, which greatly lessened the danger of damage to the photoreceptor by overcharging. Walkup and Wise invented cascade development. In this development system, two powders, toner and much larger sized carrier, were mixed together. By carefully choosing the surface material of the carrier and toner, most of the toner would be charged with one sign (i.e. either all positive or all negative) and would electrostatically adhere to the carrier. The carrier with its attached toner would then be literally cascaded over the latent image, depositing toner in the process. Schaffert invented the electrostatic transfer method in which the back of plain paper was corona charged, electrostatically attracting toner in the process. These inventions formed the heart of the first automatic copier introduced by Haloid (Xerox) in 1959, the Model 914. The 914 was so named because it could reproduce documents up to 9 by 14 inches in size. It made seven copies a minute (cpm) and produced a revolution in the office. It is difficult to imagine an office today without a copier.

During those early years, the top priority had to be to produce a marketable product. Much of the information obtained was highly empirical and not suited for reporting in scientific journals. Status reports on the state of the technology appeared in 1965 in the form of two books, one by *Roland Schaffert* [1.1] entitled *Electrophotography* and one edited by *John Dessauer* and *Harold Clark* [1.9] called *Xerography and Related Processes*. These books demonstrate two interesting aspects of the technology: First, small groups of people (sometimes called "electrophotographers") were emerging with physical intuition about which parameters were important in each subsystem. Second, it was very difficult for people to accept that this complicated process was the best way to copy documents. Extensive searches for alternative, simpler and less expensive methods were (and continue to be) actively pursued, obviously so far unsuccessfully. Those who think they have new ideas for simplifying the process would be well-advised to read these books or the summary listed in Sect. 1.4; the early pioneers were very thorough.

In one sense the 914 copier failed to meet Carlson's vision: it was not an inexpensive device. This problem was solved by a marketing decision: copiers were not sold; instead, customers paid for each copy. That brilliant marketing decision brought the price of copying down to an affordable level for the typical office environment.

The corporations primarily involved in electrophotographic research during the 1950s and 1960s were Xerox, Eastman Kodak, IBM and RCA [1.3]. For Xerox, there was the obvious need to improve the technology. Slowly, more scientific approaches were applied to subsystem work. Eastman Kodak was motivated by the concern that electrophotography was a potential threat to its Verifax copying process and perhaps even to silver halide photography. Apparently, work was begun in the mid 1950s at Eastman Kodak, although the size of the effort has not been made public. Eastman Kodak had a variety of engineering models working during the 1960s, but it was not until 1976 that a product was actually introduced into the market. Part of the delay was due apparently to the fact that a simple, low cost natural follow-on to the Verifax process was not found. IBM's interest in electrophotography came, in part, from the desire to have faster computer printers, and IBM was licensed by Xerox for such applications. IBM also began work in the 1950s and early 1960s; some of the earliest work on novel monocomponent development systems and organic photoreceptors was begun during this period. RCA evolved an alternative copying process called Electrofax, in which the photoreceptor was built into the top layer of paper, eliminating the need for the transfer and cleaning steps and avoiding the selenium patents. This was the coated paper copying process in which photoconductive materials such as ZnO mixed in a binder was thinly coated onto paper. Copiers based on Electrofax were introduced by the Charles Bruning Company (Copytron 1000, the first machine to use magnetic brush development, which produced enlargements from microfilm), American Photocopy Equipment Corporation (Apeco Electrostat copier, in 1961, the first office copier), SCM Corporation (Model 33 using liquid development), and the Dennison Manufacturing Company and Savin Corporation (Dennison Copier and the Savin Sahara copier, in 1964, which both used liquid development). However, customer preference for plain paper slowly eroded the coated paper copying market.

During the 1960s Xerox expanded its product line, introducing the 2400 series of copiers (40–60 cpm). Speed limitations and the inability of cascade development to reproduce solid areas brought about work on new development systems, including electroded cascade (used in the 2400 series) and later the magnetic brush development systems, invented during the late 1950s at RCA. IBM introduced its first copiers in 1970. The second model, introduced in 1972, used the magnetic brush development system. IBM was the first to use an organic photoreceptor coated onto aluminized Mylar in place of amorphous selenium. Being much softer, it had a substantially shorter life. To overcome this problem, the photoreceptor was made in the form of a long belt which was unrolled slowly from the inside of a drum; now the (rolled up) organic photoreceptor belt had an overall longer life than an amorphous selenium drum.

By 1975, IBM was able to introduce its first computer printer based on electrophotography, the model 3800. It runs at 215 prints per minute (ppm), and is still one of the fastest electrophotographic engines available commercially. The technology inside an electrophotographic printer is virtually identical to a copier with the exception of the exposure system. In place of lamps whose light is reflected from a document to the photoreceptor, a laser beam is scanned across the photoreceptor surface in a printer. The laser chosen by IBM was a HeNe laser. An acousto-optic modulator was used to turn the laser beam on and off corresponding to each picture element. The scanning of the laser beam was accomplished by reflecting it from the mirror facets of a spinning polygon spaced approximately 1 m from the photoreceptor. The technical problem which required invention in order to be solved was maintaining the laser position from line to line. To maintain line scan accuracy of, say, 25% for 10 lines per millimeter, requires 25 μm placement accuracy. At 1 m throw distance, a facet-to-facet angular accuracy of the polygon of 2.5×10^{-5} radians is required (equivalent to a pointing accuracy of 2.5 cm at 1 km distance)! The solution involved a clever optical trick, focusing the laser to a line on the facet face.

In 1975 an amazing coincidence occurred. Both Xerox and IBM upper management decided to discontinue their research efforts on electrophotography in their research divisions, at least partially because it was thought that electrophotography was a "mature" technology. At Xerox the majority of the electrophotographers from Research were transferred to the product development organization. Remaining applied research was directed to alternative copying and printing technologies including ink jet, magnetography, thermal printing, etc. At IBM, the primary basic research effort had been in photoreceptor development; the work was regarded as successful and transferred. The engineering solution of unrolling an organic belt solved the life problem. The sensitivity of the photoreceptor is close to the theoretical maximum. What more needed to be done? These decisions were later recognized as mistakes for several reasons. First, it opened up the technology to innovation by competitors. Second, no new technology was being developed for future machines. Third, research experts, who could be called upon to assist with problems in products being engineered, were no longer available. Fourth, future technology development had to be done in parallel with current product development, an extremely inefficient process. Fifth, rebuilding a research effort in electrophotography is a long term effort, with on-the-job training, aimed at learning the details of one subsystem and an overview of all the subsystems, taking several years.

Within one year, Eastman Kodak introduced its Model 150 copier. It was immediately recognized by the public and electrophotographers as a major advance in copy quality. The copier made the blacks blacker and the background cleaner. This was done by introducing a new development system: conductive magnetic brush development. The copier used an organic photoreceptor belt and the first recirculating automatic document feeder, which produced complete copies of reports ready for stapling by recirculating the originals each time a copy of the document was required.

Meanwhile, at Xerox during the late 1970s work was proceeding on the next mid-range copier, the 1075. The copy quality of the new Eastman Kodak copiers obviously had to be matched. As electrophotographic research was no longer being carried out in the research organization, the Xerox engineering team working on the 1075 had to shoulder the responsibility of both evaluating and bringing this technology into the new copier, no doubt contributing to the delays associated with the engineering of this product. It was finally introduced in 1982. With the 1075, Xerox introduced its first organic photoreceptor and a new charging device, the dicorotron, in which the corona wire is glass enclosed and subjected to a biased ac voltage, making it more resistant to contamination-induced nonuniformities.

By the late 1970s it was becoming increasingly clear that semiconductor lasers, which had significant advantages over gas lasers since they could be packaged in transistor-size containers and could be modulated by simply controlling the current, were going to have adequate power and life to be used in laser printers. Unfortunately their output wavelength is in the infrared, near 800 nm, where commercially available photoreceptors had little sensitivity. The search for an infrared sensitive photoreceptor was initiated at all major electrophotographic companies. At IBM, interest in using semiconductor lasers re-initiated the involvement of research in electrophotography.

A significant negative development also occurred in the late 1970s. Only three printing technologies could potentially challenge the speed and quality of laser-electrophotography: continuous ink jet, ionography, and magnetography. During the late 1970s, it became increasingly clear from work throughout the world that these alternatives would not replace laser-electrophotographic printing. To date only a few multiple nozzle continuous ink jet printers are marketed; ionographic printers are manufactured by one company, Delphax; and only one company, Bull Peripheriques, continues work on magnetography. This information, coupled with the rapidly increasing need for non-impact printers as output devices to the many computer systems becoming available, significantly heightened the potential commercial importance of the electrophotographic technology. (A fourth printing technology, bubblejet, a new form of ink jets, emerged during the early 1980s from the laboratories of Canon and Hewlett-Packard, also having the capacity to challenge electrophotography.)

Major developments in the technology came from Japan [1.3] during the period 1970–1980. Japanese companies, relying on their manufacturing strength, introduced low speed, low cost copiers, a segment of the market ignored by Xerox because it was felt that it was not possible to reduce the cost of the technology enough to make a viable product in this low speed and low volume range. The most successful during this time was Ricoh Corporation, which introduced copiers using a liquid development system (which were sold in the United States by Savin). Liquid development eliminated the need for a fuser, one of the most energy-intensive parts of the dry toner electrophotographic process; it also allowed the design of a smaller box requiring one-third of the number of components used in dry toner copiers, thus significantly improving reliability. These copiers were very successful, taking a significant share of the low end of the market.

Canon also made significant new discoveries. First, to avoid the selenium patents, Canon developed a new photoreceptor. It consisted of two layers, an upper insulating layer and a bottom photosensitive layer made from cadmium sulfide. A latent image was produced by simultaneously exposing and charging, causing the charge of the latent image to reside at the interface between the two layers. This was called by Canon the "New Process" (NP) and it forced the coining of a new name, "Carlson Xerography," for the usual latent image formation process. Canon began marketing NP machines in 1970 after eight years of development. These copiers also had liquid development. Second, in 1980, Canon announced the first copier with a monocomponent development system using insulative magnetic toner, the NP-200, a 20 cpm desktop copier. This system eliminated the carrier used in the cascade and magnetic brush development systems. Instead, magnetic material was put inside the toner, allowing magnetic forces to transport the toner into the development zone. There the toner developed across a gap in response to the dc electric fields of the latent image and a superimposed ac field. The development characteristics were excellent (good blacks, low background), the dry toner allowed the use of true plain paper, and the small size allowed the design of a small, relatively inexpensive tabletop copier. Within two years, Canon produced a whole line of copiers based on this monocomponent development system, from 12 to 30 cpm. The next big advance also came from Canon and addressed the Achilles' heel of electrophotography: reliability. In 1983 Canon introduced the cartridge concept for personal copiers. Many of the less reliable electrophotographic steps, including charging, monocomponent development, and cleaning, were incorporated into a throwaway cartridge in the PC-10 copier (8 cpm), which sold for the incredibly low price of $995 (plus $65 for the disposable cartridge, good for about 2000 copies). This copier was so inexpensive and the perceived reliability (actually availability) was so significantly increased that it opened up new markets in the low end of the copier business. Canon introduced yet another advance in 1985: the first amorphous silicon photoreceptor, which was put into its NP-7000 (50 cpm). The hardness of amorphous silicon is expected to significantly extend the life of the photoreceptors: 10^6 copies per drum and higher have been reported; 0.5×10^6 is guaranteed by Canon.

In 1982, Siemens introduced a new fuser, one based on chemical vapors, in its ND-3 printer. While vapor fusing was used on hand-operated copying equipment made by Xerox (Haloid) during the 1950s, this was the first time it was introduced in a high speed machine, 103 ppm. The challenge is to contain the organic vapors. This was achieved by bringing the roll paper after fusing into a refrigerated area that lowered the vapor pressure, condensing and capturing most of the organic solvent.

In the years 1984, 1985, several companies demonstrated that the laser-spinning polygon system used to convert a copier into a printer can be replaced with an all-solid-state device. Epson and Casio devised printers that use an array of liquid crystal shutters to address the photoreceptor. Behind the 10 liquid crystal shutters per millimeter is a uniform light source. In front is a Selfoc lens array which images the liquid crystals on the photoreceptor with a spacing of ≈ 1 cm. The Epson printer made 7 cpm; Casio's, 9 cpm. Neither is generally available to the public. Oki Electric Co. and NEC demonstrated printers with LED (light emitting diode) arrays. The Oki printer has a resolution of 10 dots/mm and a speed of 10 or 20 ppm; the NEC printer has a resolution of 12 dots/mm and a speed of 8 ppm. Two new LED printers were introduced by IBM (12 ppm, 12 dots/mm) and Eastman Kodak (92 ppm, 12 dots/mm). These "image bar" technologies present solutions to two manufacturing challenges: the maintenance of sufficient light uniformity among the elements and with time (as the elements degrade) and the interconnection of all of the elements to driver electronics at a reasonable cost (so it can compete with the laser-polygon system).

Another contribution, again from Japanese companies, was announced at the 1985 IEEE-IAS (Industrial Application Society) annual meeting in Toronto. Both Ricoh and Toshiba recognized the potential benefits of monocomponent development with nonmagnetic toner, such as lower toner manufacturing cost and the potential for better colors than could be obtained

Table 1.1. Commercially available copiers $\left(\frac{\text{Model}}{\text{Development technology}^{a}}\right)$

Copies/ min.	Xerox	Kodak	IBM	Canon	Ricoh	Oce	Konica (Royal)
120	9900 **DI**						
90	1090 **DC**	250 **DC**					
70	1075 **DC**	150, 225 **DC**	Series III **DI**	NP 8070 **MIM**			
60	3600 **DI**						
						1825 **MCM**	
							5503 **DI**
50	1050 **DI**			NP 7050, 500 **MIM**	FT 6085 **DI**		5003 **DI**
	5400, 5600 **DI**					1725 **MCM**	
40	2400 **DI**			NP 4540 **MIM**			4003 **DI**
	1040 **DI**				FT 5070 **DI**		
				NP 4035 **MIM**			
30					FT 5000 **DI**		
							2803 **DI**
				NP 270, 3525 **MIM**			
	3400 **DI**			NP 3025 **MIM**			
	1038 **DI**						
	1025 **DI**				FT 4065 **DI**		
20							
							1803 **DI**
				NP 150 **MIM**	FT 2050 **MIM**		
							1503 **DI**
	1012 **DI**						1200 **DI**
	1020 **DI**			NP 115 **MIM**			
10					RiPRO **MIM**		
8				PC 10, 20, 30 **MIM**	RiPRO-jr. **MIM** (black) **MIN** (color)		
6				PC 3, 5 **MIM**			

Table 1.1 (cont.) Commercially available copiers $\left(\frac{\text{Model}}{\text{Development technology}^{a}}\right)$

Copies/ min.	Sharp	Toshiba	Sanyo	Panasonic	Minolta	Mita	Savin (Ricoh)
120							
90							
70							
60	SF9600 **DI**						
50	SF9500 **DI**					DC-513Z **DI**	7050 **L**
				FP-4520 **DI**	EP650Z **DIμ**		
40							
					EP550Z **DIμ**		
							7040, 7350 **L**
	SF9300 **DI**		Z122 **DI**				
	SF8600 **DI**						
30		BD7816 **DI**		FP3030 **DI**	EP470Z **DIμ**	DC-313Z **DI**	870, 5030 **L**
				FP2625 **DI**	EP450Z **DIμ**		
		BD5610 **DI**			EP410Z **DIμ**		
						DC-213RE **DI**	
20	SF8100 **DI**						
			Z116 **DI**				
						DC-152Z **DI**	
				FP1530 **DI**	EP350Z **DIμ**		
		BD4121 **DI**					
	SF7200 **DI**		SFT70 **DI**	FP1300 **DI**			
						DC-1001, 111 **DI**	
10	SF7100 **DI**	BD-3110 **DI**			EP50 **DIμ**		
8	Z-50 **DI**		SFT600 **DI**				
6							

[a] D: DUAL
I Insulating
C Conducting
μ: Micro-carrier

M: MONOCOMPONENT
I Insulating
C Conducting
M Magnetic
N Nonmagnetic

L: LIQUID

with toner loaded with magnetic material. They simultaneously announced nonmagnetic monocomponent development systems; so far only Ricoh has manufactured products with this system.

It would seem, with the observed rate of new ideas, that Chester Carlson's electrophotography after 50 years still has a way to go before "maturing." Who could have ever guessed in the early days that such a complicated process would even work, let alone be the dominant copying and printing technology in the last two decades of this century?

1.2 Copier Market

The number of copiers and their features and speeds has grown enormously in the 29 years since Xerox introduced the 914 copier. Listed in Table 1.1 are most of the commercially available copiers from companies that manufacture them, along with their development technology. Clearly a large choice is available to the consumer. Plotting a table such as this as a function of time shows that Japanese copiers are dominant in the low end and are slowly challenging the high speed market dominated today by Xerox, Eastman Kodak, and IBM.

Under the model number is listed the development technology. The first letter (D, M, L) distinguishes between dual or two component (magnetic brush), monocomponent and liquid; subsequent letters define which variant is used, as discussed in succeeding chapters. A study of this table reveals that most copiers use insulative magnetic brush development (Chap. 6); only Eastman Kodak and Xerox use conductive magnetic brush development (Chap. 7). Monocomponent development systems are used by Canon, Ricoh, and Océ. Canon uses magnetic, insulating; Océ, magnetic, conducting; and Ricoh, both magnetic and nonmagnetic, insulating (Chap. 9) systems. Only Savin (in the United States) sells copiers using liquid development systems, which are manufactured by Ricoh.

The reason for the large number of commercially available copiers is the huge market, 14 billion dollars in the United States in 1986. The 1986 market and estimates of the 1991 U.S. market by segment are shown in Table 1.2 taken from information provided by Dataquest [1.10]. They divide the market into seven segments.

Segment PC (personal copier) includes copiers that have speeds up to 12 cpm, have moving platens, sell for an average of $1100 and make about 400 copies per month. These types of copiers are easy to install, have minimal features, superior reliability, and are compact and lightweight. An example is the Canon PC-5. This is predicted to be the fastest growing segment of the market, going from $294M in 1986 to $600M in 1991, or a 15.3% CAGR (compound annual growth rate).

Table 1.2. U.S. copier market (Dataquest, September 1987) [1.10]

	Segment definition			Market[a]		
		Speed [cpm]	Average retail price [$]	1986 [Millions of U.S. dollars]	1991	Compound annual growth rate [%]
	PC	up to 12	1100	294	600	15.3
	1	up to 20	2500	2320	2170	−1.3
	2	21−30	4700	3030	2840	−1.2
	3	31−44	7400	2440	3230	5.8
	4	45−69	11400	2000	3820	13.9
	5	70−90	16000−75000	2020	1710	−2.3
	6	91+	78300−129775	1890	1910	0.2
Total				13994	16280	3.1

[a] Includes hardware, service and supplies

Segment 1 includes tabletop copiers up to 20 cpm, with an average monthly volume of 3500, and an average price of $2500. These copiers have some features such as reduction or enlargement, optional input/output devices, and 11 × 17 inches maximum copy size. An example is the Minolta EP-350Z. This market segment is expected to shrink from $2320M to $2170M, a CAGR of −1.3% over the period 1986−1991.

Segment 2 copiers have speeds of 21−30 cpm, make 7200 per month, and sell for $4700 on the average. They have features similar to those in Segment 1. An example is the Toshiba BD-7816. This market segment is also expected to shrink from $3030M to $2840M, a CAGR of −1.2% over the same period.

Segment 3 category copiers are increasingly offered as systems with standard features of reduction/enlargement, feeders and sorter. They typically have speeds of 31−44 cpm, make 14 000 copies per month, and sell for an average price of $7400. An example is the Sharp SF-9300. The market will grow by 5.8%, from $2440M to $3230M.

Segment 4 copiers are highly featured machines such as the Ricoh FT-6085. They typically have a speed of 45−69 cpm, make 24 000 copies per month, and sell for an average price of $11 400. This segment has the second highest predicted CAGR of 13.9%, from $2000M in 1986 to $3820M in 1991.

Segment 5 includes highly featured, fast copiers such as the Ektaprint 225, Xerox 1075, and IBM Series III. They typically have a speed of 70−90 cpm, have a monthly volume of 63 000 copies, and sell for $16 000 – $75 000. They feature modular options including finishing, input/output options and magnification. The segment is expected to shrink by 2.3% from $2020M to $1710M by 1991.

Segment 6 includes the fastest copiers such as the Xerox 9900 and the Ektaprint 250. They sell in the range $78 000 – $130 000, have a speed of

Table 1.3. Price per copy

Segment	Speed [cpm]	Copies/ month [thousands]	Purchase price	Price per copy [¢/copy]					Total price
				Purchase price	Maintenance	Supplies			
						Cartridge	Toner	Developer	
PC	up to 12	1.5	$ 1145 $ 95 cartridge/ 2000 copies	1.27	0	4.75			6.02
2	21 – 30	8	$ 5000	1.04	2.72		0.29	0.13	4.18
3	31 – 44	20	$ 9500	0.79	1.41		0.21	0.09	2.50
5	70 – 90	60	$ 30000	0.83	0.88		0.11	Included in maintenance	1.82
6	91 +	300	$ 130000	0.72	0.66		0.08	0.04	1.50

Assumptions: Five-year amortization, no financing, supplies for one year purchased, fuser oil ignored (small), paper price not included. Numbers are only representative of published values

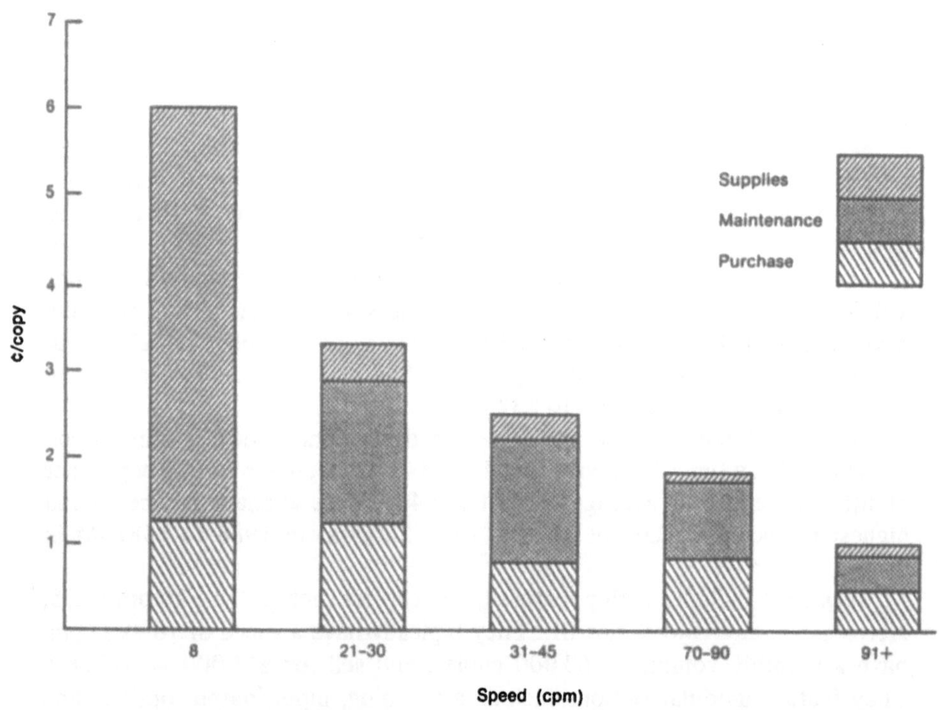

Fig. 1.4. The price a customer pays per copy for typical machines in several of the market segments. The price per copy goes up as the speed goes down, from 1.5¢ at 91+ cpm to 6.02¢ for the slowest copier (see Table 1.3)

91+ cpm, and make an average of 210 000 copies per month. This segment is expected to grow slightly, 0.2%. Its revenue in 1986 was $1890M.

The total copier market is anticipated to grow from $14 000M to $16 000M, or a CAGR of 3.1% from 1986 to 1991.

It is of interest to calculate the price the customer pays per copy across the speed range. The assumptions and numbers are shown in Table 1.3 and the data are plotted in Fig. 1.4. The purchase price has been amortized over five years and typical published prices of maintenance and supplies have been used. As can be seen, the price per copy (excluding paper) dramatically increases as the speed decreases, from 1.5¢/copy for the highest speed copiers to 6.02¢/copy for the slowest copiers. Most of the price per copy for the slowest speed copiers is due to the high cost of the replaceable cartridge, the invention that led to the increased reliability (actually availability) of these copiers. Among Segments 2–6, the increasing price per copy as speed is reduced is due primarily to the increasing price of maintenance.

1.3 Printer Market

The need for printers, which are primarily output devices for computers, has grown with the computer industry. The recent enormous proliferation of computers, availability of document preparation software, and large-scale access to data bases have led to a rapidly expanding electrophotographic printer business. This is because printers based on electrophotography are quiet, can handle multiple fonts, and can produce pictorial information over a wide speed range. Reliability concerns (relative to impact printing technologies) are being addressed by innovative manufacturing and engineering approaches. For example, some printers allow customers to throw away used-up parts, such as the Canon customer replacement cartridge concept. Others use internal microprocessors to diagnose and aid customers in fixing their own machines. Still others even use these microprocessors to call, by themselves, a central office requesting service and specifying parts required.

In 1973, Xerox introduced the first modern electrophotographic printer, the Xerox 1200, which was based on the Xerox 2400–3600 series of copiers. The copier optics were replaced by spinning character masks in front of xenon flash lamps. Each mask with all the characters for a font set spun in front of xenon flash lamps at each character location across a line. Obviously, font flexibility was limited to the character set on the spinning mask. By 1975, 1976, IBM and Canon had introduced the first laser-based electrophotographic printers, the IBM 3800, operating at 215 ppm, and the Canon LBP 2000 C1 operating at 31 ppm. Both used HeNe gas lasers. By 1977, many more products had appeared, including the Xerox 9700 (based on the 9200 copier) operating at 120 ppm, the Hitachi 8196–20 operating at 112 ppm, the

Siemens ND-2 operating at 206 ppm, and the NEC 7370 operating at 112 ppm. A significant product, announced in 1983, that opened up the low speed, low cost market was the Canon LBP-CX, an 8 ppm printer, built by adding a semiconductor laser to its PC-10 low cost copier.

At present there are a large number of manufacturers offering electrophotographic printers. A list of the major manufacturers is given in Table 1.4. Other corporations offering such printers include Hitachi, Fujitsu, NEC, Burroughs, Hewlett-Packard, Minolta, Sharp, and Eastman Kodak.

The reason many corporations are manufacturing electrophotographic printers is the expected market growth. U.S. market forecasts provided by

Table 1.4. Some commercially available printers $\left(\frac{\text{Model}}{\text{Development Technology}^{a}}\right)$

Prints/min.	Xerox	IBM	Siemens	Canon	Ricoh
220		3800 **DI**			
			ND-2 **DI**		
120	9700 **DI**	3800-6 **DI**			
			ND-3 **DI**		
70	8700 **DI**				
50	4050 **DC**			LBP-3400 **MIM**	
	5700 **DI**				
40					LP-4400 **DI**
30		6670 **DI**			
	3700 **DI**				
20		3820 **DIμ**		LBP-20 **MIM**	
					LP-3150 **MIM**
12	2700 **DI**	3812 **DI**			LP-4120 **MIM**
10	4045 **DI**			LBP-CX **MIM**	
8				LBP-CX **MIM**	LP-4080 **MIM**
6					PC-6000 **MIN**

[a] D: DUAL
I Insulating
C Conducting
μ: Micro-carrier

M: MONOCOMPONENT
I Insulating
C Conducting
M Magnetic
N Nonmagnetic

L: LIQUID

Table 1.5. U.S. printer market (Dataquest, August 1987) [after 1.11]

	Segment definition		Market[a]		
		Speed [ppm]	1986 [Millions of U.S. dollars]	1991	Compound annual growth rate [%]
	1	up to 10	1220	4630	30.6
	2	11 – 20	379	2800	49.2
	3	21 – 30	106	578	40.3
	4	31 – 50	387	950	19.7
	5	51 – 80	432	640	8.2
	6	81 – 150	583	845	7.7
	7	151 +	1220	1200	– 0.3
Total			4325	11643	21.9

[a] Includes hardware, service and supplies

Dataquest [1.11] are shown in Table 1.5: the overall market is expected to experience a compound annual growth rate of 21.9% from $4300M in 1986 to $12 000M in 1991. Very similar printer market predictions have been made by CAP International [1.12]. Dataquest divides the printer market into seven segments.

Segment 1 covers printers with speeds up to 10 ppm and includes the Hewlett-Packard Laserjet and the Apple Laserwriter, both built on the Canon LBP-CX engine. This segment is predicted to grow by 30.6% per year from $1220M to $4630M from 1986 to 1991.

Segment 2 has the largest predicted CAGR, 49.2%. These printers have a speed range of 11–20 ppm. Examples include the IBM 3812 and the Texas Instruments 2015. The market is expected to grow from $379M to $2800M from 1986 to 1991.

Segment 3 has the second largest predicted growth rate: 40.3%. Printers in this segment make 21–30 ppm. An example is the Xerox 3700. Here the market is predicted to grow from $106M to $578M.

Segment 4 printers can print at speeds of 31–50 ppm, and include the Xerox 4050 and the Ricoh LP4400. This segment is expected to grow at a CAGR of 19.7%, from $387M to $950M.

Segment 5 printers have speeds of 51–80 ppm. The market is expected to grow from $432M in 1986 to $640M in 1991, a CAGR of 8.2%. Examples are the Xerox 4060 and 8700.

Segment 6 printers have speeds of 81–150 ppm. Examples include the Xerox 9700, Siemens 2200, and the IBM 3800-6. The market is predicted to grow modestly, by 7.7%, from $583M to $845M.

Segment 7 is presently the largest market, $1200M, but is anticipated to shrink slightly by 1991 with CAGR of –0.3%. Examples include the Siemens 2300 and IBM 3800-3.

1.4 Alternative Powder Marking Technologies

There have been extensive searches for copying or printing technologies simpler than electrophotography. The motivation should be obvious to the reader after reading Sect. 1.1: electrophotography is very complicated, involving six separate process steps with significant interactions between the steps. Alternative powder marking technologies were summarized by *Weigl* in 1977 [1.13] and in [1.1] and [1.9] in 1965. We shall briefly describe them below. The most successful alternative technologies so far invented are ionography and magnetography, both of which are printing technologies. In both, the latent image formation process has been changed, allowing replacement of the photoreceptor with a hard dielectric surface. These technologies are discussed more extensively later.

In direct electrophotography or "Electrofax" the photoreceptor is placed on the surface of the paper. This is accomplished by coating the paper with a thin photoconductive coating such as dye-sensitized zinc oxide in a binder [1.14,15]. In this process the paper (with the coating) is charged and exposed, forming a latent image. Development is done directly on the paper surface, eliminating the transfer step. As the photoreceptor (the coated paper) is used only once, there is no need to clean. The price paid for this simpler process, is the need for "special" paper. Electrofax copying was widely used for low volume black-and-white copying in the 1950s and 1960s. The need for "special" paper and the emergence of low cost electrophotographic ("plain" paper) copiers have slowly eroded its market.

Persistent internal polarization was invented independently by *Kallmann* et al. [1.16–19] and *Fridkin* and *Zheludev* [1.20,21]. It is an alternative method of forming the latent image. Instead of charging in the dark and exposing with light, a special photoreceptor is uniformly illuminated in the presence of an applied field. This separates charges which are deeply trapped in the vicinity of both electrodes (the need for deep traps distinguishes this photoreceptor). Image-wise exposure then releases the carriers from the deep traps, creating the latent image. Unfortunately, the presence of the deep traps inherently limits the carrier transport and photosensitivity.

In the "Magnedynamic" process [1.22] the charging step is completely eliminated. Here a "photoreceptor," upon illumination, maintains its conductivity. By developing with conductive toner (see Sects. 8.1 and 9.4), an image is formed. This occurs because in the areas of illumination where the "photoreceptor" is conductive, the conductive toner is charged by injection to the same sign as the photoreceptor and is repelled. In the uncharged regions, the toner is charged by induction to the sign opposite to that of the photoreceptor and adheres electrostatically.

A process which eliminates the toner transfer and photoreceptor cleaning steps is called TESI, for transfer of electrostatic image [1.23–28]. In this process, the electrostatic latent image itself is transferred to a special paper prior to development. The special paper is a dielectric-coated paper which is pressed into intimate contact with the photoreceptor, usually in a suitable bias field. As with Electrofax, the special paper requirement is a handicap; in addition the process is sensitive to contact geometry and ambient humidity.

Processes have been devised in which the image-forming material is part of the photoreceptor, called "Frost" [1.29]. By overcoating the photoreceptor with a thermoplastic layer, the electric field of the latent image can be impressed on the plastic layer. When the thermoplastic is softened, the surface is deformed into a random wrinkle pattern which scatters light, making it useful as a display. The deformations are caused by mutual repulsion of the trapped surface charges (which tend to increase the total surface area and produce the wrinkle pattern) and mutual attraction of surface charges and the counter charges in the photoreceptor ground plane (which produce hydrodynamic instabilities) [1.30,31]. Images may be erased by softening the film and discharging the field across it.

In photocontrolled ion flow electrophotography [1.32,33], the photoreceptor is built into a corona device, allowing the use of a hard dielectric receptor in place of the photoreceptor. Here the photoreceptor is coated onto a metal screen interposed between an active corona source and a dielectric receiver. Illuminated areas of the screen freely pass corona ions, while the rims of holes in dark areas quickly become charged and block additional corona ions from reaching the dielectric receptor.

The electric fields associated with the latent image can be used to do electrochemistry [1.34]. Work in this area actually began about 1900 with attempts to use these fields to control the electrolytic coloration of contiguous layers such as paper and gelatin. In use, the dye-sensitized light-gray ZnO binder coating is imaged, then developed by passing it under a sponge wetted by an aqueous electrolyte containing reducible silver ions and biased about +100 V relative to the Al coating base. A color version was suggested by *Tulagin* et. al. [1.35] in which electrolytically reducible ionic dye precursors were substituted for the metallic cations. Approaches to provide photoreceptor gain were suggested [1.35–40] in order to speed up the process. As recently as 1986, IBM discussed "molecular matrix technology", in which currents from styli cause the creation of color on paper by direct electrolytic effects on colorless oxidation-sensitive leuco dyes included in the paper matrix [1.41].

A great deal of work has been devoted to photoconductive pigment electrophotography (PAPE). In this technology the latent image is created as a variable toner charge (which contains the "photoconductor"), which then moves in a constant electric field. The general theory has been treated by *Schmidlin* [1.42], *Scharfe* and *Schmidlin* [1.43], *Hartmann* et al. [1.44] and *Cheng* and *Hartmann* [1.45]. The first practical embodiment was invented by *Gundlach* [1.46], who deposited a uniform layer of loose photoconductive toner on a flat electrode, charged the deposit, exposed it, and removed the loosely adherent powder from the illuminated areas. The remaining toner could be fused on the electrode. *Goffe* [1.47] invented another form in which photosensitive particles initially placed on top of thermoplastic migrate image-wise through the thin, temporarily softened thermoplastic. A solvent wash can be used to remove the thermoplastic and its residue of incompletely migrated pigments. Recent work is summarized by *Vincett* et al. [1.48]. By using spectrally selective pigments, *Tulagin* [1.49] invented an elegant solution to the challenge of single-step full-color electrophotography. Feasibility was soon demonstrated [1.50,51] and problems identified [1.52,53]. Work on this technology was recently discussed [1.54]. Other variants of PAPE have been reported [1.55–63].

An alternative printing process called "Magnestylus" was reported by *Kotz* and *Nelson* [1.64]. In this process conductive magnetic toner is chained up under conductive styli in the development zone by a magnetic field. If voltage is applied to the styli, current flows through the conductive toner chain, charging the toner and creating an electrostatic force causing development onto a dielectric receptor. An advantage of this system is that very low voltages (5–40 V) are reported sufficient to create and develop the image, making the electronics compatible with integrated circuits.

1.4.1 Magnetography

Magnetography, the magnetic analogue of electrophotography, has attracted considerable attention from at least five different corporations. In this printing process, a magnetic latent image is created on a ferromagnetic material using an array of magnetic inductive heads which is then developed with magnetic toner. In 1972 a serial magnetographic printer was offered by Data Interface, Inc.; in 1979 a line printer was offered by General Electric [1.65]; and in 1982 Iwatsu offered magnetographic products [1.66]. Ferix and Xerox have worked on the technology. Honeywell-Bull (now Bull Peripheriques) [1.67] currently is the only company which manufactures a magnetographic product. The technology has been discussed at the Fourth International Electrophotography Conference [1.68,69].

The magnetic latent image can be created in a variety of ways. A vertical magnetization (perpendicular to the ferromagnetic film) can be created (see Fig. 1.5) by a pin-shaped magnetic pole aligned perpendicular to the film. The return path for the flux is through a magnetic sublayer under the film and a large counter pole. Horizontal recording can be achieved with a ring or horseshoe-head placed in close proximity to the film. An unusual thin film head which records horizontally was suggested by *Springer* [1.70]: it has a flat geometry which has Permalloy layers both above and below a spiral current coil. The magnetic gap is annular at the edge of the two Permalloy layers creating a radial magnetic latent image. Some of these technologies allow a

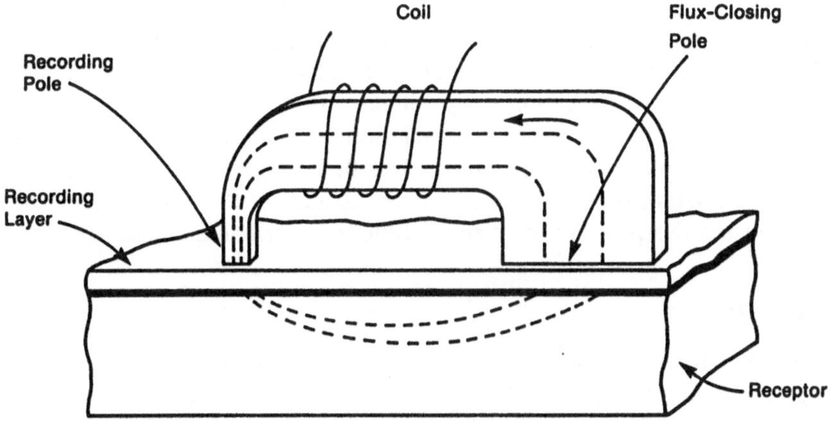

Fig. 1.5. Energizing the recording head's coil creates a magnetic flux in the circuit composed of the head and drum cores. Because field intensity is below the coercivity of the recording layer under the longer flux-closing pole, a dot is recorded only under the higher-intensity recording pole

choice of recording on the front or behind the film with a belt configuration. A printhead in front of the film is exposed to toner contamination; a printhead behind the film loses resolution because of field spreading through the thickness of the substrate and magnetic film.

Development is accomplished with magnetic toner. The development forces have been studied by *Parker* et al. [1.71], *Eltgen* [1.69], and *Schlömann* [1.72]. The force attracting toner containing soft magnetic material is proportional to the product of the magnetic field and its gradient. Predicting toner development onto a latent image is complicated by the fact that the toner becomes magnetized, attracting more toner. The force of attraction is approximately an order of magnitude smaller than in electrophotography. Among several consequences, this necessitates a means for eliminating contact electrostatic charging of the magnetic toner, which would then be attracted by image forces to the background. The small forces also necessitate in some machines an air-knife or magnetic scavenger after development to remove toner loosely attracted to the background regions of the magnetic media.

If the toner is insulating, it can be corona charged and then transferred using normal electrophotographic transfer methods, such as corona or roll transfer. If the toner is conducting, other means are necessary to deal with the well-known (see Sect. 9.4) high humidity conductive toner problem (in which the toner charge, but not the toner, transfers to moist, conductive paper).

Printing up to 170 m/min has been reported [1.67] by Bull Peripherique; it offers products operating at 50 and 90 ppm.

At this time it is unclear how successfully magnetography will compete with laser-electrophotography. Its advantages include a harder receptor, which should improve the reliability of the system, and demonstrated speeds of 170 m/min or approximately 500 ppm, approximately two times faster than any electrophotographic engine. The primary disadvantages of the system are a direct result of the printhead: its complexity and the short range magnetic forces. The complexity, with one printhead required per picture element, will probably limit the resolution capabilities as compared with laser-electrophotography where increased resolution only requires a more tightly focused laser spot on the photoreceptor. The short range magnetic forces require close subsystem-receptor spacing. Close spacing of the printhead, on the toner side of the receptor, makes it susceptible to toner contamination.

1.4.2 Ionography

The concept for this printing process goes back to *Selenyi's* [1.7] electrographic recording system (Fig. 1.2). In this technology a latent electrostatic image is created by selectively depositing ions directly on either dielectric paper or a reuseable dielectric medium. By eliminating the photoreceptor (and its charging and exposure) this represents a simpler process. Several techniques for creating the latent image have been devised. The first uses multistylus arrays, with one stylus per pixel element. A voltage is applied to each stylus (greater than 350 V to cause air breakdown). Devices with 8–12 styli per millimeter have been reported. Versatec, a Xerox subsidiary, manufactures printers using this technology coupled with liquid development. A recent discussion of advances in this form of ionography was presented at the 1981 Conference on Advances in Non-Impact Printing Technologies [1.73], including discussion of resolution limitations imposed by the air breakdown mechanism of generating charge.

The second technique for creating the latent image also uses air breakdown to create the ions but then shutters them with a third electrode. This concept has been reported by *Rumsey* and *Bennewitz* [1.74]. Current embodiments operate at 30–90 ppm, are produced commercially by Delphax Systems and are marketed by a few other corporations. The patented print cartridge for creating the ions is shown in Fig. 1.6. Voltages high enough to ionize the air

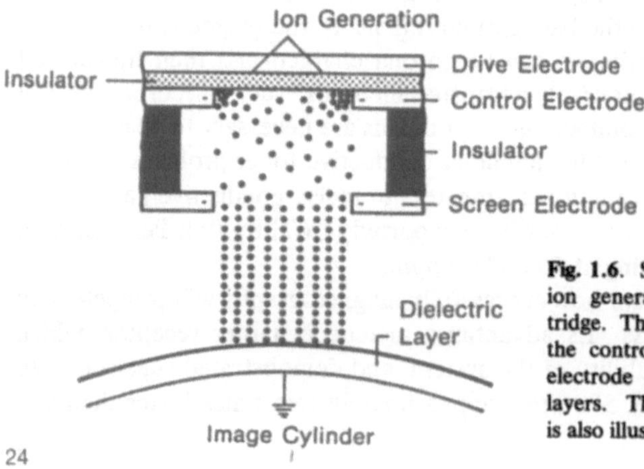

Fig. 1.6. Schematic cross section of an ion generation site in the print cartridge. The three electrodes, the drive, the control electrode and the screen electrode are separated by insulating layers. The dielectric cylinder surface is also illustrated [1.74]

are multiplexed (to reduce the number of high voltage drivers required) between the drive and control electrode; the screen electrode is used to gate the ion flow. Commercially available equipment has 240 holes per inch.

Given an electrostatic latent image on a dielectric, one could, in principle, use familiar electrophotographic subsystems for the subsequent steps to develop, transfer, fuse, and clean. However, characteristics of the print head plus the desire for high reliability determine Delphax's subsystem choice.

First, as the ions stream toward the dielectric, they are repelled by already present ions, causing the latent image to bloom. This phenomena is discussed by *Omodani* et al. [1.75,76]. It can be minimized by using a low voltage latent image, which therefore requires a development system sensitive to low voltages. A conductive magnetic monocomponent development system, which develops high optical densities at 120 V, is described by *Rumsey* and *Bennewitz* [1.74].

The conductive toner transfer problems at high relative humidity (see Sect. 9.4) were solved by combining the transfer and fusing steps into one: a high pressure cold transfix step (transfer and fix occur simultaneously), an obviously simpler and potentially more reliable process. Transfer efficiencies of 99.6% are typically observed.

Cleaning against a hard dielectric as compared to a soft photoreceptor is obviously easier to implement. Delphax physically scrapes off the remaining toner with a steel(!) doctor blade. The erasure of the residual electrostatic image is accomplished with an ac corona.

The advantages and disadvantages of this system are similar to those in magnetography. The hard receptor surface should lead to improved reliability (which is in fact one of the primary design goals of the Delphax engine). But the more complex printhead will probably limit resolution and is currently a source of unreliability caused by paper or toner dust plugging the holes in the screen electrode. This form of ionography has a unique problem: a trade-off between resolution and the magnitude of the latent image caused by the repulsion of incoming ions by those already deposited on the receptor. This may also limit the resolution capabilities of this form of ionography.

2. The Electrophotographic Process

Chapter 1 deals with the history and market of electrophotography and discusses alternative powder marking technologies. In this chapter we begin to discuss the actual technology. Section 2.1 defines the six process steps of electrophotography and provides the reader with a description of the physics within each step (or subsystem) and a guide to the literature. Section 2.2 compares the implementation of 70 cpm copiers by three different corporations; here the pervasive issue of subsystem interactions becomes immediately obvious. Section 2.3 lists and compares some of the technology choices for each subsystem.

2.1 The Six Steps of Electrophotography

Electrophotography is a complex process (see [2.1–18] for other books and reviews on electrophotography) involving, in most embodiments, six distinct

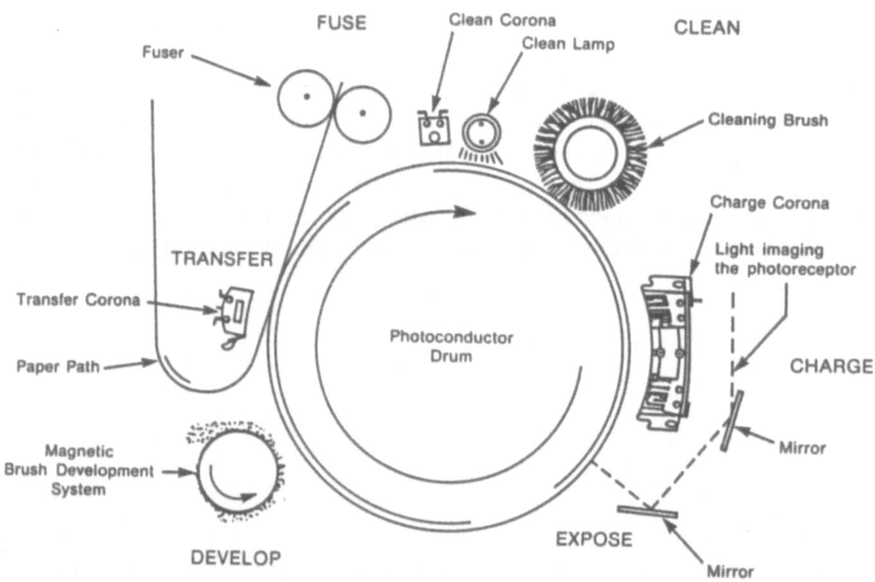

Fig. 2.1. The six steps of the electrophotographic process are charge, expose, develop, transfer, fuse and clean

steps, which are shown schematically in Fig. 1.1 and in one possible configuration in Fig. 2.1. Below, each step is defined in bold print and then discussed, with an emphasis on the relevant physics.

2.1.1 Charge

A corona discharge caused by air breakdown uniformly charges the surface of the photoreceptor which, in the absence of light, is an insulator.

The charge per unit area σ_p that the corona device is required to place on the photoreceptor is determined to first order by the potential V required for development, (≈600V) and the photoreceptor thickness. The photoreceptor is made as thin as possible, for manufacturing cost and charge transport reasons, but thick enough to avoid dielectric breakdown. For organic photoreceptors whose dielectric strengths are ≈ 30 V/μ m, 20 μm thicknesses are necessary. The required photoreceptor charge per unit area is then obtained by a straightforward application of Gauss's Law (Fig. 2.2a).

$$V = E_p d_s = \frac{\sigma_p}{\varepsilon_0} \frac{d_s}{K_s}, \tag{2.1}$$

where E_p is the electric field internal to the photoreceptor, d_s is the thickness

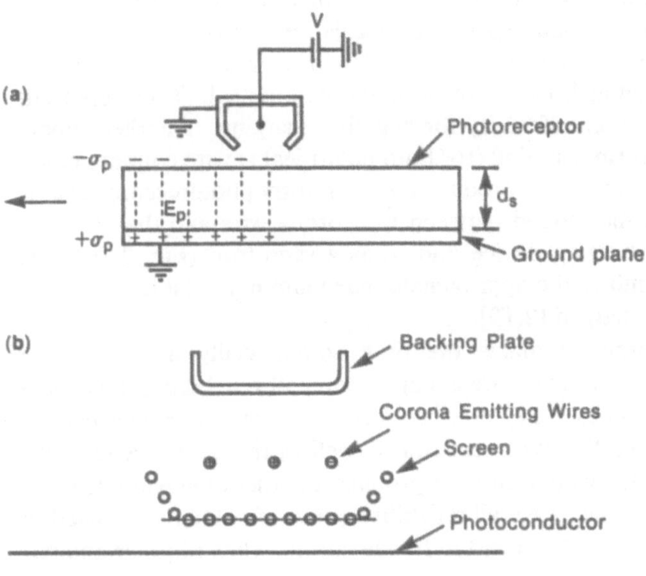

Fig. 2.2a,b. Schematic of devices used for charging a photoconductor. In **a** a corotron is shown; in **b** a scorotron is shown in which a screen is placed between the corona wires and the photoreceptor. The potential of the screen determines the approximate maximum potential to which the photoreceptor will be charged

of the photoreceptor, K_s is the photoreceptor dielectric constant ($\approx$3 for organic photoreceptors), and ε_0 is the permittivity of free space. This gives for organic photoreceptors $\sigma_p = 80\ nC/cm^2$; for amorphous Se photoreceptors (with dielectric strength of 10 V/μm and $K_s = 6.6$) $d_s = 60\ \mu m$ and $\sigma_p = 60\ nC/cm^2$.

A device used to place this charge on the photoreceptor is called a corotron [2.19–21] (Fig. 2.2a). If a sufficiently high voltage is applied between a fine wire and a nearby grounded shield, the air near the wire will become ionized [2.21–24]. Ions of the same polarity as the wire will be swept by the electric field towards the photoreceptor (and the shield). While the shield current is a source of inefficiency, it provides stability to the corona by forcing the operating condition to be far from threshold. The nature of the ionized molecules resulting from the corona has been shown by *Shahin* [2.25–27] to be primarily CO_3^- and $(H_2O)_nH^+$ with n = 4 to 8 for negative and positive coronas, respectively. The nature of the charge on the photoreceptor surface is unknown.

The voltage necessary to create the ionized air is determined [2.21] by the electric field required to accelerate electrons to sufficient velocities to ionize air molecules (Paschen's breakdown) and the distance from the wire at which the average electric field equals the Paschen sparking breakdown voltage. This voltage depends primarily on the wire diameter. Typically, a 50 μm diameter wire spaced 1 cm from the ground plane, operating at a potential of 8000 V, is sufficient to charge a photoreceptor moving at a speed of 5 cm/s [2.19].

The corona, as the discharge is called [2.19,21], appears as a uniform blue-white sheath around the wire for a positive polarity; for a negative wire there are glowing bluish points spaced at regular intervals along the wire. These nonuniformities are due to current avalanches caused by the slow velocity positive ion cloud as it moves towards the wire [2.28]. They represent a significant source of nonuniformity for negative charging. Another important source of nonuniform charging (for both polarities) is wire contamination due to toner and paper dust. To produce more-uniform photoreceptor charging, a screen is sometimes added between the corona wire and the photoreceptor to produce a charging device known as a scorotron (Fig. 2.2b); the screen potential determines the approximate maximum potential to which the photoreceptor will be charged [2.19].

While corona charging remains the only commercially used charging method, corotrons and scorotrons are a major source of reliability problems in electrophotographic engines. They produce corrosive ions and reactive neutral species that can damage the photoreceptor as well as the corona wire itself. Ozone is also produced; negative coronas produce an order of magnitude more ozone than positive coronas. Overall reliability also is affected by the need to clean periodically, without breaking, the fragile corona wires of toner and paper dust.

In addition to dc coronas, ac coronas are sometimes used to lower the toner charge on the photoreceptor (prior to cleaning) and a biased ac corona with a glass enclosed wire [2.28] was recently introduced in the Xerox 1075 copier. This new charging device appears to charge more uniformly and to be more resistant to contamination-induced nonuniformities than standard dc and ac coronas. More detailed discussions of corona devices can be found in the literature [2.2,4,19–31].

2.1.2 Expose

Light, reflected from the image (in a copier) or produced by a laser (in a printer), discharges the normally insulating photoreceptor producing a latent image—a charge pattern on the photoreceptor that mirrors the information to be transformed into a real image.

For a copier, light from a lamp is reflected from the document and imaged onto the photoreceptor. For a printer, light from a modulated laser or a linear array of light sources converts the digital information into a latent image. The energy density and wavelength of the light source must be matched to the sensitivity of the photoreceptor.

Photoreceptor properties are stringent. They must maintain high surface charge densities and correspondingly large internal fields. They must have high quantum efficiency for the conversion of light to electron-hole pairs. They need relatively high mobility for at least one charge carrier so that the charge can traverse the thickness of the photoreceptor within a time short compared to the electrophotographic process time. In addition, they must have low trap densities so that trapped charges due to previous images do not distort subsequent latent images. Finally, they must be durable, nonhazardous, and manufacturable at low cost.

Photoreceptors were originally made by evaporating amorphous selenium onto metallic drums [2.1,2]. Later, organic photoconductors coated onto webs of aluminized Mylar were introduced. These could be made into belts or unrolled from a supply inside a drum. Amorphous selenium [2.32] will transport both holes and electrons, although usually only hole transport is used (with the top surface charged positively). At first single-layer organic photoreceptors were developed [2.33–36]. A wider choice of materials became possible with the introduction of a two-layer organic photoreceptor structure (Fig. 2.3) in which a thin ($<1\,\mu$m) photogeneration layer is separated from a much thicker ($\approx 20\,\mu$m) transport layer [2.37–40]. (A third barrier layer is sometimes added to reduce hole injection from the substrate.) With such a structure the photogeneration and transport layers could be optimized independently. Since most organic transport layers transport only holes and the photogeneration layer is usually placed under the transport layer, the top surface must be charged negatively. While the organics are less expensive to manufacture, they generally have shorter life, due to the softer polymeric material wearing more quickly or the buildup of trapped charge due to photochemistry.

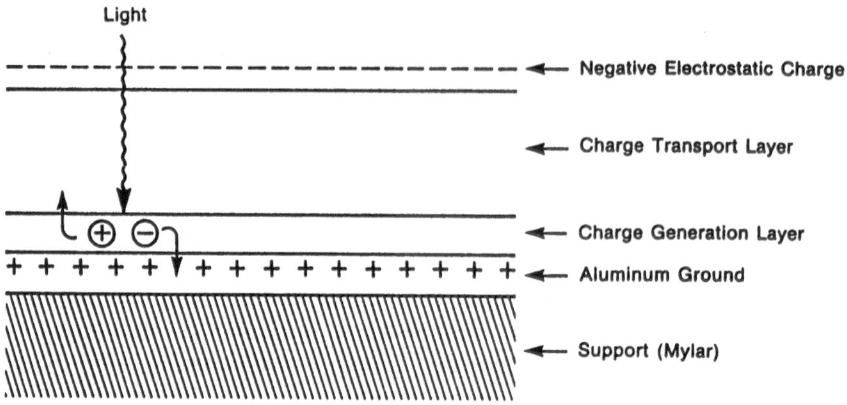

Fig. 2.3. Organic photoreceptors usually have two layers, a lower photogeneration layer less than 1 μm in thickness and a thick (20 μm) upper transport layer

Much of the scientific literature of electrophotography has focused on the microscopic processes occurring inside photoreceptors [2.32–45]. Photons of light produce electron-hole pairs in the photogeneration layer. The electrons and holes are then separated by the internal electric field and the holes drift across the thickness of the photoreceptor (charge transport) in the presence of the electric field. Several aspects of these processes are both remarkable and unexplained. Remarkably, all photoconductors, either inorganic (amorphous selenium alloys) or organic appear to have similar photogeneration and charge transport characteristics. Photogeneration is usually pictured [2.42] as the escape, by random diffusion, of a charge carrier from a Coulomb well. The source of the Coulomb well is the counter charge. The escape process is aided by temperature and the electric field, which lowers the well height. However, comparison of theory to experiments requires a fitting parameter: the initial distance between the charge carriers when diffusion begins. This parameter has yet to be calculated for any system. A list of charge generation materials is given in [2.37,38,46]. Charge transport is usually pictured as a hopping process [2.42], but the field dependence is not understood, the temperature dependence appears to be more complicated than that associated with a simple activated process [2.45], and no association of the measured parameters and the molecules in the system has been made for any photoreceptor. The field dependence of the hole mobility for several important molecules doped into polycarbonate are shown in Fig. 2.4. Another remarkable feature involves trapped charges. After repeated cycling, it usually occurs that some amount of charge becomes trapped in the photoreceptor which light cannot dissipate. The trapped charge per unit volume ρ_T causes a "residual" potential V_R

$$V_R = \frac{e\rho_T d_s^2}{2\varepsilon_0 K_s}, \tag{2.2}$$

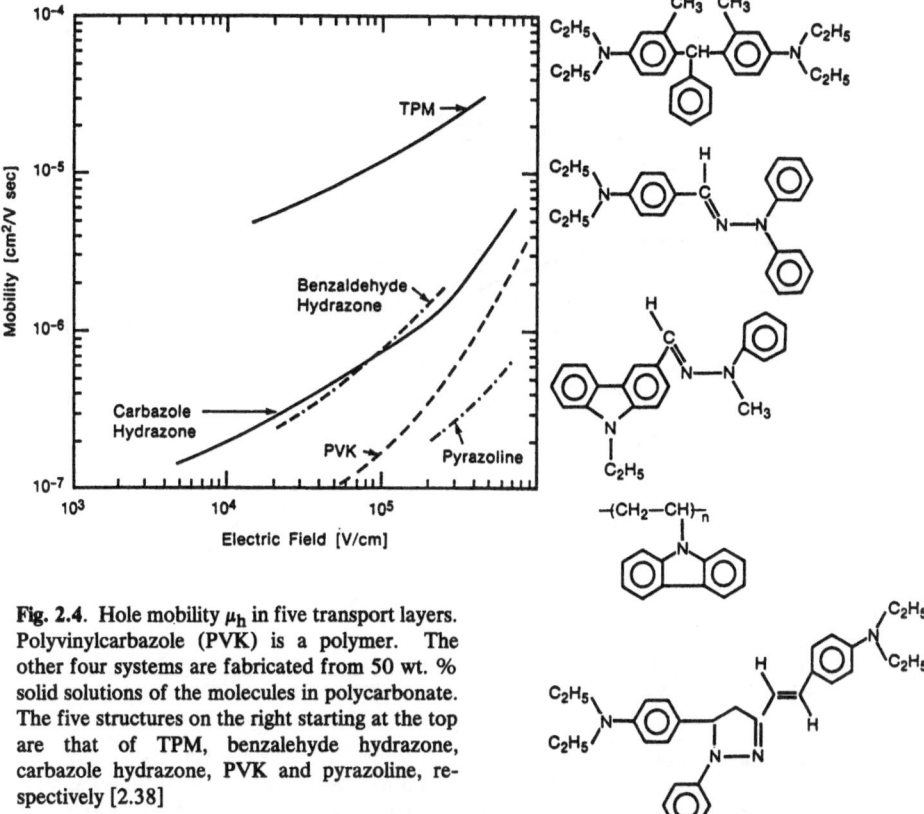

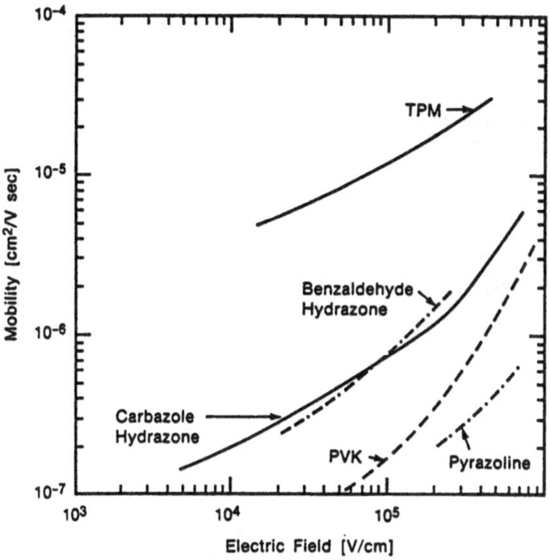

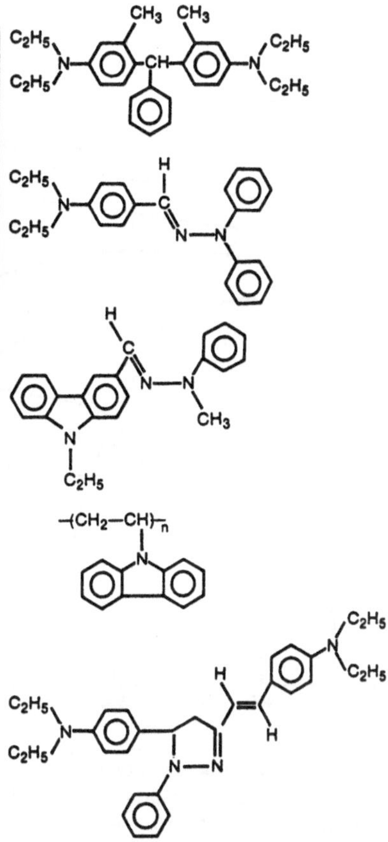

Fig. 2.4. Hole mobility μ_h in five transport layers. Polyvinylcarbazole (PVK) is a polymer. The other four systems are fabricated from 50 wt. % solid solutions of the molecules in polycarbonate. The five structures on the right starting at the top are that of TPM, benzalehyde hydrazone, carbazole hydrazone, PVK and pyrazoline, respectively [2.38]

which must remain small compared to the potential associated with the latent image (approximately 600 V, which is determined by development system requirements). For an estimate, we will assume V_R must remain less than 100 V. Using (2.2), this corresponds to $\rho_T = 8 \times 10^{13}$ cm^{-3} (for $d_s = 20$ μm, $K_s = 3$) for an organic photoconductor and $\rho_T = 2 \times 10^{13}$ cm^{-3} (for $d_s = 60$ μm, $K_s = 6.6$) for an amorphous selenium based photoreceptor. This is an incredibly low level of traps, approximately 1 in 10^8 molecules (for 10^{22} molecules cm^3). That any such material exists in nature is remarkable; that it is a glassy or amorphous material is even more surprising given the solid state physics discussions of bandtail states that are postulated to exist in amorphous materials. Despite this, many photoreceptors have been made with low trap concentrations which, in addition, have quantum efficiencies between 10^{-1} and 1 (Fig. 2.5) all the way into the near infrared (800 nm) [2.6,46], and have reasonable mobilities (Fig. 2.4), which allows charge to move across the thickness of the photoreceptor in times $d_s/\mu_h E_p$ small compared to the electrophotographic cycle time ($\approx$1 s), where μ_h is the hole mobility.

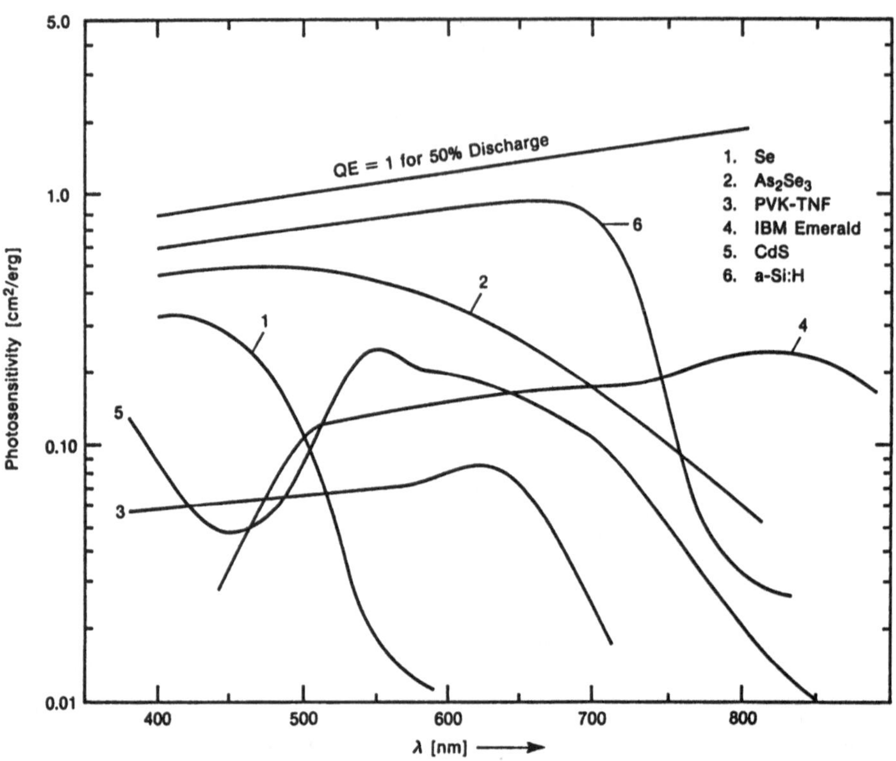

Fig. 2.5. Spectral response of typical xerographic photoreceptors. (1) Selenium; (2) As_2Se_3; (3) PVK:TNF, 1:1 molar complex of polyvinyl carbazole and trinitrofluorenone; (4) IBM Emerald; (5) CdS-binder coating; (6) amorphous Si:H. Curves 1–3 and 5 are taken from [2.6]; 6 is taken from [2.47]. The absolute value of the sensitivities of these curves may differ by a factor of 2 due to the different experimental definitions of sensitivity (to 50% or 90% discharge) Also shown is the curve for unity quantum efficiency, assuming 50% discharge.

2.1.3 Develop

Electrostatically charged and pigmented polymer particles, called toner, ≈10 μm in diameter, are brought into the vicinity of the latent image. By virtue of the electric field created by the charges on the photoreceptor, the toner adheres to the latent image, transforming it into a real image.

The known development systems are described in Chap. 3; the physics of these development systems are described in detail in the remainder of this book.

All of the development systems used in electrophotography have in common the fact that the deposition of charged toner onto the photoreceptor is controlled by a variation in the electric field (whose source is the variation in the local photoreceptor charge density produced during exposure). Other de-

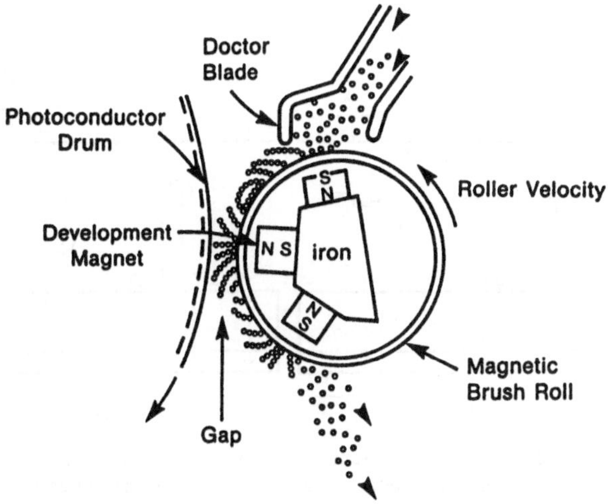

Fig. 2.6. A schematic of the magnetic brush development system. Carrier (with attached toner) is brought near a rotating roller within which are stationary magnets. Friction forces, caused by the magnetic attraction of the carrier to the roller, transport the carrier particles around the roller into the development zone

velopment technologies exist in principle: the latent image could be created as a variation of toner charge in a constant electric field, the basis of photoelectrophoresis imaging (Sect. 1.4 and [2.2,6,48–50]). Development systems based on the fact that uncharged polarizable toners are attracted to nonuniform electric fields have also been suggested [2.51,52].

One important example of a development system is the magnetic brush development system, which is used today in almost all copiers and printers operating above 30 cpm. This system is shown schematically in Fig. 2.6. In this development system, the toner is electrostatically charged by mixing it with another powder, called carrier. The toner has an average diameter of 10 μm and is a polymer loaded with carbon black as a colorant. The carrier is an approximately 200 μm diameter magnetically soft sphere coated with a polymer chosen to correctly charge the toner on contact. By virtue of the magnetic properties of the carrier beads they are transported by friction forces around the roller, which has stationary magnets within it. In the development zone or gap, the radial magnetic field causes the carrier beads to form a chain; at the end of the chain, toner contacts the photoconductor and "develops," i.e., transfers from the carrier to the photoreceptor, if conditions are "correct". The "correct" conditions are defined quantitatively in the following chapters. To enhance toner development, multiple rollers can be used, with the developer mix passed magnetically from one roller to another across inter-roll gaps.

The latent image, consisting of surface charge patterns on the photoreceptor, produces electric field lines connecting the surface charges to image charges, which reside in the ground plane of the photoreceptor in a config-

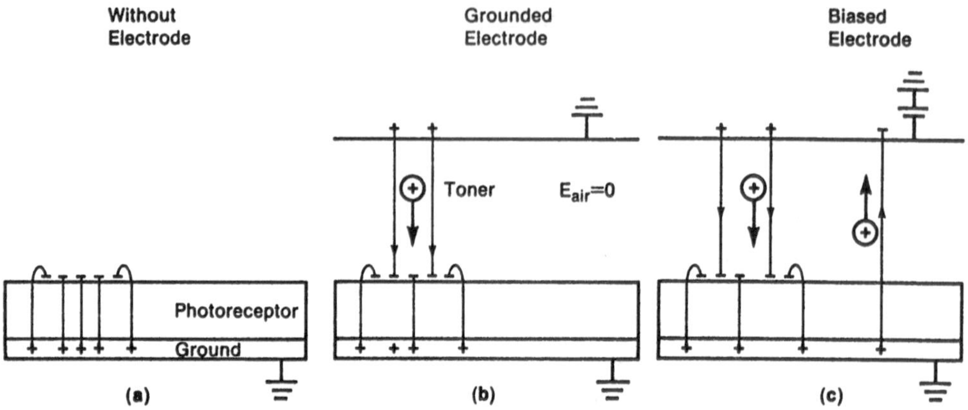

Fig. 2.7. a The external electric field associated with a latent image only appears in the form of fringe fields around the edges of the charges in the absence of an electrode. **b** In the presence of an electrode some electric field is capacitively coupled to the electrode. **c** A bias on the electrode creates a gray scale control and can be used to lower background development

uration without an electrode (Fig. 2.7a). External electric fields appear only in the form of fringe fields around the edges of the charged areas which extend above the photoreceptor a distance comparable to the thickness of the photoreceptor (Sect. 3.2). Close approach of a grounded counter electrode (Fig. 2.7b), such as the roller in the magnetic brush development system, draws field lines outward in proportion to the ratio of the capacitance per unit area of the photoreceptor and the counter electrode. Charged toner introduced into this region (during development) is attracted by the electric field to the photoreceptor. The electrode allows additional functions. For solid area and background latent images (Fig. 2.7c), a bias on the electrode can (1) create a force drawing toner away from background regions, leading to lower background development and (2) decrease the electric field above solid area latent images, leading to a gray level control. A bias control on the development electrode is very commonly used on copiers as a gray level control. Note that these two effects occur together: changing the bias to give lighter (darker) images will usually give lower (higher) background.

To minimize cost, toner usually has a wide size distribution (Fig. 2.8a). Size-classified toner (Fig. 2.8b) and spherical toner (Fig. 2.8c) are also available for experimental purposes. A higher magnification SEM of 10 μm size-classified toner is shown in Fig. 2.9a. After appropriate mixing with the carrier, the toner electrostatically adheres to the carrier beads, as shown in Fig. 2.9b.

Once the polarity of the photoreceptor charging is determined (negative for organic photoreceptor, positive for amorphous selenium), the polarity of the toner charge is determined for a copier (opposite to the photoreceptor charge). Since the latent image comes from the originally charged areas of the photoreceptor, this is called charged area development, or CAD (Fig. 2.10). In a

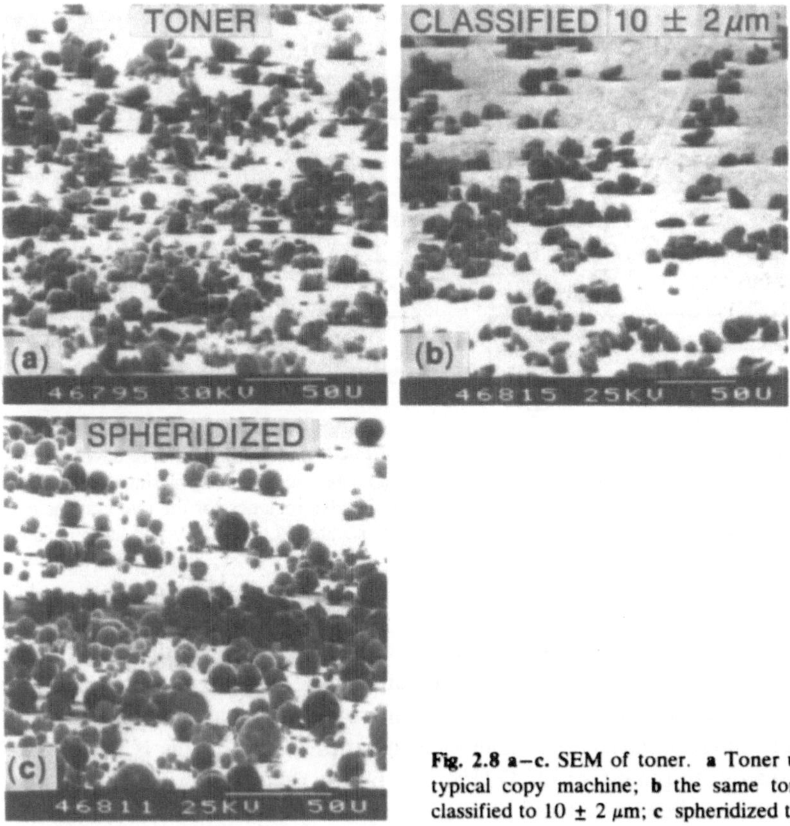

Fig. 2.8 a–c. SEM of toner. **a** Toner used in a typical copy machine; **b** the same toner size-classified to 10 ± 2 μm; **c** spheridized toner

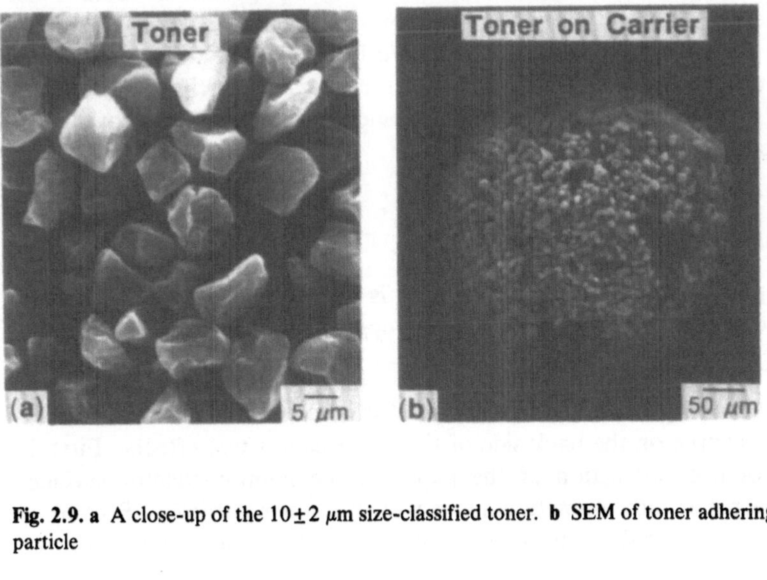

Fig. 2.9. a A close-up of the 10±2 μm size-classified toner. **b** SEM of toner adhering to a carrier particle

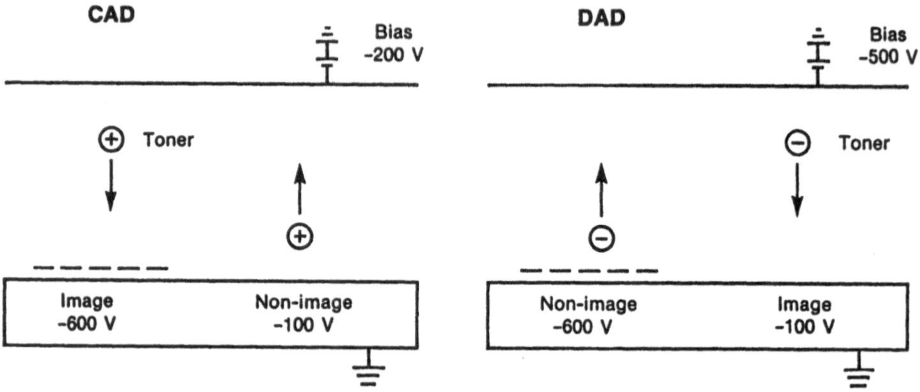

Fig. 2.10. A comparison of charged area development, CAD, and discharged area development, DAD. The direction of force on the toner particles is indicated. By properly biasing the magnetic brush roller and changing the sign of the toner charge the electrostatics for the two systems can be made identical. It is assumed that the photoreceptor is charged to −600 V and has a residual potential of −100 V.

printer, the laser (or LED array) can discharge either the white areas (emulating a copier) or the image areas. In the latter case, the latent image is in the discharged areas. By changing the bias of the roller in the development system and toner charge polarity, the same electrostatics results. This is called discharge area development, or DAD (Fig 2.10). The relative merits of these two approaches have been discussed [2.53]; usually they come from secondary effects on other subsystems. For example, in DAD the photoreceptor dielectric strength uniformity is critical; otherwise the small fringe field associated with breakdowns are developed by the toner. In CAD, the light source lifetime can be an issue because it is on approximately ten times longer than in DAD, as most of a page is usually white. It is shown in [2.53] that wider surface potential latitudes exist in DAD. When a printer is made by changing the light source on a copier, CAD is easier to implement, since toner and corona polarities need not be changed.

2.1.4 Transfer

The developed toner on the photoconductor is transferred to paper by corona charging the back of the paper with a charge opposite to that of the toner particles.

A qualitative description of the transfer process is given by *Andrus* and *Hudson* [2.54]. The charge on the back side of the paper causes two effects. First, it creates a force of attraction of the paper to the photoconductor surface, bringing the paper into intimate contact with the toner particles. Second, it creates a force pulling the toner towards the paper. Normally, some toner size classification is observed during the transfer process; the average transferred toner diameter is slightly larger than the average developed toner diameter. It is suggested this is due to the larger adhesion force of smaller diameter toner to the photoreceptor. This subsystem can actually improve copy quality because it rejects wrong-sign toner present either in the background or attracted to the reverse fringe fields near character edges.

The physics governing transfer is probably related to the competition between the force on toner due to the electric field caused by charge on the paper, and the adhesion force of toner to photoconductor. This adhesion problem has been the subject of numerous investigations [2.55–64] with controversy over whether the adhesion is due primarily to image forces or van der Waals short range forces. The arguments seem to come out in favor of the image forces with the caveat that nonuniform charge on the toner surface probably needs to be taken into account [2.57, 62, 63].

Unfortunately, very little work has been directed towards a detailed analysis of the transfer step. A good start has been made by *Yang* and *Hartmann* [2.65] in which they present a transfer model in which toner cohesion is regarded as a constant, i.e., the presence of the photoreceptor is ignored. They predict the minimum transfer voltage needed to move the interface of transferred to untransferred toner to the photoconductor surface (in their model, when the predicted transfer efficiency is 100%). As might be expected, this voltage is related to the total toner charge and the ratio of the paper to toner dielectric thicknesses. Under the condition of low transfer efficiency, there is experimental support for their model. Characterization of transfer systems also has been carried out by *Hida* et al. [2.66]. They point out that since the Coulomb force causing transfer is linear in the toner charge, and the adhesive force is quadratic in toner charge, there is predicted, and observed, an optimum toner charge for transfer. The optimum toner charge-to-mass ratio Q/M equals

$$Q/M = \sigma_{tr}/2\rho_t p_t d_t, \qquad (2.3)$$

where σ_{tr} is the charge per unit area on the paper back, ρ_t, p_t, d_t are the developed toner density, packing, and thickness, respectively. For $|\sigma_{tr}| = 32$ nC/cm^2, independent of the sign of toner or whether the toner originated from a two component or monocomponent development system, it is predicted and observed that the optimum Q/M is between 14 and 20 $\mu C/g$, consistent with (2.3) for a toner layer thickness of 30 μm.

The maximum transfer efficiency observed by *Hida* et al. is 80%, consistent with generally observed values of typically 80%–85%. It is not known whether the observed 85% transfer efficiency represents a fundamental limit.

2.1.5 Fuse

The image is permanently fixed to the paper by melting the toner into the paper surface.

The various methods of fusing usually require the use of heat to soften the toner and allow it to melt and flow into the paper fibers. While engineering analyses of the thermodynamics of these processes are available, the central problems here are material problems. Some examples: the toner's thermal properties are a compromise between being soft for the fuser, but hard enough so that they are not "fused" in the development system, on the photoreceptor, or in storage in a hot environment. Roll fusing (squashing the toner on paper between two rollers at least one of which is hot) requires polymeric materials to which the toner will not stick and which can withstand high temperatures (180°C) and stress for many hundreds of thousands of copies, a nontrivial requirement for polymers. The toner rheological properties must be optimized to allow coalescence, spreading, and finally penetration into the paper [2.67].

After fusing, the black toner decreases the reflectance of the paper, giving the perceived blackness of the image. The relationship of the reflectance of toner on paper to the deposited toner mass per unit area has been studied from first principles by *Castro* and *Lu* [2.68]. They take into account light spreading through the paper bulk and optical characteristics of the equipment (which usually focuses a collimated beam of light onto paper at 45° relative to the paper normal and collects reflected light close to normal). They show that the optical reflection density D is

$$D = -\log_{10}(R_p \exp[-2.88\,(M/A) < A/M >] + R_T\{1 - \exp[-(M/A) < A/M >]\}), \tag{2.4}$$

where the number 2.88 results from the analysis of the enlargement of the toner area by shadowing, M/A is the toner mass per unit area, $<A/M>$ is the mean toner projected area divided by the mean toner mass, R_p is the reflectance of the paper and R_T is the reflectance of the toner. Under the condition that the second term can be neglected, this can be simplified to

$$D = -\log_{10} R_p + 1.25\,(M/A) < A/M >, \tag{2.5}$$

demonstrating the linearity of D with M/A, which is generally observed (Fig. 2.11) for M/A below 0.8 mg/cm^2.

2.1.6 Clean

The photoconductor is discharged and cleaned of any excess toner using coronas, lamps, brushes and/or scraper blades.

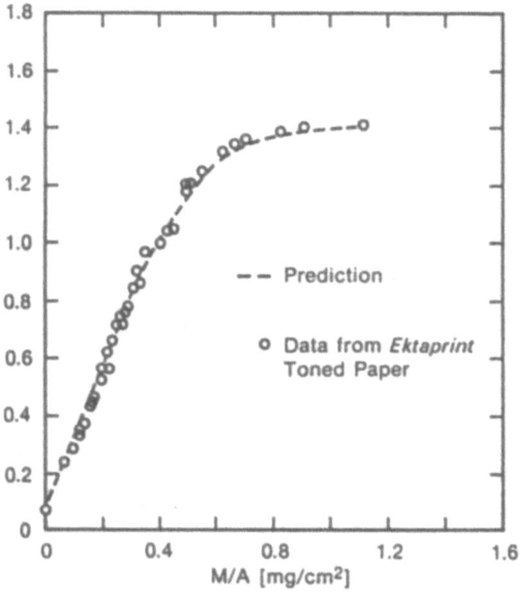

Fig. 2.11. Reflection density versus mass per unit area. Data are represented by o's. Solid curve calculated from (2.4) [2.68]

All untransferred toner and all excess charge must be removed from the photoreceptor before the process starts again. Several subsystems are generally used: light to discharge the photoconductor (decreasing the toner's electrostatic adhesion), ac coronas to bring the toner charge near zero, and finally a brush to wipe off the excess toner. A system to clean the cleaner is also necessary. For a brush cleaner this usually involves a vacuum system plus filters. As occurs in the transfer step, the physics of the cleaning process is probably related to a competition between toner adhesion to the photoreceptor and the forces on the toner caused by the cleaner. Systematic studies of the physics of cleaning are limited [2.69–74]. *De Palma* [2.70] has shown that the cleaning ability of a rotating brush depends primarily on the number of fibers that collide with the photoreceptor. *Harpavat* [2.71] has analyzed the mechanics of blade cleaning and *Abowitz* [2.73] and *McMillen* and *Salamida* [2.74] have dealt with the material requirements. It has been shown that the magnetic brush development system, biased appropriately, can be used as a cleaner, i.e., the toner can be "developed" from the photoreceptor to the carrier beads [2.75]. Presumably the physics of magnetic brush cleaning is related to the physics of magnetic brush development, although no discussions have yet appeared in the literature.

2.2 Implementation–Interactions

The six process steps of electrophotography are illustrated in Figs. 2.12–14 for three commercially available copiers, an IBM Series III, an Eastman Kodak

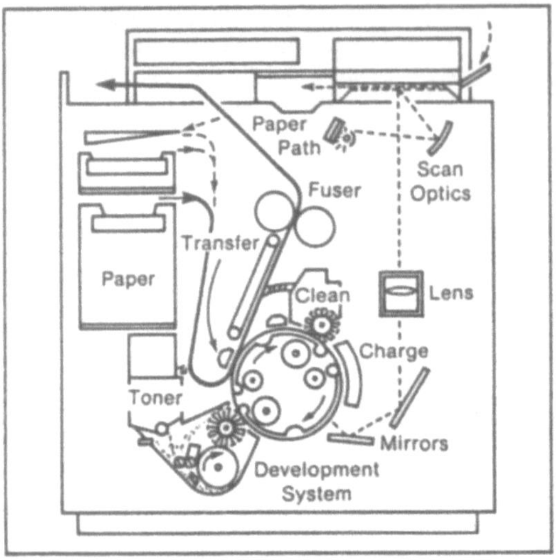

Fig. 2.12. Schematic of the IBM Series III copier

Ektaprint 150, and a Xerox 1075, all of which run at 70 copies per minute. A comparison of these three machines reveals interesting architectural design choices.

First, note the position of the six steps in each machine. The IBM Series III copier charges a photoreceptor drum at the 3 o'clock position. Exposure is achieved with scan optics at 5 o'clock. Development is at 7 o'clock, the transfer corona is at 9 o'clock, fusing is above the drum and cleaning is done with three elements near 12 o'clock. In the Eastman Kodak Ektaprint 150 copier, a photoreceptor belt is used. Charging is at 5 o'clock and exposure is achieved with a flash at the bottom of the belt, followed by development. Transfer is at 12 o'clock and cleaning is at 2 o'clock. The Xerox 1075 copier also uses a belt photoreceptor. Charging is at the top right followed by flash exposure. Development is at the bottom, transfer is on the right side and cleaning is at the top right.

Consider the photoreceptor. Note that the photoreceptor is a drum in the IBM Series III, but a belt in the other two machines. The IBM machine has a roll-up feature inside the drum so that, when new photoreceptor is desired, the push of a button allows new photoreceptor to unroll. The Xerox and Eastman Kodak machines require a belt change, i.e., the storage of fresh belts and the services of a trained technician to change it. Note the IBM system uses scan

Fig. 2.13. Schematic of the Eastman Kodak Ektaprint 150 copier [2.76] (© 1977 IEEE)

Fig. 2.14. Schematic of the Xerox 1075 copier [2.77].

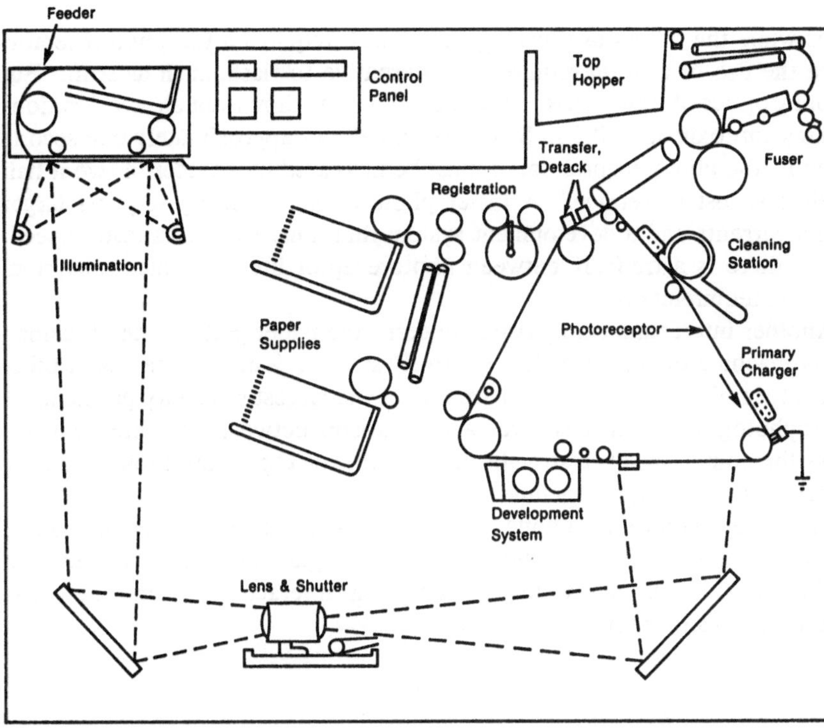

Fig. 2.13

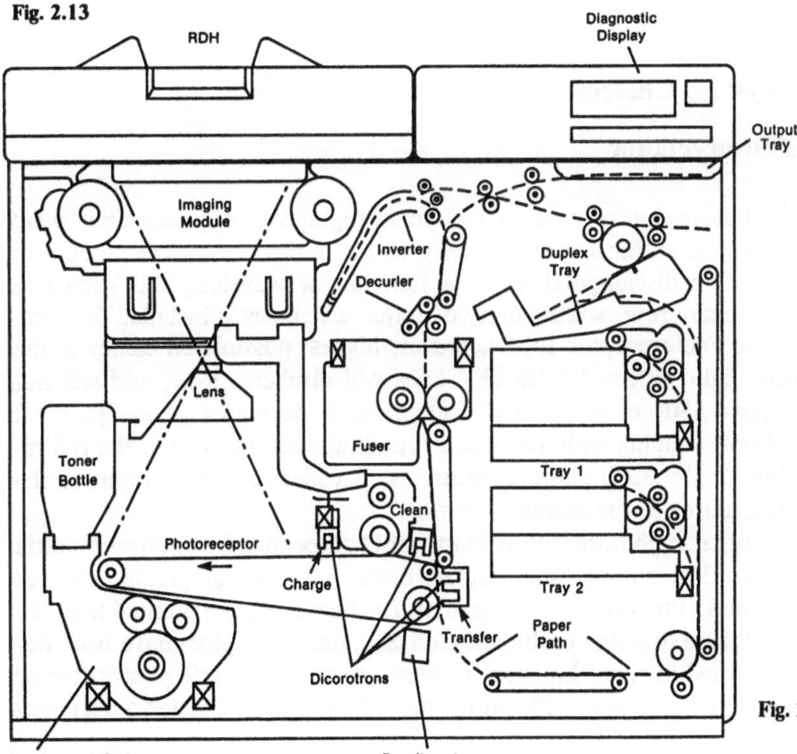

Fig. 2.14

optics while the others use flash optics. Flash requires a flat photoreceptor; hence the use of a drum requires an alternative to flash, such as scan. But scanning optics of finite mass across a document with reasonable accelerations requires approximately 0.5 s. Since 70 cpm allows approximately one second per copy, the photoreceptor speed must be increased by a factor of two if half the time is lost to retrace! A faster photoreceptor velocity requires larger corona currents and a development system that must work at a faster speed. Hence, there is a tradeoff between photoreceptor life and other subsystem operating requirements.

Another interesting comparison concerns the paper path. In the Eastman Kodak machine the paper path is relatively straight, a design one would think is ideal for reliability (paper jams are one of the largest reliability problems of electrophotographic engines), yet a comparison between the three engines shows the compromise: a straight paper path, for top optics, requires a long optical path and a larger machine.

This interrelationship among subsystems is a pervasive phenomenon in electrophotography; it limits choices of technologies for each subsystem and requires considerable unravelling of subsystem interactions when a new machine is put together for the first time.

2.3 Subsystem Choices

2.3.1 Photoreceptor

The first photoreceptors used in electrophotographic copy machines were amorphous selenium films approximately 60 μm in thickness. The amount of light required to discharge them as a function of wavelength is shown in Fig. 2.5. A curve representing unity quantum efficiency is included for comparison. The photoreceptor approaches the highest possible efficiency in the blue (hence light reflected from blue lines will discharge a-Se, and will not copy). Above 550 nm its sensitivity significantly decreases. Later [2.78] it was found that alloying with Te or compounding with As extends the red response (Fig. 2.5). Such photoreceptors use a wider spectral content of the illumination source for photogeneration.

Other inorganic photoreceptors exist. ZnO suspended in a polymer matrix [2.17] was used in special-paper copiers before electrophotographic engines were developed. Photoreceptors based on ZnO have never exhibited long life because of charge trapping problems. CdS_nSe_m photoreceptors have been developed and used in some Japanese copiers. A new inorganic photoreceptor development that may be significant for the future is hydrogenated amorphous silicon (a-Si:H) [2.47,79–81], a material produced by the rf glow discharge decomposition of SiH_4. The reported advantages of this photoreceptor include excellent photosensitivity over a broad spectral range (450–750 nm), which can be extended into the infrared by alloying with Ge, high surface hardness, mechanical strength, and thermal stability. Because of its hardness it is hoped that amorphous silicon photoreceptors will have useful lives of 10^6 copies or higher, ten times longer than the life of amorphous Se photoreceptors. At this time, the manufacturing cost of amorphous Si appears to be its major problem: the process is a plasma discharge of silane (SiH_44), which requires several hours using capital-intensive equipment.

Organic photoreceptors were first introduced into electrophotographic engines by IBM. At the time of writing, virtually all manufacturers use the same two-layer structure but have their own materials system. The photoreceptor has a thin ($< 1\ \mu m$) photogeneration layer which is coated onto aluminized Mylar. The upper layer is a $\approx 20\ \mu m$ transport layer, made from a molecularly doped polymer. Several examples of the materials used in the transport layer are shown in Fig. 2.4; a comparison of the photosensitivity (energy density required for discharge) of several photoreceptors is presented in Fig. 2.5. Again, quantum efficiencies approaching 1 are achieved.

The use of semiconductor diode lasers with output wavelength near 800 nm in electrophotographic printers has required the development of infrared-sensitive photoreceptors. The search has centered around two technical approaches: (1) extending the spectral response of inorganic photoreceptors through alloying, and (2) finding new photogeneration molecules for organic photoreceptors. Several investigators have reported success in producing adequate sensitivity near 800 nm in selenium [2.82,83], cadmium selenide [2.84], and arsenic triselenide [2.85] by addition of tellurium. Examples of infrared-sensitive organic photoreceptors include the new Ricoh azo-based charge generation material [2.86], the IBM [2.87], Xerox [2.88,89], and Pitney Bowes [2.90] work on squarylium dye systems, and the Kodak trimethine thiopyrylium dye [2.37,91]. Novel organometallic photoreceptor systems have also been described [2.92–95]. These include photoreceptors based on sublimed chloroaluminum phthalocyanine chloride [2.92], chloroindium phthalocyanine [2.93], and a complete two-layer infrared photoreceptor with ε-type copper phthalocyanine as the charge generation material [2.95].

2.3.2 Charge

Charging is almost always done with a corona, as described above. However, several novel approaches have been discussed. For example, a conductive brush could be biased appropriately and rotated against the photoreceptor [2.19,96]. Even the magnetic brush development system is known [2.97] to charge the photoreceptor during contact. The challenge is to develop an alternative to a wire which has better life, charging uniformity, and cost.

2.3.3 Light Source

The light source for exposure in a copier is an electric flash tube for flash exposure or a tungsten-halogen lamp or fluorescent lamp (for low speed copiers) for scan exposure.

A printer requires a bright light source such as a laser to convert the data bit stream into the latent image. The bit stream controls a modulator (for a gas laser) or the current (for a semiconductor laser). A schematic of the optics used for a gas laser is shown in Fig. 2.15. The acousto-optic modulator [2.4,98], required for the HeNe laser, consists of a piezoelectric transducer bonded to a transparent material whose refractive index is sensitive to pres-

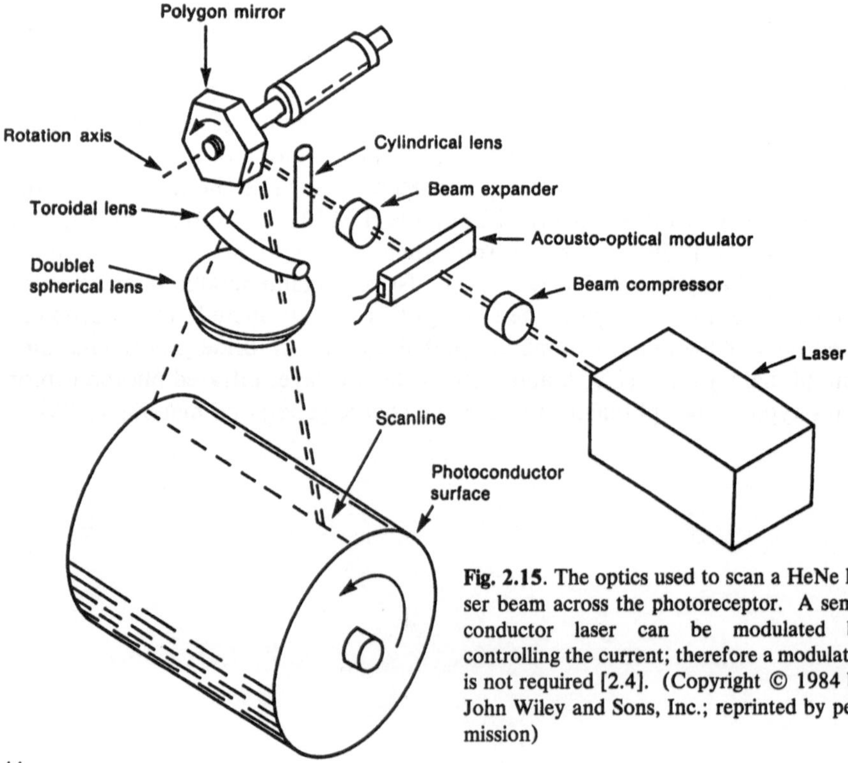

Fig. 2.15. The optics used to scan a HeNe laser beam across the photoreceptor. A semiconductor laser can be modulated by controlling the current; therefore a modulator is not required [2.4]. (Copyright © 1984 by John Wiley and Sons, Inc.; reprinted by permission)

sure. The acoustic waves launched by the transducer set up a spatial periodic variation of the index of refraction, which acts as a diffraction grating, deflecting the laser beam past a stop. The laser is scanned across the photoreceptor with a spinning polygon mirror. The polygon mirror typically has 8–36 facets and rotates at speeds of 2 000–50 000 rpm. Usually a single-beam system [2.98] is used, but multibeam configurations have been discussed [2.99]. The beam leaving the facet has constant angular velocity which must be transformed into a constant linear velocity across the photoreceptor, sometimes accomplished with an f-θ lens. Sources of error [2.100] include nonplanarity of the facet surface, solved with anamorphic optics [2.101], variations of polygon rotational velocity, up and down wobble, and lack of synchronization between scan rate and pixel clock.

The laser choice and photoreceptor choice are another example of a coupled electrophotographic problem. Table 2.1 lists the photoreceptors and lasers used in several electrophotographic printers. Because of the sensitivity of amorphous selenium, a HeCd or argon laser has been used in the Xerox 9700 printer. Most other printers use either a HeNe laser (632.8 nm) or a semiconductor GaAlAs laser (≈800 nm). The advantages of the GaAlAs laser, including its small size and elimination of the modulator, suggest increased use in future printers.

A printer also can be made by exposing the photoreceptor with a linear array of light elements. $GaAs_{1-x} P_x$ light emitting diode (LED) arrays arranged in one or two stitched rows to form one entire print line have been discussed [2.102–107]. IBM (Model 3812) based on Kentek K-2, NEC (Model LC-800) and Eastman Kodak (Model 1392) have recently announced printers using LED arrays. Printers using linear shutter arrays based on liquid crystals [2.108,109] have been built by Casio (Model LCS-2400) and Epson (Model GQ-3000). Magneto-optic [2.110-112], electro-optic [2.113], and electroluminescent [2.114] arrays also have been described. Critical to all linear array technologies are solutions to the interconnect and packaging

Table 2.1. Well-known laser printers

	Speed [ppm]	Photoreceptor	Laser	Resolution [dpi]	Laser Power [mW]
IBM 3800	215	Organic	HeNe	240	30
Xerox 9700	120	a-Se	HeCd/Ar	300	10
Siemens ND-3	103	a-As_2Se_e	HeNe	240	7
Hewlett-Packard 2680	45	CdS	HeNe	180	5
Canon LBP-CX	8	Organic	GaAlAs	300	3

problems and output uniformity among elements sufficient to eliminate any signature of the individual elements in the final copy.

2.3.4 Develop

There are three types of development systems used in electrophotography, two component, single component, and liquid.

The two component systems usually are used for high speed machines. They are called two component because the developer mix has two components, toner and carrier. The toner particles are "developed" onto the latent image on the photoreceptor and are then transferred and fused on paper. The purposes of the carrier beads are twofold: to charge the toner by static electrification and to transport the toner into the vicinity of the latent image. Examples of two component systems include cascade development (Chap. 5), insulative magnetic brush development (Chap. 6), and conductive magnetic brush development (Chap. 7). Cascade development was used in the first commercial automatic copier (the Xerox 914). In this system, the carrier beads are literally cascaded with the aid of gravity down the photoreceptor surface. Magnetic brush development, described above, is used in one of the fastest electrophotographic laser printers (the IBM 3800). Such systems require that the toner concentration (fractional mass of toner to carrier) be maintained nearly constant. This requires sensing of depleted toner, and the addition and mixing of fresh toner.

Conceptually simpler are monocomponent development systems in which carrier particles are eliminated. There are three independent choices to be made in the systems. The toner could: (1) be conducting or insulating, (2) be magnetic or nonmagnetic, and (3) contact or jump across a gap to the photoreceptor. Conducting toner monocomponent systems have been found to be unsatisfactory because at high humidity paper becomes conductive enough to bleed off the toner charge, decreasing the transfer efficiency (Sect. 9.4). The most important moncomponent system today is insulating, magnetic, and jumping, used by Canon in their whole line of copiers (Sect. 9.5). With magnetic material in the toner, a magnetic transport, similar to the two component system, can be used. Under active development at several laboratories are nonmagnetic systems (Sect. 9.6). They offer the advantage of less expensive toner that can be made in different colors.

The most important liquid development process is based on electrophoretic particle migration (Chap. 10 and [2.47–49]) in which submicron-sized toner particles are dispersed in a dielectric liquid such as kerosene, and charged by selective adsorption of ionic species. The dispersion is applied to the photoreceptor and the toner particles are attracted by the electric field of the latent image. The hardware necessary is simple since no mechanical transport device is needed and no fusing is necessary. The process also has high resolution capability since much finer toner particles can be used (less than 1 μm as compared to 10 μm in a powder system). Its principal drawback is retention in the paper of solvent, which is slowly released into the environment. This liquid development system is the one used by Ricoh (and others) in low speed, low cost copiers first introduced in the 1970s. Its high resolution and color capability has led Kodak [2.115,116] recently to introduce a pre-offset proofing system based on liquid development. Another liquid development process has been described by *Gundlach* [2.117]. Conductive ink is drawn out of the recesses of a finely patterned conductive applicator, e.g., a gravure roll, onto the photoreceptor in response to the electric field of a latent image. *Gesierich* et al. [2.118] have described selective wetting development: liquid developer is applied uniformly to a low surface energy photoreceptor, but retained selectively only in charged areas. The effect may be understood in terms of an electrostatic enhancement of the effective photoreceptor surface energy. *Moradzadeh* [2.119] has described a water-based system in which the ink is applied to a photoreceptor covered with a thin oil layer; the electric fields between the latent image and the aqueous ink create electrohydrodynamic instabilities that displace the oil, bringing the ink to the latent image. The residual ink is wiped off.

2.3.5 Transfer

Transfer of the developed image to paper is usually accomplished by corona charging the back of the paper (Sect. 2.1.4). A biased roller has sometimes been used to create an electric field across the paper-toner-photoreceptor sandwich. This requires careful control of the resistivity of the roller; if it is too low, air breakdown will occur between the roller and the back of the paper; if it is too high, the potential on the roller will not reach the applied potential during transfer and the transfer efficiency will not be optimized. Recently, *Butler* et al. [2.120] theoretically studied transfer under the conditions of an infinitely conducting roller; under such conditions toner can become airborne as the roller-photoreceptor gap decreases and the electric field becomes large enough to overcome toner adhesion to the photoreceptor. Roll transfer is sometimes used in magnetography [2.121].

2.3.6 Fuse

Following *Lee* et al.'s discussion [2.10], the toner image transferred to paper from the photoconductor must be permanently bonded or fused by inducing the powder to coalesce and adhere to the substrate [2.67]. Most common materials having these characteristics are based on thermoplastic resins. They can be fused using contact methods such as wicked or dry hot roll and cold pressure. Noncontact methods include radiant heat, chemical vapor, and flash.

In the most prevalent technology, hot roll fusing, toned paper passes between a pair of rotating rollers, at least one of which is heated. The heat melts the toner and roller pressure pushes it into the paper. Although the technique is straightforward, care must be exercised to prevent the toner from sticking to the roller and consequently creating an offset or ghost image on the next sheet of paper. The problem can be addressed by either using a special oil wicked onto the surface of the roll [2.122,123] or choosing the roll material to have the desired release properties [2.124,125].

Toner can be fused without heat by supplying sufficient pressure between a pair of metal rollers. Although cold pressure fusing requires very little energy, it has not been widely accepted because of some copy quality drawbacks. Foremost are the glossy appearance of the print and the background paper from the high compression. This has led to innovations in textured coatings on the roll to simulate the appearance of hot roll fused copy [2.126].

Hot roll and cold pressure require physical contact between the substrate and the roller, and thus these approaches limit the range of media that the machine can handle. This has sparked interest in reviving and improving older noncontact methods. The Hewlett-Packard 2680 Laser Page Printer [2.127] uses a two-stage system. The toned paper is first pre-heated conductively as it passes over a hot platen. Final fusing is accomplished by radiant heat transfer from a quartz halogen lamp. The hot platen insures that the paper is warm so that it absorbs little lamp radiation compared to the toner during the second stage. A serious safety problem with radiant heating in the past has been the ignition of paper when jams occur. This deficiency has largely been eliminated by advances in paper sensing and control techniques [2.128]. Siemens [2.129] recently introduced a laser printer with a fusing technology similar to some of the first electrophotographic systems. In operation the paper web passes through a vapor that melts the toner. The principal concern with this technique is similar to that in liquid development: absorption of solvent into the paper and its subsequent evaporation. This has apparently been reduced by bringing the roller paper after fusing into a refrigerated area that lowers the vapor pressure, condensing and capturing most of the organic solvent [2.130]. Another noncontact method is flash fusing, which uses short pulses from a high-power lamp [2.131–133] to deliver radiant energy to the toner. High hardware cost has limited this method to high-end applications.

One important consideration in selecting the fuser technology is its effect on paper properties. In roll-type fusing, the pressure deforms the paper fibers, leading to changes in the paper dimensions and surface smoothness. Heat drives off the natural moisture. These factors cause many reliability problems in the paper path through the fuser and into the subsequent output or duplexing operation. In the usual approach, paper transport is managed by a judicious combination of hard rolls and those with uniform deformable coatings [2.134]. An interesting alternative is found in the Kodak Ektaprint 250 duplicator, which introduced automatic single-pass duplex copying. In this machine the toned image is transferred to both sides of the paper before it goes through the fuser. The station consists of two wicked hot rolls, each of which is covered with a sophisticated multiple-layer material to fuse the toner on both sides simultaneously without smearing or offsetting the powder [2.135,136].

2.3.7 Clean

The simplest method of cleaning the photoreceptor of untransferred toner is to scrape the photoconductor with a blade [2.71]. In low-end applications, a blade cleaner is ideal. It is simple, inexpensive, and very compact. If usage is high, the blade and the photoconductor must be changed more frequently. Thus at the mid range and above, a mechanical brush with vacuum assist is most common [2.70]. Xerox recently introduced in their Xerox 1075 copier a magnetic brush cleaner [2.75] similar in principle to the developer. This cleaner is much quieter and more compact than a fiber brush cleaner with vacuum assist.

3. The Development Step

Mastery of the development step in the electrophotographic process is challenging for three reasons: (1) this step usually determines the best image quality the copier or printer will produce, (2) significant aspects of the physics of development are not well understood, and (3) as a result, significant empirical material and hardware parameter searches are standard procedure in optimizing a development system for a new copier. Reducing the "black magic" of the development system should result in a better, more efficiently optimized system. Therefore the focus of the reminder of this book is on a thorough description of our current, but limited, understanding of the physics of the development process. We describe and compare known development systems in this chapter. Detailed discussions of the physics occurring in each development system and their associated toner charging mechanism are the subjects of succeeding chapters.

3.1 Challenges

The copy quality challenge results from the fact that the development step determines the best image quality a copier produces. The blackness of line copy and the blackness and smoothness of solid areas are determined by the amount and uniformity of toner developed onto the latent image. The whiteness of the nonimaged areas (the background) is determined by the amount of toner developed onto the nonimage areas of the photoreceptor. A casual glance by the reader will reveal that almost all copies made by electrophotography have a light gray appearance in the nonimage areas caused by unwanted toner. Comparison with offset print quality (from a book or catalog) dramatically reveals the difference.

Two quantum leaps in copy quality have been the direct result of the introduction of new development systems. The cascade development system (Chap. 5) was used in the first automatic copier, the Xerox 914. Only the edges of solids were reproduced and background development was high by today's standards. A test pattern reproduced on a copier with a cascade de-

Fig. 3.1. A test pattern including solid area, typical characters and background reproduced on copiers using: (a) cascade development, (b) insulative magnetic brush development, (c) conductive magnetic brush development

Cascade

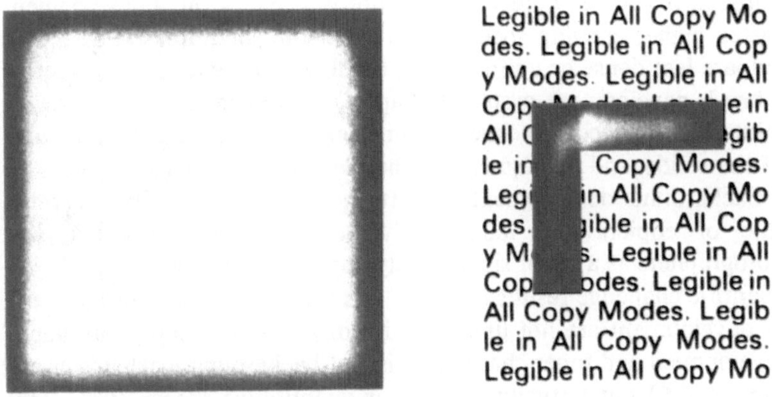

Insulative Magnetic Brush

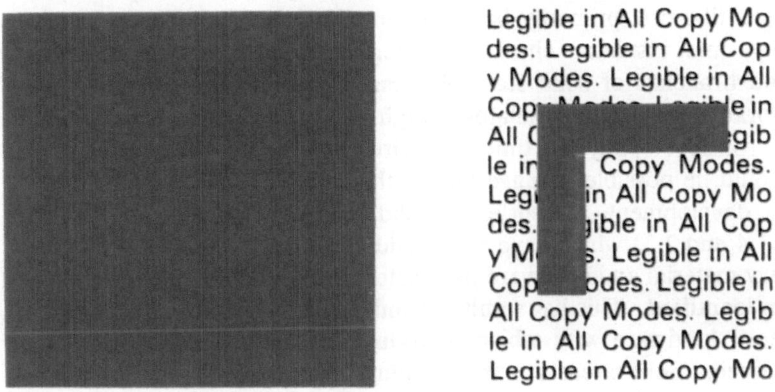

Conductive Magnetic Brush

velopment system is shown in Fig. 3.1a (the original test patterns look like Fig. 3.1c). The cascade development system was replaced with insulative magnetic brush development (Chap. 6) in the early 1970s primarily to improve this solid area development defect; it also improved background development and permitted higher copying speeds. Note that the solid areas (Fig. 3.1b), while filled, are less dense or uniform than the lines. In 1975, the conductive magnetic brush development system (Chap. 7) was introduced by Eastman Kodak. Note the enhanced blackness of lines and solids (Fig. 3.1c), the uniformity of the solids, and the equality of the blackness of lines and solids. In addition, background development was further decreased. The Eastman Kodak line of copiers was immediately perceived by the public and electrophotographers as a set of significantly improved products.

The scientific challenge results from the fact that significant aspects of the physics of development are not understood today. Two examples are background development and toner charging. Lists of background mechanisms are available (Sect. 6.6) but virtually no data or quantitative theory exist. Our understanding of toner charging, which is critical to solid area as well as line copy development and is probably critical to background development, is in the pre-scientific era, primarily based on empirical studies. Toner charging, specifically, and insulator charging, generally, remain one of the least-understood branches of solid state physics. Enormous disagreement exists among workers who study insulator charging (Sect. 4.2) and, consequently, guidelines for developing new toners are virtually absent. That a background theory should be possible is suggested by the status of our understanding of solid area development. Theories of solid area development are available (Sects. 6.3.4 and 7.3) which have been validated experimentally. As a result, the primary material and hardware parameters driving solid area development have been identified. This has resulted from carefully controlled experiments using special hardware with which reproducible, highly accurate data have been obtained. Solid areas are the obvious first aspect of development to study because the "simple" electric field makes both experiments and theory easier.

Our lack of knowledge of the physics of background development and toner charging creates the third technological challenge. Without hardware or material guidelines, mastery of the development step is very costly in terms of manpower and time because extensive empirical hardware and material searches are standard procedure in optimizing a development system for a new copier. With no predictive ability or off-line tests, one can only proceed by building actual hardware, making copies, and running lifetests. This is an expensive and wasteful manpower-intensive procedure. Even worse, after a failed life test, it is unclear whether to make a hardware or material change.

An example of the procedure followed in optimizing a new development system illustrates the point that our lack of knowledge is costly. It is known, i.e., in the electrophotographic folklore, that a tradeoff exists between line copy and background development. A lower average toner charge-to-mass ratio Q/M increases line copy development but also increases background development. The standard hardware approach (after trying to get the background specification raised!) is to set Q/M to achieve the background specification and then to change other variables, such as number of rollers in the magnetic brush development system (Sect. 2.1.3), to achieve the line copy specification. This can become expensive when the number of rollers approaches 3 or 4. It can also have a deleterious effect since mechanical stress is placed on the developer mix at each roller-photoreceptor gap, causing damage to the surface of the toner and carrier particles. This can lower Q/M, which increases the background! A better approach might be to try to understand the basic reason why a tradeoff seems to occur. Perhaps the reason lower Q/M causes increased background is that such mixtures have more wrong-sign toner. If this is the case, then narrowing the toner charge distribution should allow one to reduce simultaneously the average toner charge (increasing line copy development), and the amount of wrong-sign toner, minimizing background development. Such an approach requires knowledge of the physics determining the toner's charge distribution and physics of background and line copy development.

3.2 Focus

The primary focus of this book is a thorough description of our current understanding of the physics of the development process. This includes solid area, line copy, and background development for all known development systems and their associated toner charging mechanism. A secondary focus is to point out areas where significant unanswered questions exist to encourage future research.

That significant issues are associated with the physics of solid area development can be demonstrated with two simple calculations. First, if development proceeded until the toner charge per unit area σ_t completely neutralized the latent image (surface charge σ_p) then after development (Fig. 3.2)

$$\sigma_t = \sigma_p \, . \tag{3.1}$$

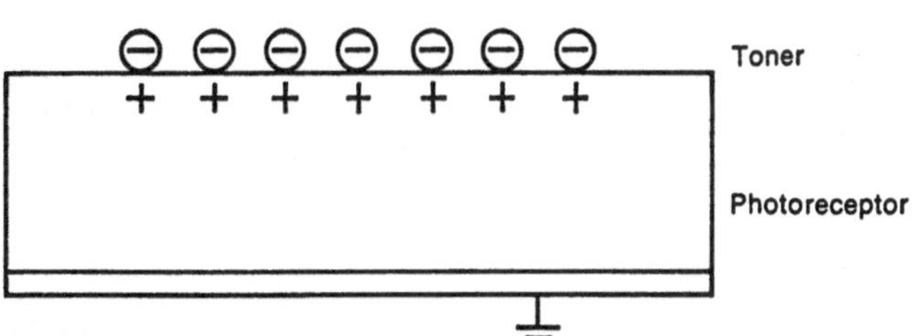

Fig. 3.2. The neutralization condition occurs when the toner charge per unit area equals the photoreceptor charge per unit area

Since

$$\sigma_t = \frac{M}{A}\frac{Q}{M} \tag{3.2}$$

and

$$V = \frac{\sigma_p d_s}{\varepsilon_0 K_s}, \tag{3.3}$$

where M/A is the developed toner's mass per unit area, Q/M is the developed toner's charge to mass ratio, V is the electrostatic potential associated with the charge of a solid area latent image a dielectric distance d_s/K_s from the ground plane, and ε_0 is the permittivity of free space, then neutralization predicts

$$\frac{M}{A} = \frac{V\varepsilon_0}{(Q/M)(d_s/K_s)}. \tag{3.4}$$

Such high M/A 's are never observed. For example, for a typical organic photoreceptor one can assume $d_s = 20\ \mu m$, $K_s = 3$ and $V = 600$ V. A single roll magnetic brush development system operating with toner having $Q/M = 20\ \mu C/g$ will produce approximately 0.45 mg/cm^2 (Sect. 6.4); (3.4) predicts 4.0 mg/cm^2, nine times higher than observed. Hence a factor other than neutralization limits development.

Second, only a small fraction of the toner available for development is used. At synchronous motion (the magnetic brush roller and photoconductor are moving at the same speed), the total amount of toner in the carrier chain per unit area of the carrier chain is

$$\frac{M}{A} = C_t \frac{M_c}{4R^2}\frac{L}{2R}, \tag{3.5}$$

where C_t is the toner concentration (ratio of toner to carrier mass), M_c is carrier mass, $4R^2$ is the area occupied by the carrier, L is the photoconductor to roller gap and $L/2R$ is the number of carrier beads per chain in the gap. Adding the speed ratio factor (v_r/v_p, roller divided by photoreceptor velocity) to account for nonsynchronous motion, we obtain (with ρ_c being the carrier density)

$$\frac{M}{A} = \frac{\pi}{6} C_t L \rho_c \frac{v_r}{v_p} \tag{3.6}$$

counting all the toner in the bead chains, or

$$\frac{M}{A} = \frac{\pi}{3} C_t R \rho_c \frac{v_r}{v_p} \tag{3.7}$$

counting only the toner in first layer of carrier beads [obtained by omitting the $L/2R$ factor in (3.5)]. This predicts (for $C_t = 2\%$, $L = 1250$ μm, $\rho_c = 5$ g/cm^3, $v_r/v_p = 2$, $R = 100$ μm) 13 mg/cm^2 total toner in the bead chains, or 2 mg/cm^2 toner in the first layer of carrier beads, again at least an order of magnitude more than is observed. Most of the toner passing by the latent image is unused.

Line copy development is actually much more important than solid area development from the point of view of usage. However, study of it has been limited for two reasons. First, it is obviously a more complicated problem because of the nonuniform electric fields. Second, it is observed that the ratio of line copy to solid area development is somewhat constant, 1.5–2, for insulative magnetic brush development and close to 1 for conductive magnetic brush development (consistent with the model discussed in Chap. 7).

The electric fields associated with lines have been studied [3.1–3] and a qualitative understanding is useful for grasping the complicated nature of this problem. *Neugebauer* [3.1] solved the electrostatic problem of a line of charges on a dielectric, i.e., photoreceptor surface. He showed that the electric field depends on the thickness of the dielectric and the width of the line, and varies rapidly in space above the line. Variations of the perpendicular field as a function of distance above a 25 μm thick dielectric with dielectric constant 6.6 charged to 100 V for a line of 10 μm half width are shown in Fig. 3.3.

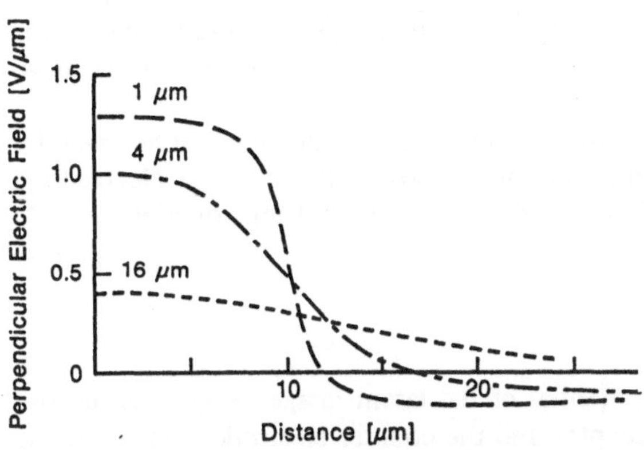

Fig. 3.3. Perpendicular electric field plotted versus the distance (1, 4, 16 μm) above a 25 μm thick photoreceptor of dielectric constant 6.6 for a line charge half width of 10 μm [3.1]

The sensitivity to line width is shown in Fig. 3.4. Note the rapid spatial variations of the electric field and that at large line width, i.e., solid areas, the electric field goes to zero. Now imagine adding to the space above the line moving 200 μm diameter metal balls (carrier beads) which must maintain an equal potential across their surfaces. This will obviously change the electric field both spatially and with time. Further, as toner develops and neutralizes some of the line charge, the electric field lines will move toward the interior of solids, and development will proceed toward central areas, increasing the thickness of fringe field development at the edge of solids. If that were not

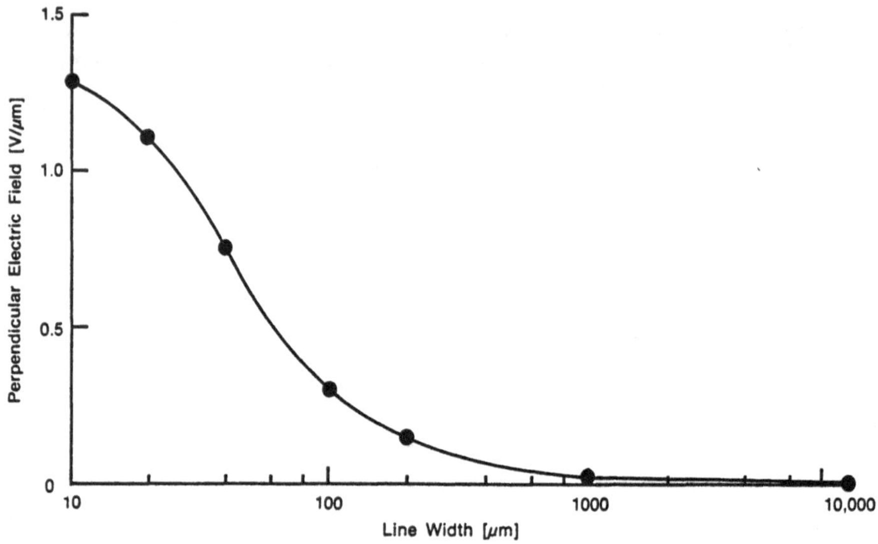

Fig. 3.4. Perpendicular electric field component at the center of a line charge plotted versus the width of line [3.1]

complicated enough, it is unclear where in space to evaluate the electric field when calculating toner development. Some workers have evaluated the field at a toner radius over the photoreceptor, but that is probably not accurate for either a powder cloud or a field stripping mechanism of development (Chaps. 5 and 6).

An estimate of the ratio of line to solid area development can be made by ignoring most of the problems mentioned above. If a counter electrode is added (Fig. 2.7), a uniform field is added to the above fringe field of

$$E_{air} = \frac{V}{d_s/K_s + L/K_E} , \tag{3.8}$$

where V is the electrostatic potential of the latent image (3.3), L is the the distance between the photoreceptor and the counter electrode, and K_E is the dielectric constant of the developer mix. For $V = 100V$, $L = 1250\ \mu m$, $K_E = 1$, this field is 0.08 V/μm, small compared to the fringe fields (Fig. 3.3). If $K_E = 7$ (approximate value for insulating magnetic brush development as shown in Sect. 6.2.2), this field becomes 0.55 V/μm, about half the fringe field value. These numbers suggest that the electric fields due to lines are approximately twice as strong as the electric fields due to solids, and 2:1 ratios of line to solid area toner mass per unit area are reasonable for insulating magnetic brush development.

The development mechanisms governing background development represent extremely important but essentially unexplored areas of development physics. One may regard the development system as a signal-to-noise discriminator, where solid areas and lines are the signal and background is the noise. Pity the poor engineer who must maximize the signal-to-noise ratio, yet has little insight into the noise sources! Where information is available on background development, it will be mentioned. It is hoped that this book will encourage research on this problem.

There are other aspects of the development technologies which will not be discussed, such as the design and mechanism of particle flow in the dual component development systems, lifetime problems, and sensing and replenishment of toner.

3.3 Descriptions

One of the first problems faced by the inventors of electrophotography was devising methods of charging the toner particles and bringing them into close proximity with the photoreceptor containing the latent image; i.e., inventing a development system. A history of the various ideas tried can be found in the books by *Schaffert* [3.4] and *Dessauer* and *Clark* [3.5]. We will focus here on those development systems which have successfully made it into products.

Cascade development, invented [3.6] during the Haloid-Battelle collaboration and used on the first electrophotographic copiers is one of the dual component development systems, so named because both toner and carrier are used. The carrier in this system is polymer-coated glass beads with diameters of several hundred micrometers. Contact between the carrier and toner surfaces causes charge exchange (via static electricity, Chap. 4). This neutral mixture of charged toner and (equal and oppositely charged) carrier is literally cascaded down the photoreceptor surface (Fig. 3.5). Bead motion is con-

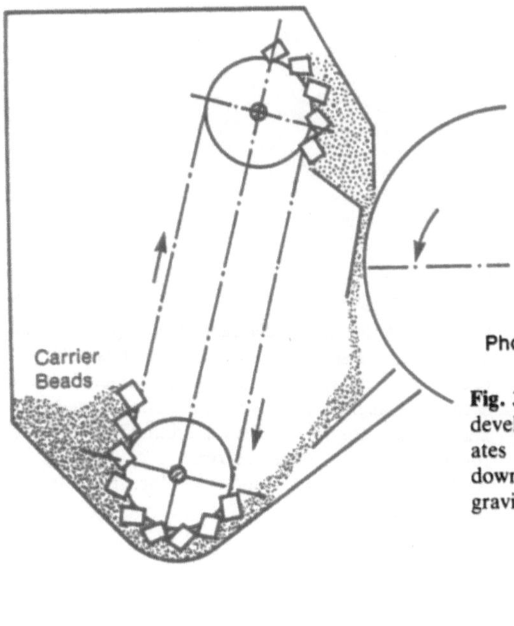

Fig. 3.5. A schematic diagram of cascade development. Cascade development operates by literally cascading carrier beads down the photoreceptor under the force of gravity [3.7]

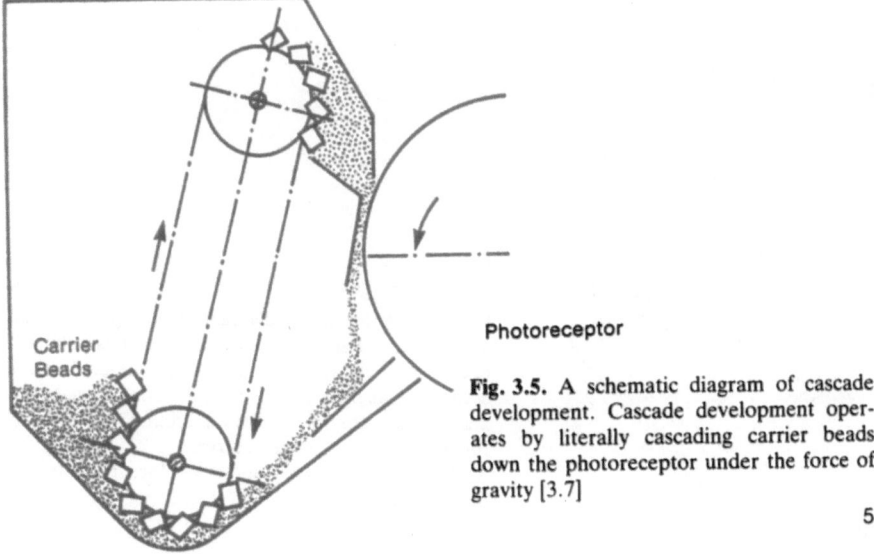

trolled by gravity, imposing an architectural design constraint. The counter electrode, capacitively coupling the electric field out of the photoreceptor, is very far away (Fig. 2.7), leading to the classic solid area washout (Fig. 3.1) associated with these copiers. The carrier particles are recirculated while the toner is used up as it is developed onto the photoreceptor. Hence, means must be provided to sense depleted toner, add toner, and mix new toner with the carrier to produce the proper charge. These functions add considerable complexity to the system. Many of the ideas for possible mechanisms of development, such as the powder cloud and field stripping models, were conceived by researchers working on this system. A discussion of the results of this work can be found in Chap. 5.

Magnetic brush development, invented in the late 1950s at RCA [3.8], has completely replaced the cascade system today. It is used in almost all copiers with speeds above 30 cpm This is also a dual component system, with toner and carrier, but the carrier is made from magnetically soft material such as iron or ferrite. At the bottom of the roller (Fig. 3.6) the carrier is attracted to the stationary magnets and, by magnetic forces and the resulting friction forces, the carrier beads are transported around the rotating roller. Because magnetic forces are used, the architectural design constraint imposed by gravity in cascade development is removed. For enhanced development several rollers can be utilized. Usually the developer mix is magnetically passed from one roller to the next. Because magnetic material is usually conductive, the counter electrode is much closer to the latent image and solid area development is now possible. Quantifying the magnitude of the distance to the counter electrode, in a region of moving conductive balls, has been a challenging electrostatic problem, as discussed in Sect. 6.2.2. The physics associated with this development system has been the most studied of all development systems, and the discussion in Chap. 6 is correspondingly the most extensive.

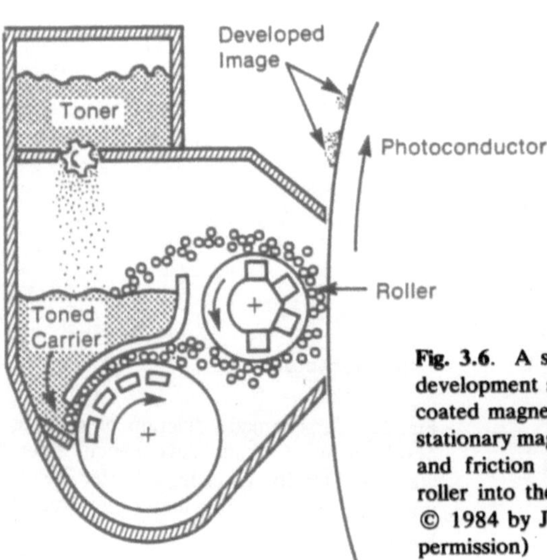

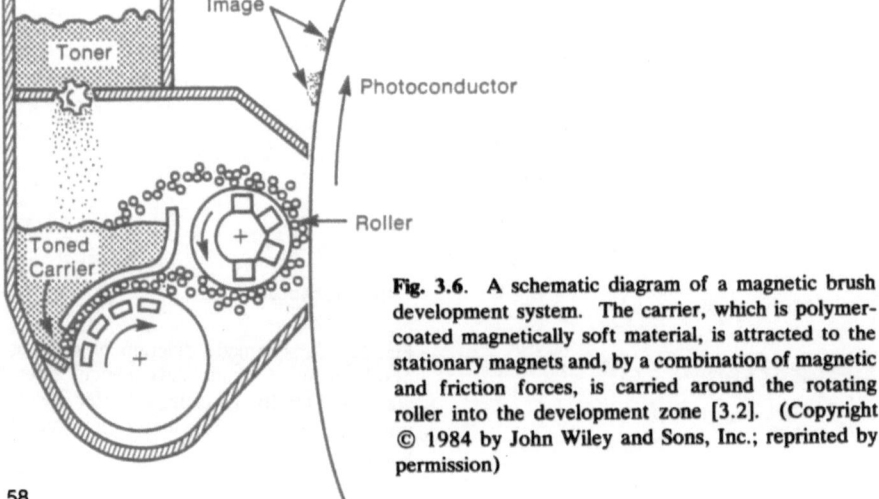

Fig. 3.6. A schematic diagram of a magnetic brush development system. The carrier, which is polymer-coated magnetically soft material, is attracted to the stationary magnets and, by a combination of magnetic and friction forces, is carried around the rotating roller into the development zone [3.2]. (Copyright © 1984 by John Wiley and Sons, Inc.; reprinted by permission)

In the mid-1970s Eastman Kodak [3.9] announced an improvement to the magnetic brush development system. They replaced the spherical carrier particles with irregularly shaped particles (called sponge carrier). To understand why this constituted an important advance, one must understand that toner development on the photoreceptor with insulative spherical carrier is limited by the buildup of charge on carrier beads adjacent to the photoreceptor (Chap. 6). The use of rough beads allows this charge to be short circuited to the roller, significantly increasing the amount of development possible (Chap. 7). The counter electrode now becomes so close, basically at the carrier adjacent to the photoreceptor, that lines and solids look the same electrically. The line to solid area electric field ratio, and consequently the line to solid mass-per-unit-area ratio, approaches one, the ideal situation, approaching the quality produced by offset printing.

All the development systems described so far are dual component, that is, they require two components, carrier plus toner. Having two components entails some nontrivial hardware complications, which involve the sensing of depleted toner and the addition and mixing of fresh toner. These complications can be avoided with a monocomponent development system in which only toner is used. Such systems have been researched over the years by almost all of the manufacturers of copiers. 3M and Canon Corporation were the first to introduce commercial versions of these systems.

There are three independent characteristics of monocomponent development systems: conducting or insulating toner, magnetic or nonmagnetic toner, and contact or noncontact between the photoconductor and the toner-loaded roller. Monocomponent development systems are used usually in low speed machines, below 20 cpm, where manufacturing cost is particularly important.

The first such system was introduced by 3M [3.10] in the early 1970s, the VHS (for very high speed) copier operating at 20 cpm. It used conducting toner, magnetic transport, and contact development. By loading the toner with magnetite, magnetic forces could be used to move the toner into the development zone (Fig. 3.7). In this system the magnets rotate and the roller is stationary. The high conductivity of the toner allowed the use of induction, an extremely simple method of charging. The field due to the latent image induced charge flow through the toner chain to the toner particles adjacent to the photoconductor, which were then attracted electrostatically to the latent image. Unfortunately this system has two inherent flaws (Sect. 9.4), monolayer development (hence, gray copy) and humidity sensitive transfer, which have caused it to be all but abandoned.

In the early 1980s Canon introduced another monocomponent development system [3.12] based on insulating toner, magnetic transport, and noncontact (they called it jumping) development (Sect. 9.5). The charging of the toner was achieved using static electrification (instead of induction), just as in dual component systems, with the other part of the tribo-couple being the roller surface (Fig. 3.8). Magnetic forces were again used for transport. In

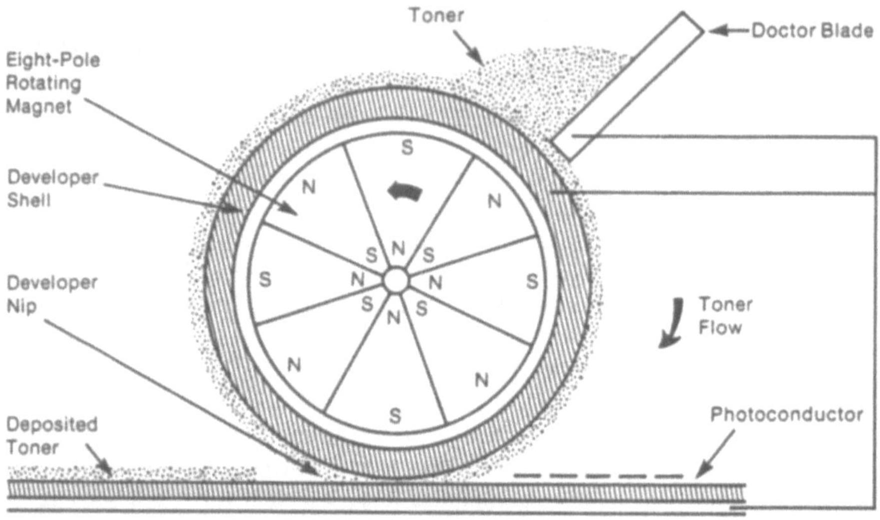

Fig. 3.7. In 3M's conductive monocomponent development system the magnets rotate inside a stationary shell, causing the toner chains to move in the opposite direction around the shell into the development zone [3.10,11] (© 1983 IEEE)

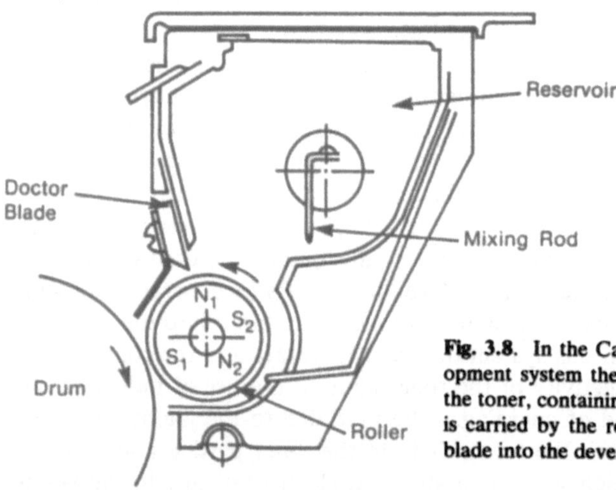

Fig. 3.8. In the Canon monocomponent development system the magnets are stationary and the toner, containing magnetically soft material, is carried by the roller past a magnetic doctor blade into the development zone [3.12]

order for the toner to jump across the gap, large (1200 V p-p) ac fields were superimposed across the development zone. This development system has been extremely successful for Canon and is used in their full range of products, from their personal copier PC10 to their higher speed versions NP-8070, covering a range from 8 to 70 cpm.

A nonmagnetic development system might also be attractive because of the potential lower cost of manufacturing toner and the ease of making colored toner. In 1985, Ricoh [3.13] and Toshiba [3.14] discussed such systems at the

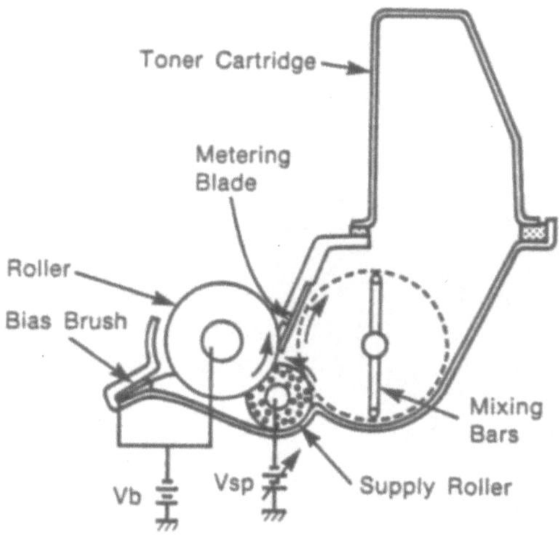

Fig. 3.9. In the Ricoh nonmagnetic monocomponent development system, the toner is brought into contact with a roller by contact with a soft supply roller; those toner particles which escape through a spring-loaded metering blade obtain a uniform charge and thickness [3.13] (© 1985 IEEE)

annual IEEE-IAS conference in Toronto. The Ricoh system (Fig. 3.9) uses contact development and has since appeared in products: a copier, the Ricoh RePRO jr. (8 cpm), and a printer, PC Laser 6000 (6 ppm). The Toshiba system uses noncontact jumping development. Both are discussed in Sect. 9.5. As shown in Fig. 3.9, in the Ricoh version the toner is still brought near to the photoconductor with a roller. The roller is loaded by flooding it via a supply roller with a bath of toner, which must escape through a spring-loaded metering blade. When it exits this region, it is of uniform thickness and charge. Clearly, sophisticated material engineering has been achieved.

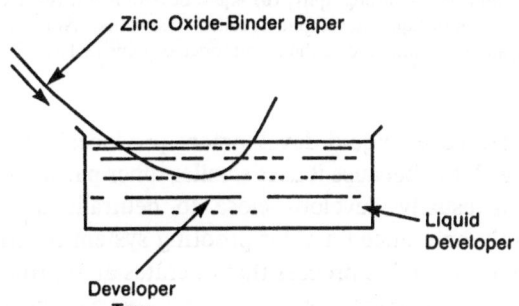

Fig. 3.10. A schematic of the immersion-type liquid development system [3.15]

All of the above systems use dry powder to develop the latent image. Liquid development systems [3.15] have also been used in copiers. Perhaps the most successful were copiers made by Ricoh, because they produced slow but small, inexpensive and reliable copiers, a market ignored by Xerox during the 1970s. In these systems (Fig. 3.10) the latent image is dipped into a liquid that contains charged toner particles. Obviously the liquid must be insulating so that it does not destroy the latent image. This requires the use of organic

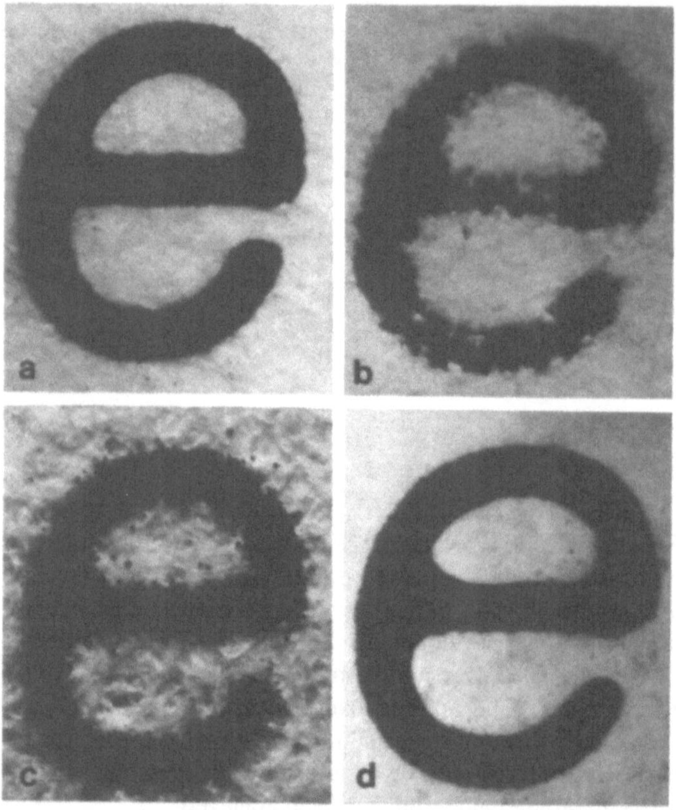

Fig. 3.11. A comparison of the print quality for (a) lithography, (b) liquid development, (c) dry toner electrophotography, and (d) a new form of liquid development called ElectroInk. Note the lack of edge crispness in the dry toner system as compared to the liquid toner systems [3.16]

solvents, a nontrivial concern. However, liquid development produces some of the highest quality images (Fig. 3.11) because it uses smaller toner particles, less than 1 μm, and because it usually develops close to neutralization. Eastman Kodak [3.17,18] recently announced a color proofing system based on a liquid-development electrophotographic process that operates at approximately one page every few minutes. A discussion of the liquid development system is found in Chap. 10.

4. Toner Charging for Two Component Development Systems

In two component development systems, two powders, toner and carrier, are mixed together (Fig. 2.9). Toner particles have diameters of approximately 10 μm and are blends of polymers and carbon black pigment. Carrier particles have diameters of approximately 200 μm and are composed of magnetically soft cores coated with a thin polymer coating. Contact between the toner and carrier causes charge to be exchanged. Depending on the materials chosen for the toner and carrier coating, the resulting charge on the toner may be positive or negative. This mixture, which has zero net charge, is introduced into cascade or magnetic brush development systems, as described in Chap. 3, where toner particles are attracted to the latent image on the photoreceptor.

The proper charge properties of the toner are crucial requirements for a good development system. The average charge-to-mass Q/M ratio determines the amount of toner developed onto solid area and character latent images; the lower Q/M, the darker the images on the page. It is believed that wrong sign toner is "developed," i.e., attracted to photoconductor, onto nonimaged areas, giving an objectionable gray color to the white paper. Zero charged toner becomes dust in the machine, leading to reliability problems.

The phenomenon of charge exchange between contacting materials is a pervasive and interesting solid-state physics problem which remains poorly understood. It is pervasive in both a negative and positive sense. Sparks generated by static electricity may cause explosions in mines, flour mills and supertankers. One merely has to walk across a rug under low relative humidity conditions and experience the shock on touching grounded metal for a demonstration of the pervasiveness of charge exchange phenomena (between shoes and rugs). In the positive sense, besides electrophotography, electrostatic charge exchange is used in electrostatic precipitators to control pollution and in electrostatic spray painting.

It is an interesting phenomena because it occurs between all materials (metals and insulators, organic and inorganic) and remains one of the few solid-state physics problems that is at such a rudimentary level of understanding. The difficulties in making progress in this field should not be underestimated and are well documented in prior reviews and books [4.1–8]. When the surfaces of two materials are brought into contact and separated, the actual area which made contact is difficult to determine. Whether pure contact or friction is required has not been determined. In fact, the terms contact electrification and triboelectrification, i.e., frictional electrification, are often used interchangeably. The precise nature of the surfaces are usually not well defined: dust particles, surface contaminants and even water layers may be the "surface." Even for "clean" surfaces the nature of intrinsic and extrinsic surface states on insulators is not well understood. The magnitude of return currents during separation remains controversial. Finally, the number of surface molecules involved in the charging process is extremely small, on the order of one molecule in 10^4 or 10^5. The determination of the nature of the charging sites by surface science tools remains an unsolved solid-state physics problem.

In order to give the reader perspective on the problem of toner-carrier charging, we will first review the literature of metal-metal, metal-insulator, and insulator-insulator charging. Then we will try to relate this information to the problem of toner-carrier charging.

The construction of contact charge exchange models requires specification of the following items: the nature of the charge carrier (electrons, ions, or mass transfer), the driving force (difference in work function, concentration gradients), the mechanism, (thermionic emission, tunneling), the energy states involved (bulk or surface, extrinsic or intrinsic), and the condition (whether the dynamic or equilibrium condition is being addressed). These parameters are agreed upon only for metal-metal contacts [4.1–3] (electrons, difference in work functions, tunneling, bulk intrinsic states, equilibrium condition). When insulators are involved, serious experimental and theoretical disagreements abound.

4.1 Metal-Metal Contact Charging

If two metals with different work functions ϕ_i are brought together (Fig. 4.1), and electrons are allowed to exchange by tunneling so that thermodynamic equilibrium is maintained, a contact potential difference V_c is created across the interface given by

$$V_c = (\phi_B - \phi_A)/e \tag{4.1}$$

and the charge Q exchanged by electron tunneling is

$$Q = C_{AB} V_c, \tag{4.2}$$

where C_{AB} is the capacitance between the two adjacent bodies. As the two bodies are separated, C_{AB} decreases (and consequently Q decreases) until charge exchange by tunneling stops. *Harper* [4.3,9] showed that the cutoff of tunneling currents with distance is so abrupt, at about 10 Å, that no velocity of separation effects should ever be observed experimentally. Hence, the final observed charge should be

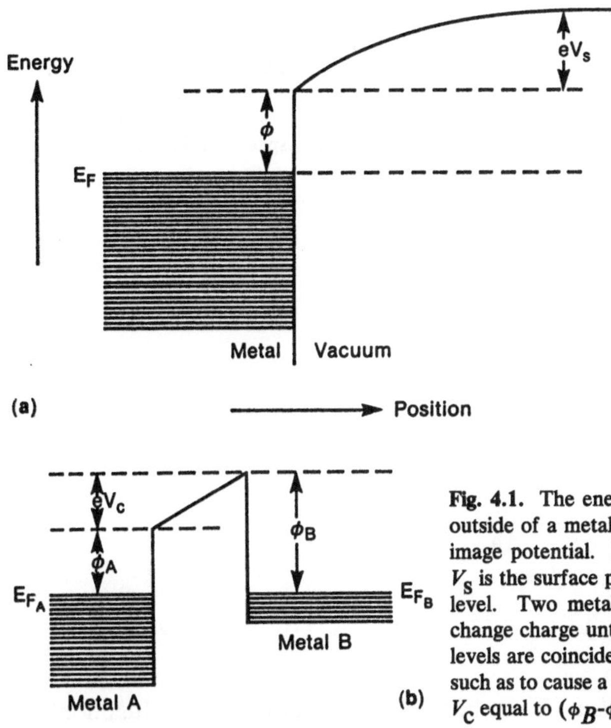

Fig. 4.1. The energy of an electron inside and outside of a metal is shown in (a) ignoring the image potential. Here ϕ is the work function, V_S is the surface potential, and E_F is the Fermi level. Two metals in close proximity (b) exchange charge until, in equilibrium, their Fermi levels are coincident. The transferred charge is such as to cause a difference in surface potential V_C equal to $(\phi_B - \phi_A)/e$ [4.3]

$$Q = C_0(\phi_B - \phi_A)/e\,, \tag{4.3}$$

where C_0 is the capacitance at a spacing of 10 Å.

The experimental results [4.3, 9] of making a single contact with a 5/32 inch diameter chromium plated ball against 1/2 inch diameter balls electroplated with different metals are shown in Fig. 4.2. The contact potential difference was measured in situ for the identical contact points used in the charge measurements. Clearly both the trend with V_c and the magnitude of the charge appear correct. The slight discrepancy between theory and experiment can be traced to the roughness of the metal surfaces, as suggested by *Harper* [4.3,9] and proved by *Lowell* [4.10].

4.2 Metal-Insulator Contact Charging

This problem has received a great deal of attention because of its practical importance and the perception that it is only one step from the understood metal-metal problem. Nonetheless, significant experimental and theoretical controversies exist.

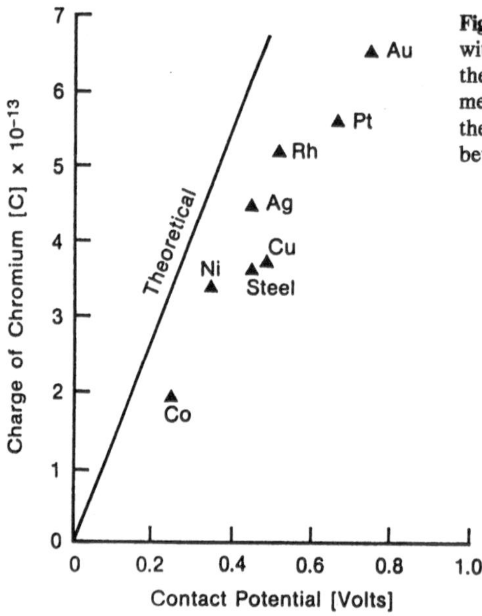

Fig. 4.2. Charge on a Cr sphere after contact with a sphere of another metal, plotted against the contact potential difference between the metal and Cr. The line marked "theoretical" is the prediction of a theory based on tunneling between closely spaced smooth metals [4.3]

4.2.1 Controversies

a) Experiments with Teflon. It has been claimed that the "work function" of Teflon is both larger [4.11] and smaller [4.12] than the work function of Au. It also has been claimed [4.3] that Teflon *does not charge* against Au. Others [4.13,14] claim that the charge exchanged is *not influenced* by the metal used for charging. It is difficult to imagine any more disagreement than this!

b) Experimental Determination of Polymer "Work Function". In 1969, *Davies* [4.12] did the logical extension of Harpers' metal-metal experiments with polymers (which has been repeated by others [4.11,15–18]). If one can define a polymer work function (see below), then the charge exchanged between a metal and a polymer would depend on the energy difference between the work function of the metal ϕ_M and the work function of the polymer ϕ_I

$$Q \propto (\phi_M - \phi_I)\,. \tag{4.4}$$

If one measures Q for a series of metals, then the value of the work function at $Q = 0$ should be ϕ_I, the insulator work function. Such a linear relationship was in fact observed by Davies (Fig. 4.3) and others. However, a tabulation of ϕ_I for a wide range of polymers (Table 4.1) reveals a curious fact: the values show very little variation. Quoting *Lowell* and *Rose-Innes* [4.1], "This remarkable constancy is unexplained. It surely provides an important clue about the mechanism of contact electrification."

c) Contact Charge "Spectroscopy". Another example of the difficulty of the field is given by the recent experimental and theoretical work of *Fabish* and

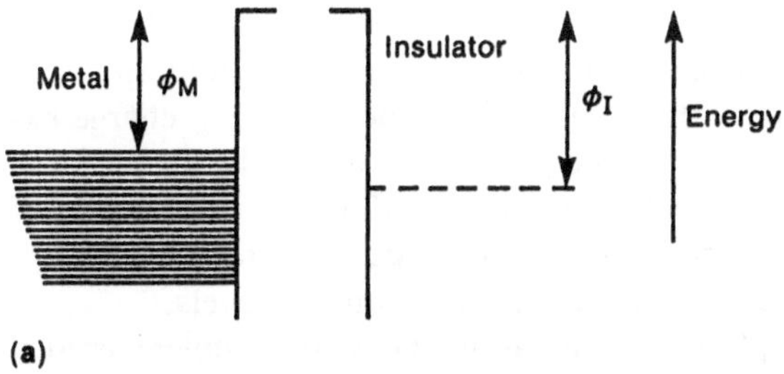

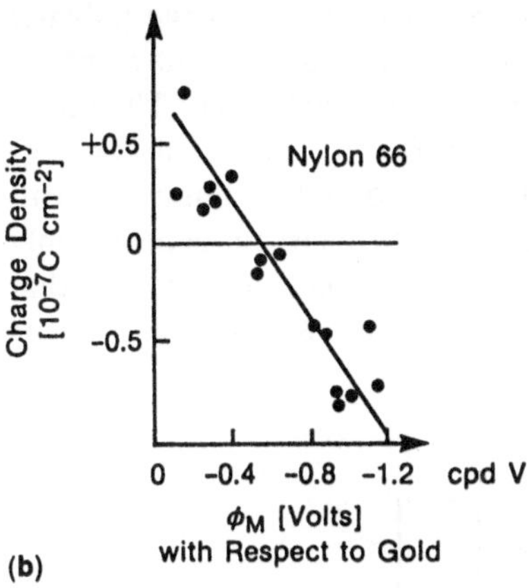

Fig. 4.3a, b. Evidence that contact charging of insulators by metals is determined by the energy difference between the metal Fermi energy and some energy *tag*ϕ_I characteristic of the insulator. **a** Energy levels in a metal and an insulator. **b** Contact charging of nylon 66 by various metals. The horizontal axis is the contact potential difference between each metal and a gold reference [4.12]

Table 4.1. Insulator work function (after [4.12])

Insulator	Work function [eV]
Polyvinylchloride	4.85 ± 0.2
Polyimide	4.36 ± 0.06
Polycarbonate	4.26 ± 0.13
Polytetrafluoroethylene	4.26 ± 0.05
Polyethyleneterephthalate	4.25 ± 0.10
Polystyrene	4.22 ± 0.07
Nylon 66	4.08 ± 0.06

co-workers [4.19] and *Cottrell* et al. [4.20] on metal-insulator contact electrification. Fabish et al.'s experimental results are presented in Fig. 4.4a. In these experiments, two smooth plates, one metal and one an insulator, are repeatedly contacted. The charge remaining on the insulator is measured periodically. Fabish et al. found that a given metal produces a specific magnitude change in insulator charge after a sufficient number of contacts. This change is independent of whether the insulator has previously been contacted by other metals. They explained their results by assuming (1) that the insulator has a range of localized bulk energy levels ("intrinsic molecular-ion states", each of which has its energy spread over 0.5 eV due to molecular vibrations and variations in the local structure) near the contacting surface, and (2) charge exchange can occur only within a narrow energy window about the metal Fermi level. Thus, each different metal depletes or fills a different energy region of localized insulator states. Therefore metal-insulator contact charge exchange could be used as a spectroscopic tool to map insulator energy levels.

Cottrell and co-workers [4.20] investigating the same problem used a slightly different experimental approach. In order to insure that the polymer was contacting the same microscopic region of metal surface, they used a hemispherical polymer surface in contact with a plane metal sheet. They tried to maintain the relative positions of the two surfaces to insure a constant contact location. Their results are shown in Fig. 4.4b and do not agree with the results of Fig. 4.4a. Their results indicate that the charging depends only on the last metal that contacts the polymer surface. Note, for example, that Pt seems to "want to" leave the surface neutral, whereas Mg charges it negative. There is no additivity.

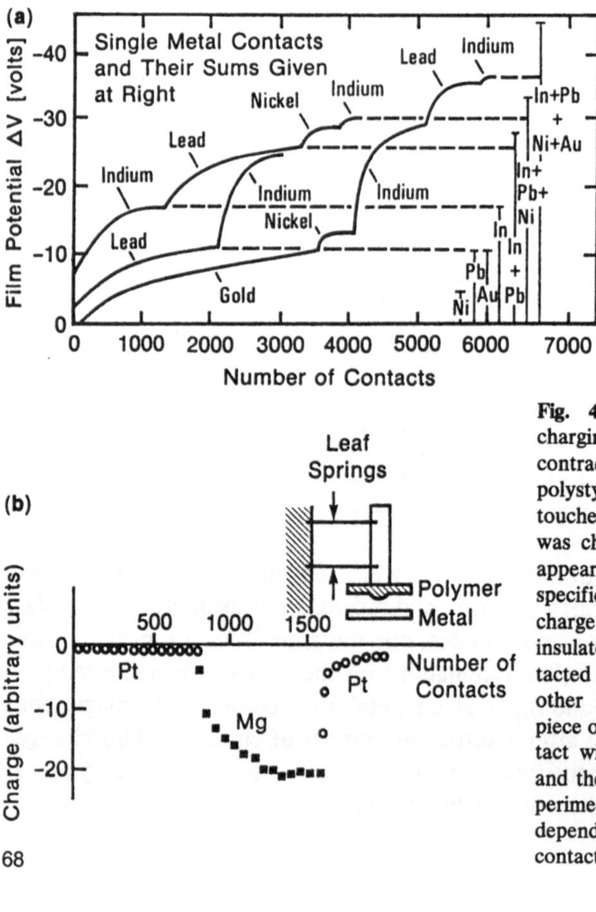

Fig. 4.4a, b. Two metal-insulator charging experiments give apparently contradictory results. In one (a) a polystyrene disk was repeatedly touched to a metal surface; the metal was changed from time to time. It appears that each metal produces a specific magnitude change in insulator charge independent of whether the insulator has previously been contacted by other metals [4.19]. The other (b) involved a hemispherical piece of PMMA polymer making contact with platinum, then magnesium and then platinum again. In this experiment it appears that the charging depends only on the last metal that contacts the polymer surface [4.20]

4.2.2 Experimental and Theoretical Difficulties

These disagreements highlight both the experimental and theoretical difficulties in this field: The extreme disagreement on the behavior of Teflon in contact with Au probably reflects differences in the surfaces of the Teflon samples. The disagreement between the experiments of Fabish et al. and Cottrell et al. may be understood, as discussed by *Lowell* and *Rose-Innes* [4.1], if the contacted areas were different for each metal in Fabish's experiment.

The experiments of Davies raise several important, unresolved questions about contact charging of polymers:

(1) A first thought, as described in solid-state physics textbooks, is to regard an insulator as a wide band gap semiconductor. A problem occurs when the band gap exceeds ≈ 2 eV. Traps in the band gap are now so deep that at room temperature they cannot come into thermodynamic equilibrium in reasonable times. The release time τ is generally written as

$$\tau = \nu_e^{-1} \exp(E/kT) \tag{4.5}$$

where ν_e is an attempt-to-escape frequency ($\sim 10^{14}$ s^{-1}) and E is the depth of a trap from the conduction or valence band. For a trap depth of 0.5 eV, τ = 5 μs; for a 1 eV trap depth, τ = 2000 s; for a trap depth of 2 eV, τ = 10^{20} s! The polymers used in Davies's experiments are believed to have forbidden gaps of ≈ 7 eV! Hence, concepts requiring thermodynamic equilibrium, such as the Fermi level (the electron energy level at which the probability of occupancy is 50%) and work function, may not be useful in such materials.

(2) It remains to be explained why so many polymers have such similar "work functions." Despite these two questions, people have taken data such as Davies's, which show a correlation of charging with metal work function, to suggest that electron transfer is responsible for metal-polymer contact charging.

(3) An ill-defined experimental technique is used in all of the above experiments: repeated contacts between the metal and polymer are required to achieve the "equilibrium" condition. (This is shown explicitly in the data shown in Fig. 4.4). What is happening during those thousands of contacts? Is more area being exposed to the metal? This is an obvious explanation that appears to fail on closer examination [Ref. 4.1; Sect. 4.4]. For example, a 1 cm diameter metal sphere contacted to PTFE with a force of 0.1 N causes a plastically deformed indentation 0.5 mm in diameter. The accuracy of replacing the metal sphere is much better than that. Of course, the true area of contact at each contact may be less than the indentation area, but there is reason to believe it is at least half the apparent area. Yet the charge upon repeated contacts increases by at least an order of magnitude. Viscoelastic flow during contact could also increase the area, but again this effect is much too small to account for the increasing charge. Is time a factor? A simple modification in which the time of contact is varied shows this not to be a factor

[4.1] although results can be time dependent if the surfaces of the polymers are sufficiently conductive [4.21]. *Seanor* [4.7] suggests slow diffusion from surface to bulk states driven by charge concentration gradients, but then points out that for polymers it is unlikely that equilibrium ever would be attained and charge distribution would depend on the detailed charging history. No clear answer to this question appears in the literature. An indication of the potential importance of this question is raised by the data of *Greason* and *Inculet* [4.16] which they obtained by repeating Davies's experiment but using only a single contact. They obtained very different results (Table 4.2). The results showed a "reasonable" variation of insulator work function with material.

Table 4.2. Insulator work functions and densities of surface states (after [4.16])

Matcrial	Vacuum tests		Air tests	
	ϕ_I[cV]	States [$cm^{-2}eV^{-1}$]	ϕ_I[eV])	States $_9$
[$cm^{-2}eV^{-1}$]				
Polypropylenc	5.43	1.2×10^9	9.1	5.3×10^8
Polystyrene	4.77	8.7×10^9	7.45	1.9×10^9
Polyvinylchloridc	4.86	2.0×10^9	8.20	1.5×10^9
Polycarbonate	3.85	2.2×10^9	4.40	2.0×10^9
Acrylic	4.30	3.7×10^9	2.90	1.7×10^9
Polytetrafluoroethylene	6.71	1.9×10^8	10.3	3.2×10^8

4.2.3 Other Metal-Insulator Experiments

Experiments on metal-insulator charging have been reported in which the bulk or surface has been systematically varied. Correlations of charging with Hammet substituent constant have been observed [4.22–24]. The Hammet substituent constant measures the shift in electron density on one side-group which is brought about by the substitution of other groups at another site on the molecule. *Cressman* et al. [4.25] have shown electron energy levels are proportional to the Hammet substituent constant. These experiments tend to reinforce the view that electron exchange governs metal-insulator charging experiments. That the surface can be involved in insulator charging has been clearly demonstrated by *Hays* [4.26], *Bauser* [4.27], and *Kittaka* and *Murata* [4.28] who altered the surfaces of organic films with exposure to ozone, oxygen and UV irradiation, respectively, and observed changed charging behavior.

On the other hand, there exist other authors who believe ion exchange is occurring. For example, *Harper* [4.3,29] found that the charging of 5/32 inch spheres of cleaned polyethylene, nylon, polystyrene, PTFE, PMMA and other insulators against 1/2 inch spheres of steel, chromium and gold was below the limits of sensitivity of his apparatus (3×10^{-8} C/cm^2). Using the same apparatus he measured charges up to 3×10^{-4} C/cm^2 on quartz spheres. Harper interpreted his results as showing that the polymers he studied do not charge in contact with metals when carefully cleaned. A possible conclusion that follows is that when charging of these materials is observed it is due to an exchange of surface contaminants. Consistent with these results, experiments on pure rare gas solids [4.30] show no charging when contacted by metals. The observation of zero charge exchange is an important result. It probably requires both the absence of surface ionic impurities and negligible bulk or surface electron states in the insulator band gaps near the metal Fermi energies.

Other experiments support the view that ion exchange is occurring in the presence of water. For example, experiments on glass [4.31] have clearly shown that charging depends on the acidic nature of the surface with basic materials charging positive and acidic materials charging negative. *Weber* [4.32] has suggested a correlation of charging with the acidic nature of groups or contaminants on polymer surfaces. *Sereda* and *Feldman* [4.33] have shown charging peaks on fabrics at one monolayer of surface water. The potential importance of water layers on surfaces during contact charge exchange has been hinted at in reviews. For example, *Lowell* and *Rose-Innes* [4.1] in their review state, "it has been suggested, though never unequivocally demonstrated, that ions in the layer of surface water may themselves cause contact electrification." *Morris* [4.6] states: "The part played by water in ionic charging is not clear. It may be essential in providing an aqueous layer between the 'contacting' surfaces in which the ions can diffuse. However, this layer would have to be extremely thin or discontinuous in order to avoid rapid surface leakage of the charge on separation of the contact." *Medley* [4.34] takes a much more positive attitude towards the role aqueous layers can play in contact charge exchange. With polar polymers, and in particular textiles and ion-exchange resins, he claims the presence of the acidic or basic groups promote water sorption and consequently ionic dissociation and mobility even at low humidities. One ion remains firmly bound to the polymer matrix and the other is free to take part in the charge exchange (Fig. 4.5). For example,

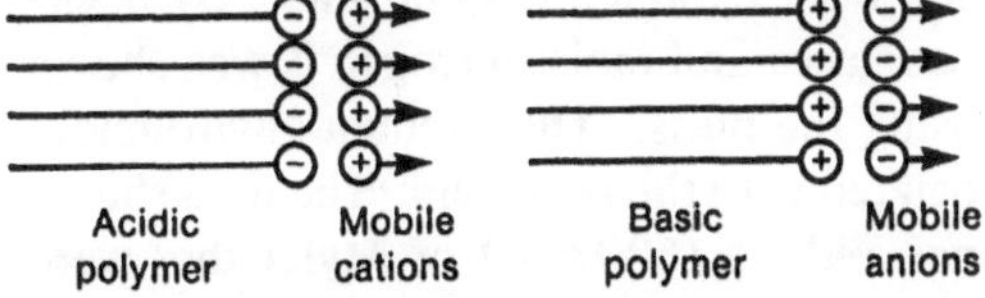

Fig. 4.5. Polymeric material with ionic components. One ion is held in place, but the opposite ion could be mobile, free to participate in contact charge exchange

consider the strongly basic resin De-acidite FF and the strongly acidic permutit sulphonated polystyrene. In the presence of water, one obtains

$$\begin{aligned} R{\cdot}\mathrm{NOH} &\rightarrow R\mathrm{N}^{+} + \mathrm{OH}^{-} \\ R{\cdot}\mathrm{SO_3H} &\rightarrow R\mathrm{SO_3^{-}} + \mathrm{H}^{+} \end{aligned} \tag{4.6}$$

i.e., free H^+ and OH^- ions are available for charging. On shaking the powders of these resins from filter paper at 30% relative humidity, ***Medley*** found that the De-acidite became consistently positive and the permutit became consistently negative, the loose ions having presumably been transferred to the filter paper. Other acids and bases were found to behave similarly. These results tend to confirm extensive experiments done by ***Rudge*** [4.35] in 1913 and *Knoblauch* [4.36] in 1902. Knoblauch slid 75 powders off glass, sulfur, platinum and paraffin wax coated plates. Both investigators found a strong correlation between the sign and magnitude of charging and the acidic or basic properties of the powders.

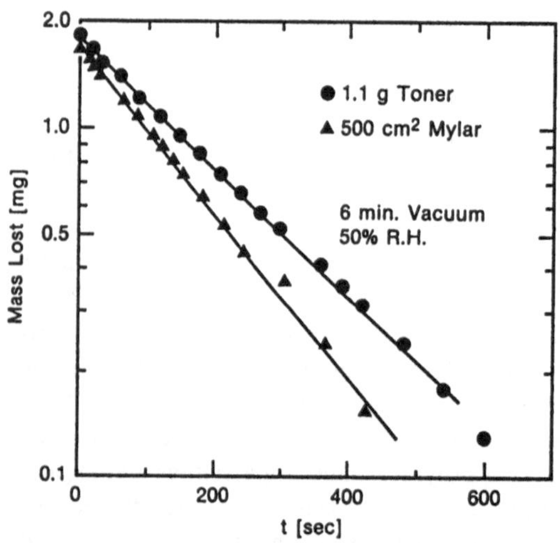

Fig. 4.6. Mass loss observed for 1.1 g of toner and 600 cm^2 of Mylar as a function of time out of the vacuum system. Initially the observed mass loss was approximately 2 mg for both samples which decreased to zero exponentially in time. It is argued in the text that the mass loss is due to water loss, which is slowly regained on exposure to normally moist air

In an effort to establish the existence of water on the surface of materials through which H^+ and OH^- could transfer, the present author carried out an experiment shown in Fig. 4.6. Toner and a sheet of Mylar were weighed with an accuracy of 10^{-2} mg after removing them from a vacuum system. Both showed a mass loss which was gained back exponentially in time. By backfilling the vacuum system with various gases and moist nitrogen it was shown that it is water uptake that is increasing the mass. The vacuum environment must have evaporated the water associated with the polymers. The data shown are for 1.1 g of 10 μm diameter toner and 1 g (600 cm^2) of Mylar that were placed in a vacuum system ($\approx$1 torr) for 6 min. Longer times in the vacuum system or higher relative humidities produced larger mass losses. Water uptake by polymers and its effect on bulk properties has been discussed [4.37]. It is usually assumed that most of the water is diffusing into the bulk polymer and the actual surface water layer is thin. However, such experiments strongly suggest some water is on the surface of polymers. Similar mass losses were found by the author for many materials, including bare aluminum and steel, although for metals a longer time in the vacuum system did not change the mass loss after a few minutes. Calculation of the water layer thickness is complicated by the existence of porous oxides and rough surface topology. However, consistent with these results, *Bowden* and *Throssel* [4.38] find an approximately linear relation between the number of monolayers and relative humidity (RH) on gold and platinum, starting at 10 monolayers at 30% RH and increasing to 40 monolayers at 80% RH. If water layers exist on the surface of polymers and metals, it is hard to ignore them when considering contact electrification under normal atmospheric conditions.

Kornfield [4.39] and earlier workers [4.32] have proposed a model in which insulators have either a net charge or a net dipole moment which is compensated by ions absorbed from the air. Charge exchange is caused by contacts which mix the surface layers, disturbing the compensation. Such a mechanism has been termed a kinetic effect by *Henry* [4.40]. Left unspecified is the nature of the "surface layer." The surface layers discussed by Kornfield may be water layers. (Kornfield states that "...air humidity considerably affected the magnitude of the charge..." exchanged). Indeed, in criticizing Kornfield's work, *Robins* et al. [4.41] showed that ions probably are not responsible for charging in metal-pyroelectric contacts by showing the same charging results in air (in which a pyroelectric material is normally covered with a layer of ions) and in vacuum (in which the freshly cleaned pyroelectric material cannot obtain the ion layer). However, in *moist* air, Robins et al. found charging increased by an order of magnitude. Perhaps two mechanisms are operative; one in vacuum and a second, larger ionic mechanism under normal, i.e. moist, atmospheric conditions. *Homewood* [4.42] also carried out experiments in high vacuum (10^{-10} torr) with cleaved PTFE samples and compared them to results obtained with samples exposed to "clean air" and then replaced in the vacuum system. He obtained the same charge exchange and also concluded that ions could not be responsible for the charge exchange. It is possible that he pumped off any ionic, i.e., water, layer that formed. Therefore, one may need to distinguish between vacuum (or dry air) and moist air conditions in order to understand metal-insulator charging.

Finally we mention the observation [4.43] of mass transfer by ESCA in a geometry similar to those used for contact electrification in which Teflon was observed to transfer to metal surfaces. While mass transfer is a possible mechanism of charge transfer, the authors themselves note that much more mass was transferred than is required to account for the charge transfer. *Williams* [4.44] has pointed out that Teflon may be a special case because its low friction has been shown to be due to drawing out of polymer lamellae from the bulk (which may then transfer to the metal surface).

4.2.4 Electron Transfer Theories

Theoretically, electron transfer theories can be divided into bulk and surface state theories. We do not discuss bulk theories because, as several authors [4.1,3] have shown, assuming equilibrium could be established and assuming reasonable band gaps and resistivities, (1) the charge densities predicted would be very much lower than observed or the thickness of the charge layer would be large compared to any typical sample, and (2) even if there are bulk states (extrinsic defect or intrinsic polaron-like [4.45]) electrons probably cannot move into the bulk of these polymeric materials (for example transient photoconductivity measurements indicate immediate deep trapping).

Surface states on insulators can be intrinsic or extrinsic. For either case, one might expect on contact with a metal that empty states below the metal Fermi level E_F will be filled and full states above it will be emptied [4.1,3,46–48]. If there are $n(E)dE$ insulator states per unit area whose energy falls in the range between E and $E + dE$, the number of electrons per unit area σ_S/e which pass from the metal to the insulator surface is (Fig. 4.7)

$$\sigma_S/e = \int_{\phi_I}^{\phi_M} n(E)dE, \quad (4.7)$$

where ϕ_I is the insulator work function, the boundary between filled and un-

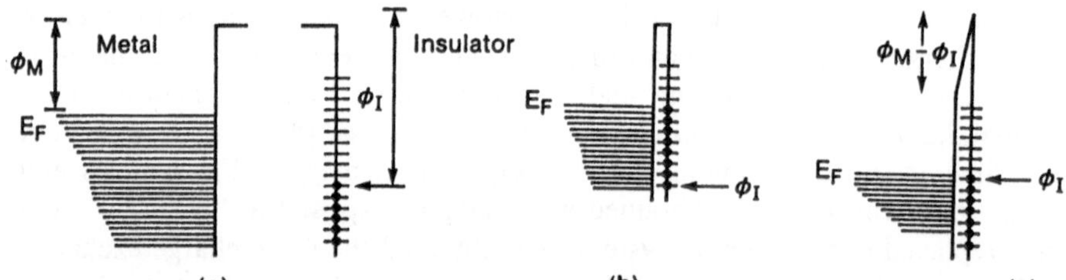

Fig. 4.7a–c. A surface states theory of contact electrification of insulators can be pictured as follows: **a** The insulator is uncharged and the surface states are filled below the *neutral level* ϕ_I and empty above. **b** When the insulator is brought into contact with a metal, insulator surface states below the metal Fermi level E_F will fill and those above will remain empty. If the density of surface states is *low* (b), the number of electrons which transfer to the insulator is equal to the number of surface states with energy between ϕ_I and E_F. The electric field between the metal and the insulator is too small to shift the insulator energy levels appreciably. **c** In the opposite limit of *high* surface state density (which should not usually occur and is not discussed in the text) the surface states need only be filled to a level slightly above ϕ_I, which creates a a strong enough electric field, raising the energy of all the insulator states; charge transfer ceases, in the first approximation, when the neutral level ϕ_I is raised to the Fermi level of the metal [4.1]

filled surface states on the insulator. This is valid if the density of insulator surface states is low enough so that the transfer of charge does not alter the energy of the states, usually a reasonable assumption. Also assumed is none-quilibrium with the bulk (probably reasonable) and thermodynamic equilibrium among the surface states (questionable).

Obviously, if $n(E)$ is constant, the insulator charge will be proportional to $\phi_M - \phi_I$, as observed by Davies. However, the next logical implication of the experiments, that $\phi_I = 4.5$ eV for a variety of polymers, is difficult to understand for either intrinsic or extrinsic surface states. The magnitude of $n(E)$ also remains unexplained. About $10^8 - 10^{11}\,\mathrm{cm}^{-2}\,\mathrm{eV}^{-1}$ states are needed to explain the results obtained by Greason and Inculet and Davies. What is the nature of these states and why is the density so low? Experiments in which the surface is altered (by ozone, oxygen or UV) and charging changes can be understood in terms of $n(E)$ changing. Experiments in which electronegativity or Hammet substituent constant are correlated with charging are perhaps understandable in terms of ϕ_I changing.

4.2.5 Ion Transfer Theories

Ion transfer theories have been considered by *Henry* [4.40] and *Ruckdeschel* and *Hunter* [4.49]. In Henry's paper, a discussion of the possible mechanisms of ion transfer is given in what he calls an "illustrative" theory. For clarity, we will separately discuss each of the mechanisms. Henry envisions the potential energy of an ion between two surfaces separated by a distance z as represented in Fig. 4.8.

The first mechanism of ion transfer depends upon the different affinities of the two surfaces for the charged particles. When applied to electrons, this is just the difference in work functions, as discussed by Harper. For ions the difference in energy is $U_1 - U_2$. The charge per unit area σ_s that must be

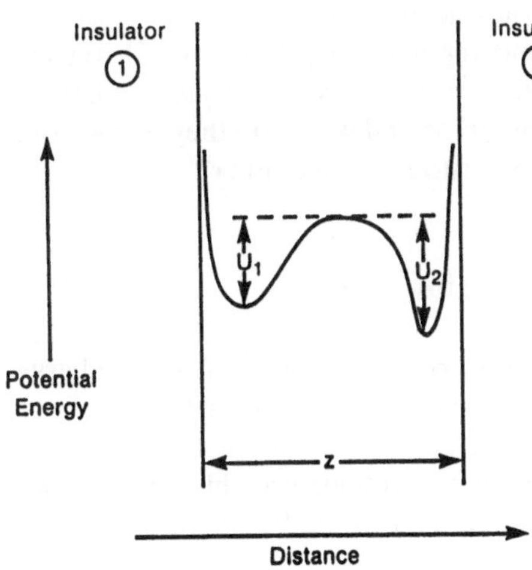

Fig. 4.8. Dependence of the potential energy of an ion on its position between two plane parallel insulator surfaces [4.40] (Reproduced by permission of Shirley Institute, Manchester, U.K.)

transferred over a distance z to equal this energy difference is

$$\frac{\sigma_S z}{K\varepsilon_0} = U_1 - U_2 \,. \tag{4.8}$$

A difficulty here is that it is unclear what value to take for z. *Lowell* and *Rose-Innes* [4.1] point out that if $U_1 - U_2$ is taken to be 1 eV and z is taken to be 3 Å, this predicts much larger charge densities (2×10^{13} charges per cm^2) than are usually observed.

The second mechanism depends on the abundance of a particular ion on one surface. Henry points out that if one ion per 10 Å^2 exists on one surface (which is actually a very large charge density, close to a monolayer), that corresponds to 10^{15} (n_i) ions per cm^2 available for charging. Even if $U_1 - U_2$ were as small as 0.2 eV, the number of ions that would transfer

$$\sigma_S = n_i \exp[-(U_1 - U_2)/kT] \tag{4.9}$$

would be $3.3 \times 10^{11}\,cm^{-2}$, well within the usually observed exchanged charge densities.

The third mechanism is based on what Henry calls a kinetic effect, basically a shearing off of charge from one surface to another as discussed by *Kornfield* [4.39] and others [4.32]. The amount of charge transferred would depend on the number of surface ions compensating the "intrinsic" electric field and the details of the shearing motion.

Any of the above three mechanisms could be operative if water layers existed between the two surfaces. Henry goes on to list other mechanisms including ones based on piezoelectric effects, pyroelectric effects and thermal gradients.

What seems to be lacking in this field are experiments that directly probe the nature of the charge carrier (ion or electron), the driving force, and the physical position of the exchanged charge (surface or bulk). The role of aqueous layers between the two materials needs to be pursued. Questions, originally proposed by *Henry* [4.40] and rephrased by *Morris* [4.6], remain unanswered: (1) What are the charge carriers–ions, electrons or charged material particles? (2) Where do they originate and where do they reside after transfer? (3) Why do they move from one surface to the other?

4.3 Insulator-Insulator Contact Charging

With such experimental and theoretical uncertainty in metal-insulator charging, it should not be surprising that our understanding of insulator-insulator charging is more limited. Traditionally, the approach that has been taken in this area is to attempt to order insulators such that a material higher up in the series will always charge positive when touched or rubbed with a material lower down. Such a "triboelectric series" should exist if one particular mechanism of charge exchange is operative. An example [4.50] is shown in Table 4.3; others can be found in [4.5, 31, 32, 51–53]. Although the positions of some materials such as glasses and natural fibers are erratic and there are even examples of triboelectric rings [4.31] (silk charges glass negative, and glass charges zinc negative, but zinc charges silk negative), there is some agreement as to the relative position of several polymers. Polytetrafluoroethylene (PTFE), polyethylene and polystyrene are invariably found at the negative end while nylon and polymethylmethacrylate are invariably found at the positive end of the series. Such uniformity tends to suggest a similar mechanism is occurring on samples prepared and handled in many different laboratories and leads to the hope that a theory of insulator-insulator contact charging could be constructed.

The next logical step is to attempt to correlate the position of a material in such a series with a property of the materials. Correlation with dielectric constant was suggested by *Coehn* [4.54]. His rationale was that a charged particle of charge Q, equidistant between two surfaces, will be attracted by its image force

$$\frac{1}{4\pi\varepsilon_0}\left(\frac{K_i-1}{K_i+1}\right)\frac{Q^2}{4r^2} \tag{4.10}$$

more strongly to the surface with a higher dielectric constant K_i, other things being equal (where r is the radius of the particle). However, it has been pointed out [4.6] that other properties order polymers similar to K_i, namely their percentage water absorption [4.53], yield strength [4.55] and their wettability [4.56]. Correlation of charging with trap densities estimated from measurements of radiation-induced conductivity was done by *Fukada* and *Fowler* [4.57]. They pictured charge exchange as the sharing of electrons between available traps. As mentioned earlier, correlations with Hammet substituent constant and with the basic and acidic nature of glass and polymers have been suggested.

Of particular interest is the work of *Davies* [4.58], and *Duke* and *Fabish* [4.59] in which they extended their measurements and theories from metal-insulator charging to insulator-insulator charging. Davies found that the charge exchange between two insulators can be predicted from a knowledge of the charge they acquired against metals. Davies interpreted this in terms of his model, that bulk charging brings the insulator Fermi levels into coincidence. A surface charging model would equally well explain the data as pointed out by *Lowell* and *Rose-Innes* [4.1]. Duke and Fabish found the same experimental result. They interpreted their result in terms of the density of states attributed to the insulators from the metal-insulator charging experiments. The only modification needed from their metal-insulator theory is that the "window" about the Fermi level from which charge exchange occurs is much wider.

Table 4.3. Triboelectric series. When two materials are contacted, the one lower on the series becomes negative (after [4.50])

Triboelectric series
Positive
Silicone elastomer with silica filler
Borosilicate glass, fire polished
Window glass
Aniline-formol resin, acid catalyzed
Polyformaldehyde
Polymethylmethacrylate
Ethylcellulose
Polyamide 11
Polyamide 6-6
Rock salt (NaCl)
Melamine formol
Wool, knitted
Silica, fire polished
Silk, woven
Polyethylene glycol succinate
Cellulose acetate
Polyethylene glycol adipate
Polydiallyl phthalate
Cellulose (regenerated) sponge
Cotton, woven
Polyurethane elastomer
Styrene-acrylonitrile copolymer
Styrene-butadiene copolymer
Polystyrene
Polyisobutylene
Polyurethane flexible sponge
Borosilicate glass, ground state
Polyethylene glycol terephthalate
Polyvinyl butyral
Formo-phenolique, hardened
Epoxide resin
Polychlorobutadiene
Butadiene-acrylonitrile copolymer
Natural rubber
Polyacrylonitrile
Sulfur
Polyethylene
Polydiphenylol propane carbonate
Chlorinated polyether
Polyvinylchloride with 25% DOP
Polyvinylchloride wihtout plasticizer
Polytrifluorochloroethylene
Polytetrafluoroethylene
Negative

The important conclusion from Davies's and Duke et al's. experiments is that it appears that insulator-insulator charging is caused by the same basic mechanism as metal-insulator charging. That is an important result. That several different models can explain the same result suggests that this conclusion is not model specific [4.1]. Another interesting conclusion that might be drawn is that if (1) metal-insulator and insulator-insulator contact charging are driven by the same mechanism and (2) if insulators do not have sufficient electrons to donate to a contact charge exchange, then perhaps neither experiment is driven by electron exchange!

4.4 Toner-Carrier Charging

The powder geometry of contact charging phenomena has some unique aspects, in addition to the fact that it represents the most commercially successful aspect of static electricity. First, the charging takes place with virtually no electric field being generated over macroscopic distances, due to the size of toner particles. This significantly reduces the probability that air discharge is affecting the results. Second, as 200 million toner particles are contained in a 10 g sample of developer, the statistics of large numbers assists in eliminating the effects of nonuniformities in materials, contact area, etc. However, thirdly, the dynamics of the contacts, the area and number of contacts, remain undetermined. We concentrate our discussion below on toner-carrier powders; there also exists literature on charging of single powders (see [4.60–65] and Chap. 8).

Toner is usually made from polymers such as polystyrene, polyacrylics, polymethyl methacrylate, etc. loaded to 10% by weight with carbon black. Carrier particles are coated with polytetrafluoroethylene, poly(vinylidene fluoride), etc. Because of manufacturing costs, there is usually a fairly broad size distribution of the toner particles from 1 to 30 μm distributed with a log normal distribution. Size-classified toner and spherical toner can be obtained for experimental purposes (Figs. 2.8 and 4.9).

The charge exchanged between particles in a powder is measured using a unique method [4.66–68]. Both powders are placed in a Faraday cage (Fig. 4.10) with screens on both ends which have holes intermediate in size between the diameter of the toner and carrier. An air jet forces the toner out of the cage. Measuring the change in charge and mass gives the average charge-to-mass ratio Q/M. It has been shown [4.68] that the measured charge is unaffected by toner interacting with the cage walls, screen, or other carrier on the way out, and air discharge does not occur upon separation. Typically observed charges are 10 μC/g on a 10 μm diameter toner particle, which corresponds to 2 nC/cm^2 or 10^{10} charges/cm^2 on the surface. This represents 3×10^4 charges per toner particle or one molecule (of size 10 Å^2) in 10^5 being charged, a very small number and consistent with charges observed in other

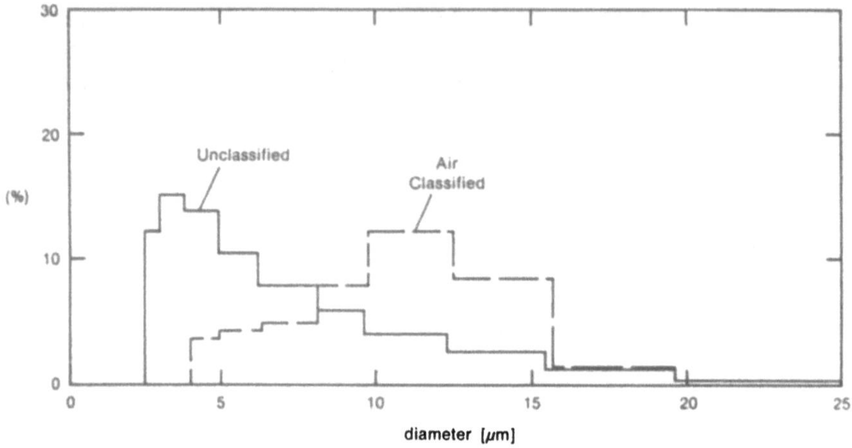

Fig. 4.9. The size distribution of the normal toner and toner size-classified to $10 \pm 2\ \mu m$. The distributions were cut off at $3.5\ \mu m$ because of inherent error in the measurement equipment [4.66]

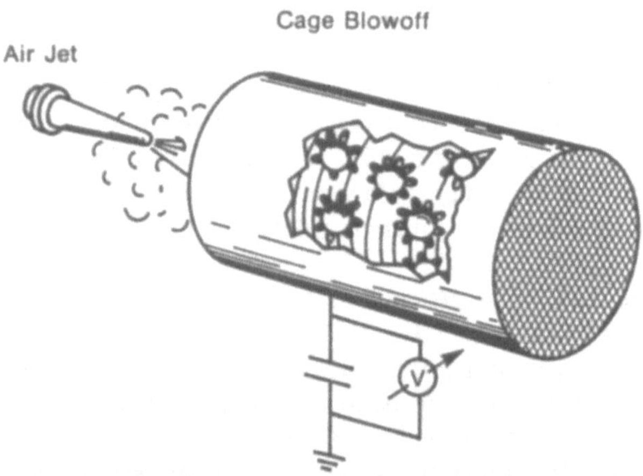

Fig. 4.10. Apparatus used in the total blowoff measurement. The stainless-steel cage has screens on both ends with holes intermediate in size between the diameters of the two powders. An air jet forces the toner out of the cage. Measurement of the charge Q and the mass left gives the Q/M ratio of the toner [4.68]

geometries. While this technique gives very limited charge characterization, it has been and remains the main charge measuring tool of electrophotographers.

A typical characterization technique [4.68] is shown in Fig. 4.11. Toner and carrier are put in a glass jar and rolled together. At various time intervals Q/M is measured. It is observed to increase for $\approx 1-2$ hours and then slowly decrease. The increase of Q/M with time is probably the same effect noted in insulator-metal charging: multiple contacts increase charge exchange. It could simply be an indication of the fractional area being contacted increasing

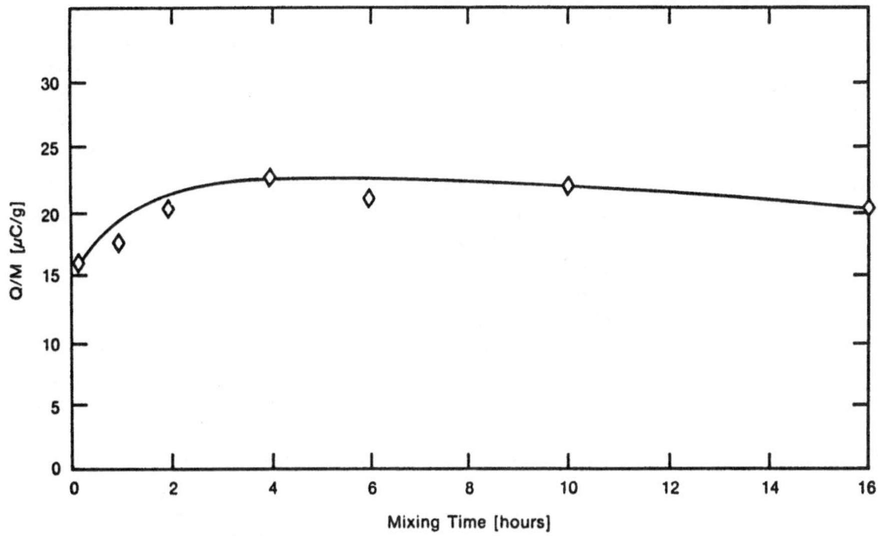

Fig. 4.11. Measurements of the average charge-to-mass ratio of toner mixed with bare Ni carrier as a function of mixing time [4.68]

as more time for interaction is allowed. The decrease of Q/M with time may reflect degradation of one of the materials. Many useful toners have Q/M of 10–25 μC/g.

Not all charging behavior in electrophotography is beneficial. *Cassiers* and *Van Engeland* [4.69] pointed out that carrier and toner also can be charged against the photoreceptor. If the triboelectric relationship is such as to produce wrong sign toner, a halo around characters will be obtained consisting of wrong sign toner. *Schein* [4.70] also has shown that photoreceptor charging occurs during development (Sect. 6.2.1). This prevents reuse of the latent image to make multiple copies.

After toner leaves the carrier during development, net charged carrier particles are returned to the developer reservoir. The discharge of these carriers has been investigated by *Hays* [4.71]. He showed that when the net charged developer is agitated, toner migrates through the carrier beads, reducing the net charge in its bulk, leaving a toner-starved bead layer on the surface. These surface beads are neutralized either by electric field assisted triboelectric charge exchange with the metallic walls of the reservoir or by net charge flow if the bead coating integrity is poor. (When charge exchange is blocked by, for example, painting the inside of the developer reservoir, nasty events occur, such as lightning bolts between the charged carrier and a nearby ground.)

Several review articles [4.7,44,72] have appeared on toner-carrier charging. As one might expect, triboelectric series of useful materials have been assembled. An example [4.73] of varying the carrier coating with a standard toner is shown in Table 4.4. Obviously the magnitude and sign of the toner charge can be controlled by properly choosing the carrier coating.

Table 4.4. Triboelectric series based on toner charge in xerographic developers (after [4.73])

Carrier coating Polymer	Polymer type	Produced by	Toner Charge [μC/g]
Kynar	Polyvinylidene fluoride	Pennwalt	+12.3
Saran F220	Vinylidene chloride-acrylonitrile copolymer (85/15)	Dow	+10.6
	Vinylidene chloride-acrylonitrile-acrylic acid (79/15/6) terpolymer		
	Cellulose nitrate		+ 9.2
Kel-F800	Chlorotrifluoroethylene vinylidene fluoride	3M	+ 8.1
Polysulfone P-1700	A diphenylene sulfone	Union Carbide	+ 8.0
Epon 828/V125	Epoxy/amine curing agent	Shell/General Mills	+ 7.0
	Cellulose acetate butyrate		+ 6.0
	Uncoated iron		+ 5.7
Cyclolac H-1000	Acrylonitrile-butadiene-styrene terpolymer	Borg Warner	+ 5.2
Hypalon 30	Chlorosulfonated polyethylene	DuPont	+ 5.0
Polysulfone P-3500	A diphenylene sulfone	Union Carbide	+ 4.6
Epolene C	Polyethylene	Eastman Chemical	+ 4.2
Polystyrene 8X	Polystyrene	Koppers	+ 3.6
Ethocel 10	Ethyl cellulose	Hercules	+ 3.1
Durez 510	Phenol formaldehyde	Durez	+ 2.7
Dupont 49000	Polyester	Dupont	+ 2.7
Estane 5740X1	Polyurethane	Goodrich	+ 1.1
Ganex V816	Alkyl-substituted polyvinyl pyrrolidone	GAF	+ 1.0
Formvar 7/70	Polyvinyl formal	Monsanto	+ 0.4
Lexan 105	A poly-bisphenol-A carbonate	GE	+ 0.3
Ganex V804	Alkyl-substituted polyvinyl pyrrolidone	GAF	+ 0.06
Ganex V904	Alkyl-substituted polyvinyl pyrrolidone	GAF	− 0.08
Dapon M	Diallyl phthalate	FMC	− 0.3
Lucite 2041	Methyl methacrylate	DuPont	− 0.4

One can be sure that a great deal of proprietary empirical knowledge has been built up on material choices for the best toners and carriers within corporations manufacturing electrophotographic toners.

4.4.1 Surface State Theory

To what extent has the study of metal and insulator contact charging been useful in increasing our understanding of toner-carrier charging? With the powder geometry one can vary the available toner area by varying the toner radius r or the toner concentration C_t (the toner mass relative to the carrier mass). One then can propose tests of models such as the surface state model discussed in Sect. 4.2.4. *Lee* [4.74] suggested the following application of the surface state theory to the toner-carrier system. Assume that after charge exchange, carrier surface states are filled between the carrier "work function" ϕ_c and energy ϕ_g and the toner surface states are filled between the toner "work function" ϕ_t and energy ϕ_g. Then on one carrier, the carrier charge Q_c is

$$Q_c = eA_cN_c(\phi_c - \phi_g) \tag{4.11}$$

and the charge on the n_0 toner particles on one carrier is

$$n_0Q_t = |Q_c| = e\,A_tN_tn_0\,(\phi_g - \phi_t), \tag{4.12}$$

where A_i are the available areas and N_i are the surface state densities per unit area per unit energy (assumed constant). Eliminating ϕ_g from the above equations gives

$$\frac{(\phi_c - \phi_t)}{n_0Q_t}e \equiv \frac{\Delta\phi e}{n_0Q_t} = \frac{1}{A_cN_c} + \frac{1}{A_tN_tn_0} \tag{4.13}$$

and, therefore,

$$\begin{aligned}\frac{M_t}{Q_t} &= \frac{M_t}{\Delta\phi e}\left[\frac{n_0}{A_cN_c} + \frac{1}{A_tN_t}\right]\\ &= RC_t\left(\frac{\rho_c}{3\Delta\phi eN_c}\right) + r\left(\frac{\rho_t}{3\Delta\phi eN_t}\right),\end{aligned} \tag{4.14}$$

where $\rho_c(\rho_t)$ is the carrier (toner) density, $R(r)$ is the carrier (toner) radius and C_t is the toner concentration,

$$C_t = \frac{n_0M_t}{M_c}. \tag{4.15}$$

This predicts M_t/Q_t is linear in toner concentration C_t with an intercept proportional to the toner radius and a slope proportional to the carrier radius (Fig. 4.12a).

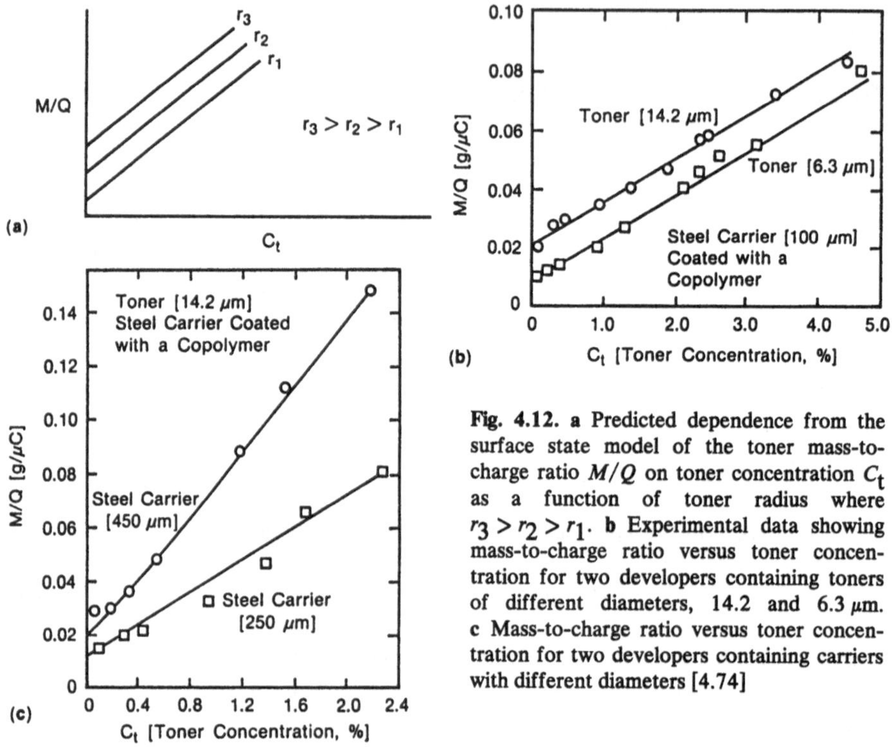

Fig. 4.12. a Predicted dependence from the surface state model of the toner mass-to-charge ratio M/Q on toner concentration C_t as a function of toner radius where $r_3 > r_2 > r_1$. **b** Experimental data showing mass-to-charge ratio versus toner concentration for two developers containing toners of different diameters, 14.2 and 6.3 μm. **c** Mass-to-charge ratio versus toner concentration for two developers containing carriers with different diameters [4.74]

The experiments presented by Lee tend to confirm this model. In Fig. 4.12b, M/Q (the average value of the mass-to-charge ratio obtained from a blowoff measurement which should approximate M_t/Q_t) is plotted versus C_t for two different toner diameters. The data do indicate M/Q is linear in C_t as predicted and the intercept appears to be linear in toner diameter. In Fig. 4.12c, the carrier diameter was varied. The slope does appear to be linear in carrier diameter as predicted. The data can be analyzed to give surface state densities ($N_c\Delta\phi$ and $N_t\Delta\phi$) of about $N_t \approx 5 \times 10^{10}\ \mathrm{cm}^{-2}$, and $N_c \approx 6 \times 10^8\ \mathrm{cm}^{-2}$ consistent with numbers obtained in insulator charging experiments. As $C_t \to 0$, Q should be proportional to r^2 (the area of a toner particle), see (4.14). This result has also been reported by *Raschke* [4.75] and *Fiedler* and *Stottmeister* [4.76] (although it is unclear whether they were actually in the low C_t region).

4.4.2 Carbon Black

Carbon black is a major constituent of toner (≈ 5%–10% by weight), with vastly higher conductivity and electron density of states than the other polymeric constituents. That has naturally led to consideration of the role that carbon black plays in the toner charging.

Daly et al. [4.77] have investigated the way in which toner preparation affects its charging and dielectric properties. They showed that a correlation exists between the dielectric constant and the charge-to-mass ratio, determined by blowoff measurements after mixing the toner and a standard carrier for a sufficient time to achieve a stable equilibrium charge. Their toner samples were prepared by varying the the milling time for a constant 10% carbon black loading. This altered the dispersion of the carbon black. It was observed that as the milling time increased, Q/M decreased and the dielectric constant decreased. They also showed that Q/M and the dielectric constant increased with carbon black loading. They concluded that the dielectric constant is due to charge polarization across chained carbon black particles (hence the dielectric measurements provide a measure of the carbon black dispersion) and high values of Q/M are associated with the existence of these chains. They suggested this association may be due to the carbon black connecting the outside surface to the internal polymer providing an increase in charging area. Charge migration into the bulk of the toner is being postulated, although one might expect mutual electrostatic repulsion would tend to keep the charges on the toner surface.

Brewington [4.78] investigated initial toner charging by drawing a monolayer of toner with a doctor blade in an applied electric field. She showed that the charging was dependent on the electric field and the carbon black on the surface of the particles. Others [4.79] have shown that acidity and oxygen content of the carbon black have an effect on toner charging. *Julien* [4.80] has observed a connection between charging and contact potential of the carbon black. *Prest* and *Mosher* [4.81] have demonstrated that various types of oxygen-containing functionalities can contribute to Q/M.

The effect of carbon black on charging is obviously complicated because of the many sources, its acidity, oxygen content, conductivity, loading, dispersion, etc. *Julien* [4.80] summarizes recent work by stating that "carbon black will have... an effect when the difference between the toner polymer and carrier polymer in a triboelectric series is small"; "when the toner polymer and carrier polymer themselves generate substantial tribo" (i.e., charge) "the effect of carbon black is muted." As Julien points out, this is a surprising result considering the conductivity and density of electron states of carbon black relative to the polymer. Of course, this result is only surprising if electron transfer theories are being considered. It is less surprising if ions are transferred.

4.4.3 Charge Control Agents

Additives called charge control agents [4.44,82] increasingly are being added to toner to control the toner charge. While virtually no scientific literature exists yet, patent claims include an increase of the rate of charging, stabilization of the charge (with mixing time), and a decrease of the amount of wrong sign toner, all potentially very useful improvements.

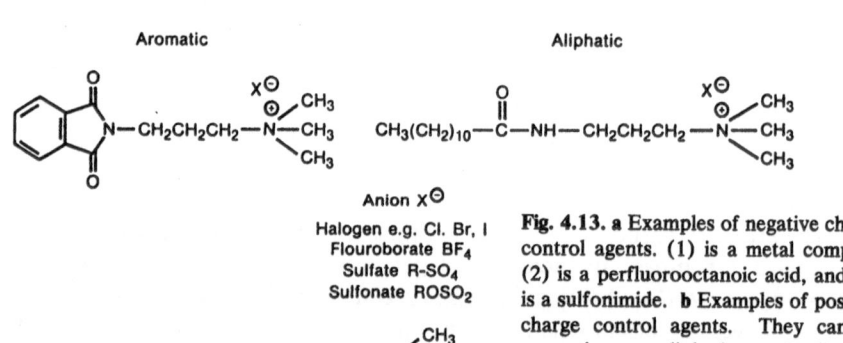

(a) Negative Charge Control Agents

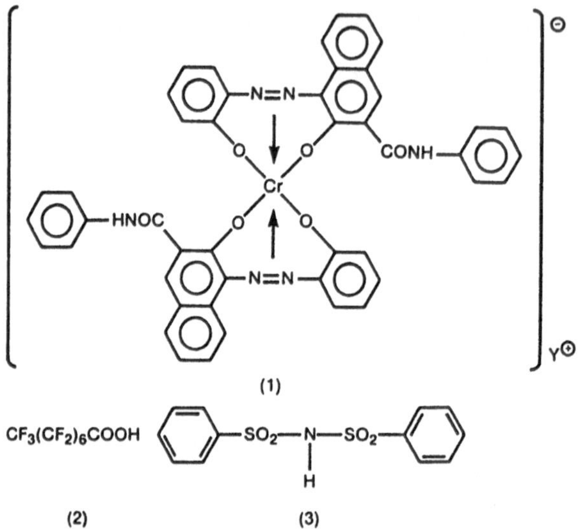

(b) Positive Charge Control Agents

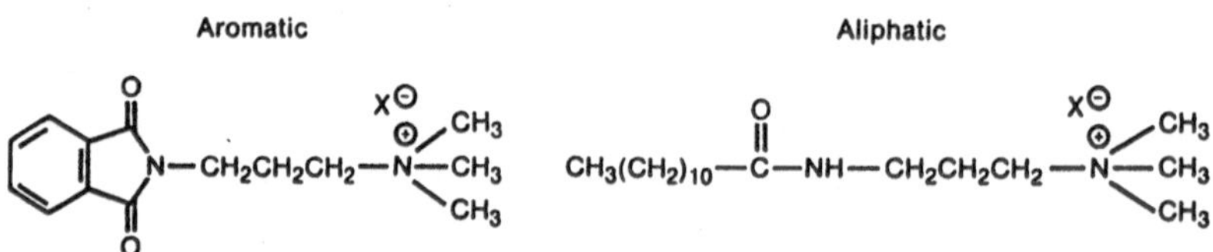

Anion $X^{\ominus}$

Halogen e.g. Cl. Br, I
Flouroborate BF_4
Sulfate R-SO_4
Sulfonate $ROSO_2$

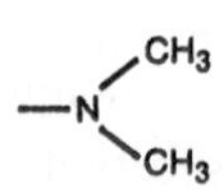

Amine Group

Fig. 4.13. a Examples of negative charge control agents. (1) is a metal complex, (2) is a perfluorooctanoic acid, and (3) is a sulfonimide. **b** Examples of positive charge control agents. They can be aromatic or aliphatic. A list of counterions is shown at the bottom of **b** [4.82] (Courtesy of Society for Information Display)

Charge control agents (sometimes known as charge directors) are added to either the toner surface or bulk [4.82]. Examples of surface charge control agents are cabosil, a fumed silica, and Kynar, a highly fluorinated polymeric material. Bulk charge control agents which are melt blended into the toner polymer differ for negative and positive toners. Metal complex dyes shown in Fig. 4.13a have been suggested as charge control agents for negative toners [4.83]. Bulk charge control agents for positive toners can be amines and quaternary ammonium salts (Fig. 4.13b); in either case they may be aliphatic or aromatic (both with long aliphatic hydrocarbon chains). The counterion can be halogens, fluoroborates, sulfates or sulfonates. These materials are ionic surfactants very similar in chemical structure to the antistatic agents used to treat fabrics [4.82]. The materials are generally colorless and therefore suited for color toner formation, in contrast to the metal complexes which are generally highly colored.

Little information appears available on how these charge control agents affect the charge distribution. *Birkett* and *Gregory* [4.83] have proposed that the charging process probably involves transfer of the counterion of the charge control agent on the toner surface to the carrier surface upon contact. They suggest this mechanism qualitatively explains why 2:1 chromium or cobalt complex azo dyes having a proton as a counterion are observed to be particularly effective control agents: (1) the proton, being the smallest counterion, is highly mobile and (2) the single negative charge in the dye anion is delocalized over an extremely large π-system resulting in an easy removal of the proton. Historically, these ideas extend the thoughts of Medley and others (Sect. 4.2.3) that charge exchange could depend on the acidic nature of the surface, with OH^- and H^+ moving between the surfaces. The observation of water layers on the surfaces of toner (Fig. 4.6) certainly suggests OH^- and H^+ could be moving in an aqueous layer between materials. At present, the number of different chemical classes of either positive or negative charge control agents is currently limited to a few basic structures. This area promises to be an exciting chemistry and physics problem with advances producing important improvements in the development subsystem.

4.4.4 Charge Measuring Tools

The above discussion deals with an average charge property, the average charge-to-mass ratio. From preceding discussions it is clear that the charge distribution within a mixture is critically important to the behavior of a development system. New tools for measuring the charge distribution are obviously needed. A refinement to the blowoff technique was introduced by *Schein* and *Cranch* [4.68] and improved by *Harpavat* and *Orr* [4.84] in which the air pressure is incremented during the blowoff measurement allowing a distribution of charge-to-mass ratios to be obtained. Further, by using size-classified toner [4.68], which fixes the mass, the actual charge distribution can be obtained. Schein and Cranch showed that size-classified toner has an extremely narrow charge distribution. Harpavat and Orr observed different distribution widths but they did not appear to correlate with the size distribution. *Collins* [4.85] demonstrated with this technique that the charge on toners within his mixes were distributed exponentially in a parameter which measured the amount of toner remaining in the mix. The data are consistent with a picture in which (1) the charge acquired by toners is proportional to the uncharged areas on the carrier beads and (2) toner particles are blown off in the reverse order in which they are attached, i.e., the first and most highly charged toners are the last to be removed.

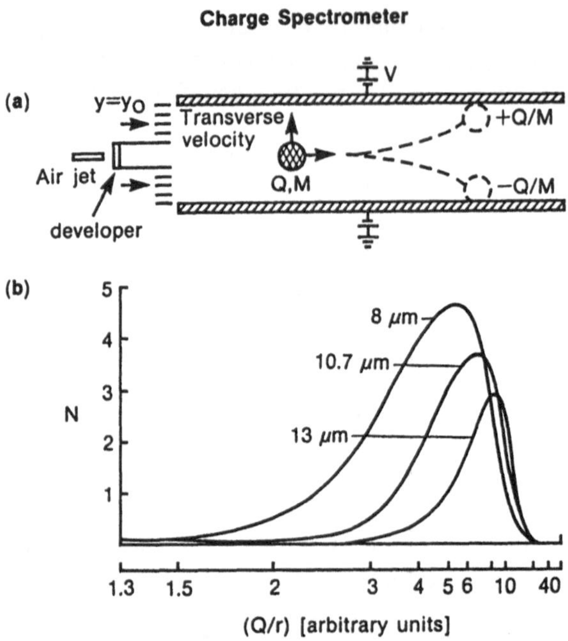

Fig. 4.14. **a** In this charge spectrometer, toner is injected into a laminar air stream, deflected by a perpendicular electric field, and deposited on the side plates. **b** Q/r distributions of toners with mean particle sizes of 8 μm, 10.7 μm and 13.0 μm. The toner concentration was held constant. Since the peak positions are nearly proportional to r, this suggests $Q \propto r^2$ for constant toner concentration (after Hölz in [4.88])

Another charge measuring technique, reported by *Simpson* (discussed by *Williams* in [4.86)]), *Stover* and *Schoonover* [4.87] and *Hölz* [4.88] and shown in Fig. 4.14, is to inject the toner into a laminar air stream flowing transverse to the electric field. This separates particles by Q/r. Results shown in Fig. 4.14b suggest that within a mix the peak $Q/r \propto r$ for constant C_t. This would appear to support the surface state theory.

Of potential importance is another charge spectrometer reported by *Lewis* et al. [4.89]. Instead of collecting toner on the side plates of Fig. 4.14, it is collected at the bottom of the apparatus on filter paper. Since the Coulomb force $Q_t E_{air}$ moves the toner against the viscous force, the toner terminal velocity v_t in the direction of the electric field is given by (Fig. 4.15)

$$Q_t E_{air} = 6\pi\eta r v_t \,, \tag{4.16}$$

where η is the viscosity of air and r is the toner radius. The displacement y for a length of time t in the spectrometer is then given by

$$y = \left(\frac{Q_t}{r}\right)\frac{E_{air}}{6\pi\eta}\,; \tag{4.17}$$

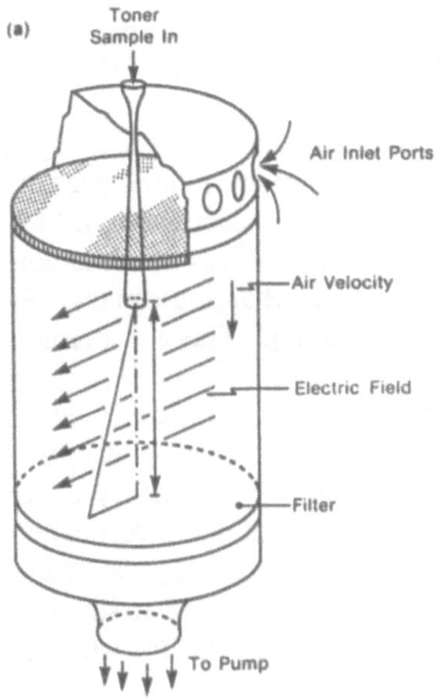

Fig. 4.15. a A different charge spectrometer in which laminar air flow and crossed electric fields are used, but toner is collected on the bottom plate. The position where a toner contacts the bottom plate depends on (Q/r). **b** A $10 \pm 0.5\,\mu$m diameter toner fraction from a working developer. Note the peak sharpness. **c** The $10 \pm 0.5\,\mu$m diameter toner fraction from a worn-out developer. Note the large numbers of uncharged and reverse-charged particles [4.89]

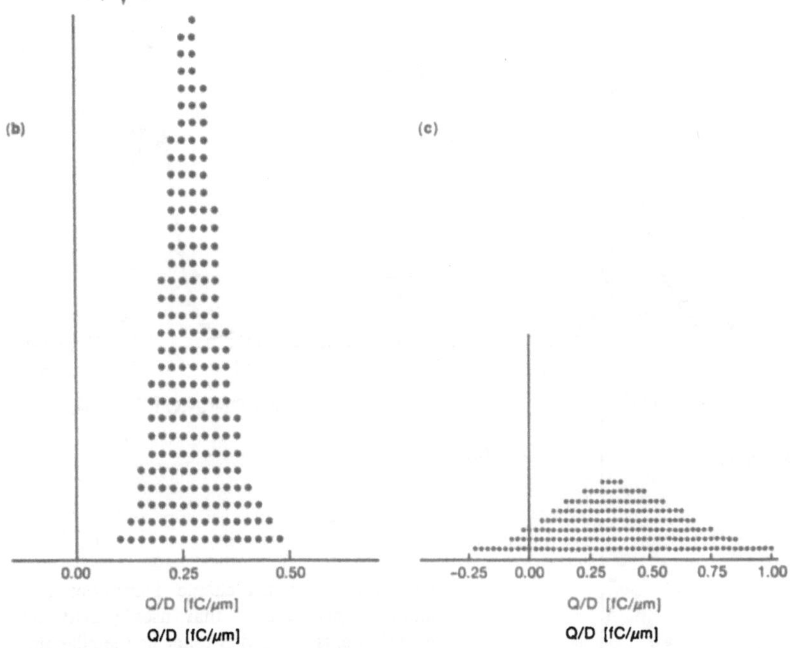

the displacement is proportional to Q_t/r. By measuring with a microscope-computer system the toner size at each displacement, the total Q_t and r distribution function can be unfolded. This spectrometer is particularly powerful because it can directly measure the amount of wrong sign and zero charged toner. Figures 4.15b and c show a comparison of fresh and worn-out developer (at the $10 \pm 0.5\,\mu$m diameter toner fraction). Clearly a rather sharp distribution has been considerably broadened to include both zero and reversed sign toner. In a similar instrument, *Terris* and *Jaffe* [4.90] observed $Q_t/r \propto r^n$ with $n = 0.4$ to 0.6 for small r, and $n = 0$ for large r. Very similar data were obtained by *Demizu* et al. [4.91] on toner from a completely different system, a nonmagnetic monocomponent development system. Terris and Jaffe suggested a transition from a surface state model at small r (which predicts $n = 1$) to a contact area model at large r in which it is assumed that the charged area resides within a fixed height above the carrier surface (which predicts $n = 0$). The contact area model requires some surface conductivity. In fact, in the limit of a "conductive" toner (with RC time constant less than the time toner sits on the carrier), one could imagine the toners charging capacitively as *Harper* [4.3] proposed for metal-metal contact charging. This also predicts Q/r is independent of r.

A modification of this technique has been suggested by *Takahashi* et al. [4.92]. Toner, released from carrier in a vibrating cell, falls under gravity in a horizontal electric field. Photographic images of the toner trajectory under chopped laser illumination gives the velocity in the horizontal and vertical directions, which can be used to obtain the toner radius and charge-to-radius ratio respectively. Charge distributions of two toners with similar size distributions and charge-to-mass ratios are shown in Fig. 4.16. The one with less

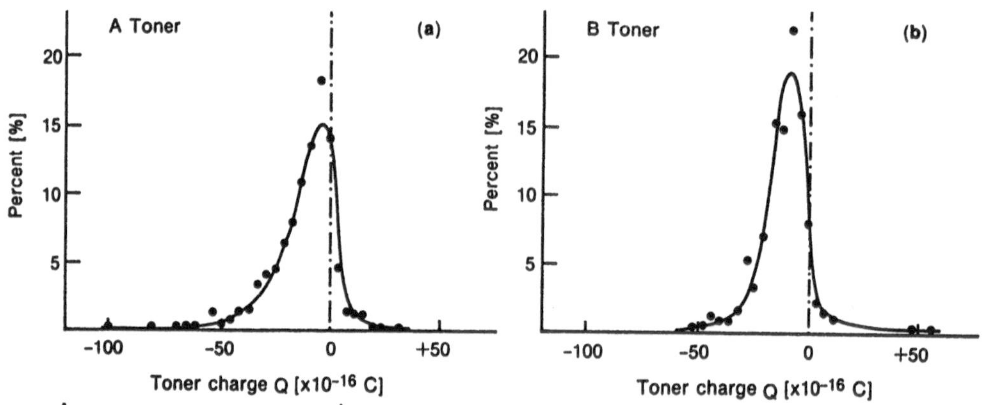

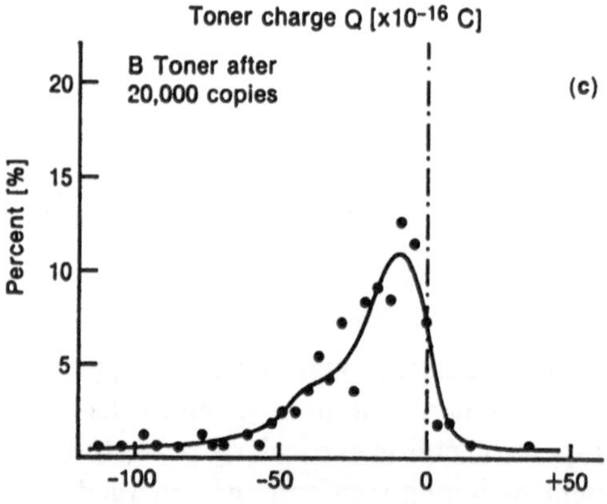

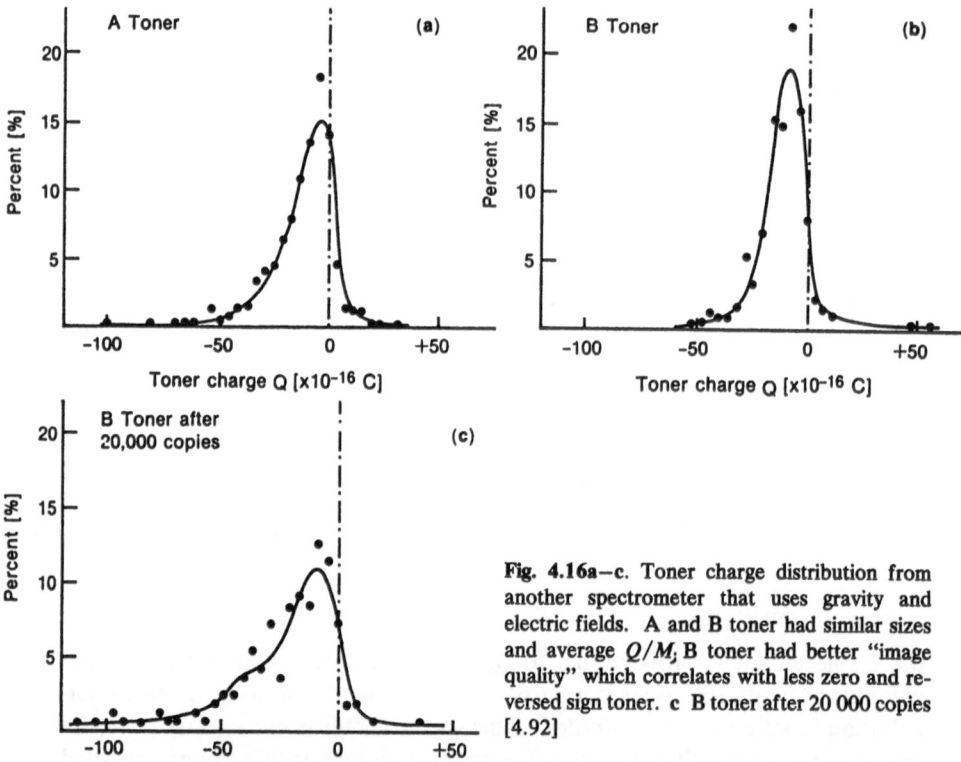

Fig. 4.16a–c. Toner charge distribution from another spectrometer that uses gravity and electric fields. A and B toner had similar sizes and average Q/M; B toner had better "image quality" which correlates with less zero and reversed sign toner. **c** B toner after 20 000 copies [4.92]

zero and wrong sign toner had "better" image quality. After running 20 000 copies, the charge distribution of the B toner was observed to change, with the peak decreased and higher charged toner appearing, suggesting continual charging is occurring in the mixing portion of the development system hardware. The authors state that the method's problems include separation of toner from carrier and reduction in measurement time.

Another charge characterization technique has been discussed [4.93] which uses laminar air flow, electric fields, and gravity to achieve a separation of the Q_t and r function (Fig. 4.17). The smaller the particle, the longer it takes to fall under gravity and the further down the plate it goes. The smaller Q_t/r, the further down the x axis it travels. Therefore, contour lines of constant charge and radius are produced by the toner particles.

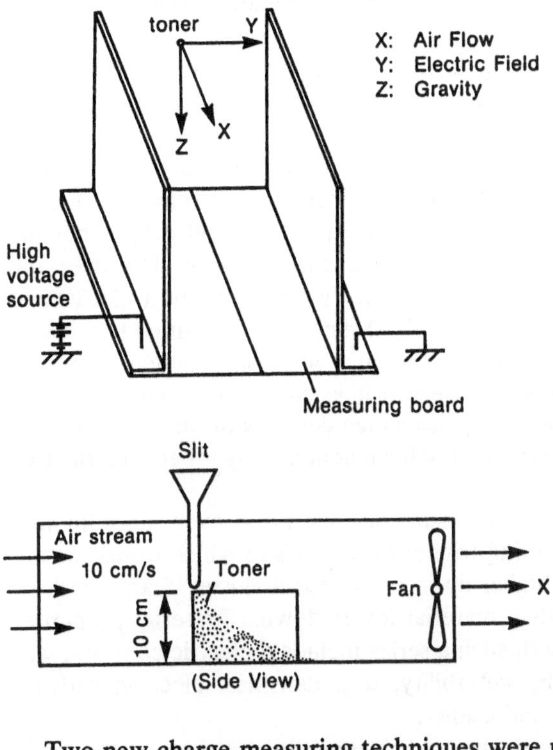

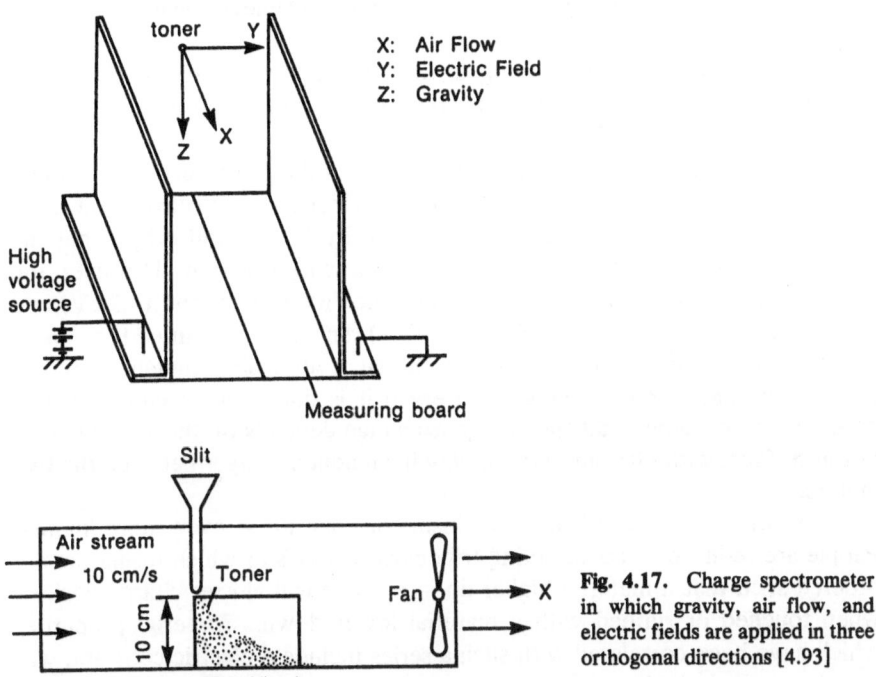

Fig. 4.17. Charge spectrometer in which gravity, air flow, and electric fields are applied in three orthogonal directions [4.93]

Two new charge measuring techniques were recently reported. It is well-known that the Millikan oil drop experiment measures the charge on particles. *Kutsuwada* and *Nakamura* [4.94] applied this technique to characterizing toner particles, one at a time. *Mazumder* et al. [4.95] described a more complicated technique in which a laser Doppler velocimeter is used to measure the velocity of a toner particle subjected to a dc electric field and an ac acoustic pulse. The phase lag between the acoustic drive and the particle motion gives the size of the toner particle; the velocity component due to the electric field give the toner's charge/diameter ratio.

While some data is given for each of these charge spectrometers, systematic experiments to understand the physics and material parameters determining toner charging behavior and its distribution have not been reported.

4.5 Summary

The toner charge distribution determines the behavior of the development system. This chapter was devoted to a detailed description of the status of our understanding of toner charging and characterization of these distributions. Toner charging is but one example of the phenomenon of contact charge exchange which occurs between all materials, organic and inorganic, metals and insulators. As such, contact charging is a pervasive solid-state physics problem. It is also a very poorly understood problem whenever insulators are involved.

Metal-metal contact charging is determined by electron exchange to maintain thermodynamic equilibrium; theory and experiment appear to be in accord.

On the other hand, there is no agreement on the mechanism determining metal-insulator contact charging. There is evidence for electron exchange. The heart of the argument is the observation by Davies and others that the observed charge exchange is linear in the metal work function. However, the observation that many polymers have the same "work function" (4.2 eV) and the extracted state density ($\approx 10^9\,\mathrm{cm^{-2}\,eV^{-1}}$) remain unaccounted for. There is evidence that clean polymers do not charge at all against metals. The implication is that when charging is observed it is due to ionic contaminants. There is also evidence that the charge exchange depends on the acidic nature of the surface, with OH^- and H^+ moving in an aqueous layer between the two materials.

Our understanding of insulator-insulator charging is even more limited. People are reduced to constructing triboelectric series in which insulators are ordered such that a material higher up in a series will always charge positive when touched or rubbed with a material lower down. Material properties which have been correlated with such a series include dielectric constant, water absorption, yield strength, wettability, trap densities, electronegativity, Hammet substituent constant and acidity.

The toner-carrier powder geometry has some unique aspects that could be exploited to study fundamental insulator charging physics, including the fact that (1) charging takes place with virtually no external electric field over macroscopic distances being generated and (2) very large numbers of particles are contained in a few grams of sample. This allows the statistics of large numbers to assist in eliminating the effects from nonuniformities in materials and contact areas. As might be expected, triboelectric series of useful materials have been assembled which serve as useful guides but have not led to an understanding of the mechanism of charge exchange. Of considerable interest is the development of new tools that allow examination of the actual toner charge distribution. None of these has yet been exploited to do systematic experiments to understand the physics or material parameters determining the charging behavior. Only one paper exists that tests the surface state theory using the standard total blowoff measurement, with some success. Also of considerable interest is the recent trend to add charge control agents to toner. An enormous patent literature is evolving but virtually no scientific studies have been published to date. The emergence of new measuring tools and new chemistry for a property as crucial to electrophotography as toner charge suggests important advances can be expected in the near future.

5. Cascade Development

It was clear to the early inventors of electrophotography that a method was required to charge and transport the fine powder that we now call toner. These powders are difficult to handle. They move like dust and settle on ev-

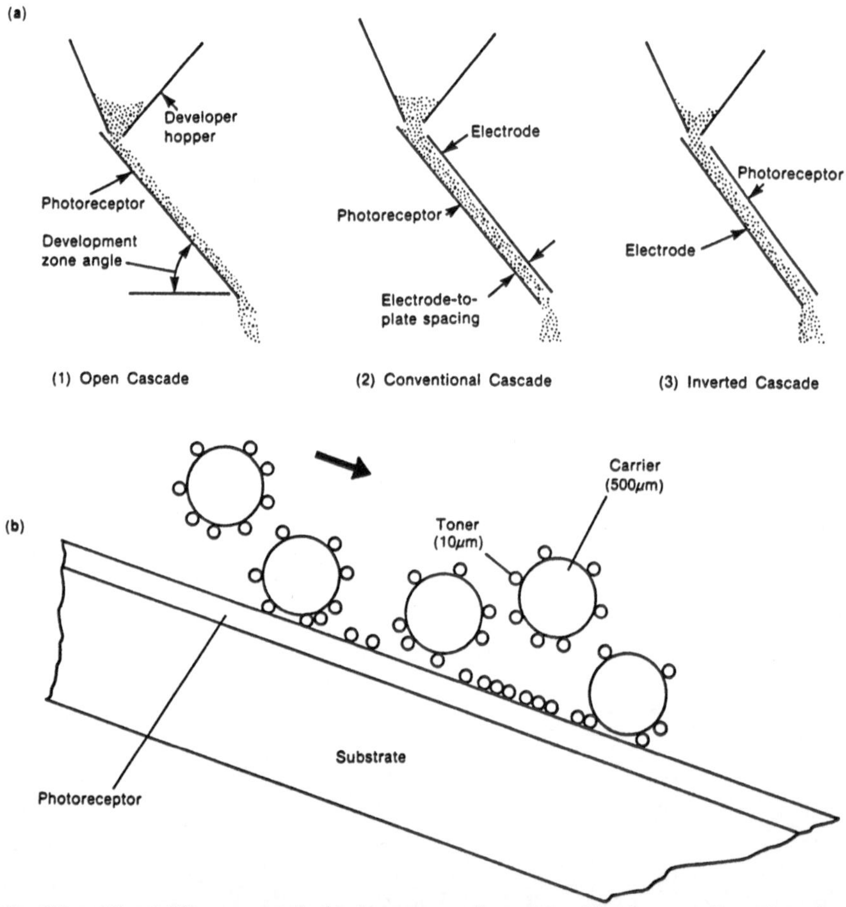

Fig. 5.1. **a** Three different cascade development configurations. The three configurations shown are (1) open, (2) conventional, and (3) inverted cascade development. Only in the last two configurations is an electrode present. **b** A blowup of the carrier beads cascading down the photoreceptor surface [5.2,3] (© 1972 IEEE)

erything from the background regions of the photoreceptor to gears and mirrors. The cascade development system invented in 1952 during the Battelle-Haloid collaboration [5.1] solved both the charging and transport problem. It is shown schematically in Figs. 3.5 and 5.1 at different magnifications. Toner (average diameter $10\,\mu m$) is mixed with carrier (diameter $\approx 200-500\,\mu m$). Charge exchange causes the toner to be electrostatically attracted to carrier beads (Chap. 4). The carrier beads are then literally cascaded down the photoreceptor under the influence of gravity. The interaction of the toner with the latent image clearly depends on the behavior of the cascading carrier beads such as bead velocity and bounce rate . These variables depend on hardware parameters such as the developer flow rate, the angle of the photoreceptor with respect to gravity, etc. The cascade development system has three forms. Open cascade has no electrode (Fig. 5.1a), conventional cascade has an electrode above the photoreceptor, and inverted cascade has an electrode underneath the photoreceptor. Figure 5.1d shows a blowup of the carrier beads cascading down the photoreceptor surface.

While cascade development was the most successful early development system which resulted from the Battelle-Haloid collaboration, many other ideas were tried [5.2,4]. (1) In aerosol or powder cloud development (Sect. 9.1), charged toner (charged by interacting with a nozzle) or zero net charged toner (with toner charge equally distributed between positive and negative polarities) is introduced as an aerosol suspended in air in a channel between the photoreceptor and an electrode. (2) In touchdown development (Sect. 9.2), a surface bearing a layer of charged toner particles is brought into actual or close contact with a photoreceptor. (3) In fur brush development, a fur brush serves to charge the the toner and transport it across the latent image. (4) In dual component liquid development (electrophoretic development) small toner particles are suspended in a high resistivity fluid in which the photoreceptor is immersed. The toner charge results from the formation of an electric double layer at the interface (Chap. 10). (5) In single component liquid development, electrohydrodynamic instabilities of a conducting liquid exposed to an electric field are used to pull liquid onto a latent image (Sect. 2.3.4). (6) In thermoplastic development the electric field of the latent image is used to produce deformations in a heated thermoplastic (Sect. 1.4).

The cascade development system was used in all of the early Xerox copiers including Model D (a hand operated copier), the famous 914 (the first completely automatic electrophotographic copier) and the 2400 series. It was finally replaced by the magnetic brush development system in the early 1970s for reasons of physical size, speed, and image quality.

Work on optimizing the cascade development system progressed in roughly three stages. At first empirical approaches were adopted, which resulted in data characterizing the system [5.3,5]. Later, qualitative concepts were introduced to explain the behavior of toner [5.2,6–8] such as airborne toner, contact development, and impact assisted electric field stripping. Finally, quantitative theories were introduced and compared with data [5.9,10]. The progression towards quantitative theories of development was driven by the need to identify and optimize the many parameters of the development system and to understand the source of copy quality defects. One example of a copy quality defect discussed in this chapter is a light area just inside the edge of a solid area. Another example which many people associate with the early copiers was the loss of solid areas (Fig. 3.1). An explanation of the source of this defect follows from an understanding of the electrostatics within the development system (Fig. 5.2). A solid black has a uniformly charged latent image (Fig. 5.2b). In the absence of a counter electrode, the only electric field above the photoreceptor is associated with the fringe field at the edges of the solid (Figs. 5.2c and d). In the inside of the solid area no electric field exists and no force on toner exists. Therefore only the edges of solids develop (Fig. 5.2e).

The difficulties in achieving quantitative theories of development, which are common to many aspects of electrophotography and which slowed

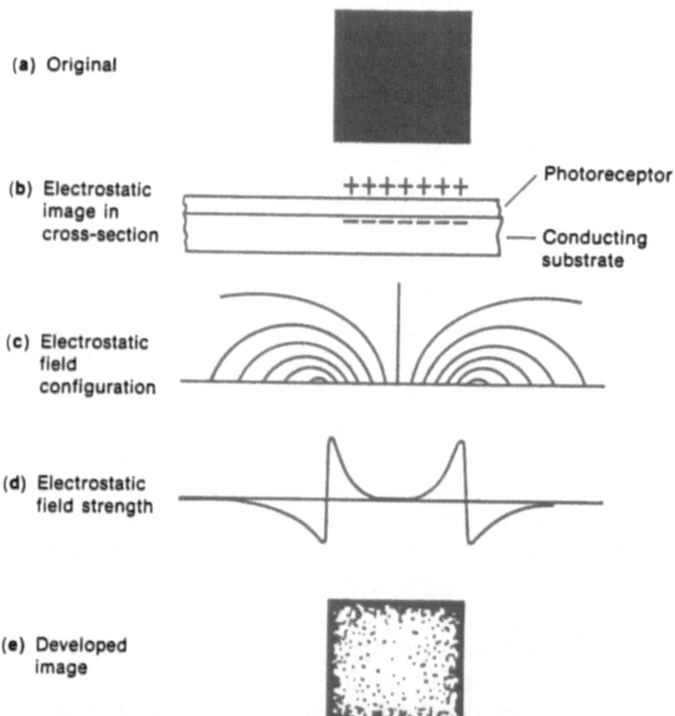

Fig. 5.2. A solid area on a copy (**a**) becomes a uniform charged density on the photoreceptor (**b**), which has only fringe fields in the absence of a counter electrode (**c,d**). Toner is therefore only attracted to the edges of the solid, creating the classic solid area image defect (**e**) [5.2] (© 1972 IEEE)

Table 5.1. List of parameters in cascade development

Development zone angles
Developer flow rate
Total developer flow
Development time
Electrode spacing
Carrier type, Diameter, Coating
Toner type, Size, Charge, Concentration
Photoreceptor image potential
Background potential
Bias on electrode

progress, are (1) controlled experiments are difficult because of the many parameters, and (2) many mechanisms of development were in principle possible. In Fig. 5.1 are shown some of the important hardware and material parameters. A list of these parameters is given in Table 5.1. Any experiment designed to test the effect of a parameter must control all other parameters, an obvious but difficult requirement. In fact, progress was made only when sufficient time was allocated (several years) to carry out such controlled experiments [5.3]. The second argument was based on estimates of the various forces on toner. In Table 5.2 and Fig. 5.3 is shown the possible range of potentially important forces on toner particles. Toner radii can vary from 1.5 to 12.5 μm; Q/M 's can vary from 5 to 40 μC/g. The image force assumes the toner's charge is at the center of the toner particle; *Jackson* [5.11] gives the force to be

$$\frac{1}{4\pi\varepsilon_0}\frac{K-1}{K+1}\frac{Q_t^2}{(2r)^2}\,, \tag{5.1}$$

where K is the dielectric constant of the material the toner contacts. We assumed $K=\infty$, i.e., the material is metallic, for this calculation. The impact force assumes an average carrier velocity change on impact of 15–200 cm/s, taken from *Stover*'s measurements [5.12] and an impact time of 10 μs (guessed, since no estimate of this number appears available). The Couloumb forces assumes a field of 0.1–1 V/μm, estimated from *Neugebauer*'s [5.13] calculations for the perpendicular electric field above the photoreceptor surface (Chap. 3), enhanced by the effective dielectric constant of the mix, 3 or 50. The viscous force assumes that relative to a toner the air velocity is 15–200 cm/s. The polarization force assumes ∇E^2 is 10^{11} V^2/cm^3 estimated from *Neugebauer*'s [5.13] calculations. Inspection of Fig. 5.3 reveals that only

Table 5.2. Parameters used to calculate the forces on toner particles

	r [μm]	Q/M [μC/g]	Voltage [V]	L (Nip spacing) [μm]	K_E (Effective dielectric constant)	Change in velocity [cm/s]
Min.	1.5	5	100		1	15
Ave.	6.5	20	400	1000	3	100
Max.	12.5	40	1000		50	200

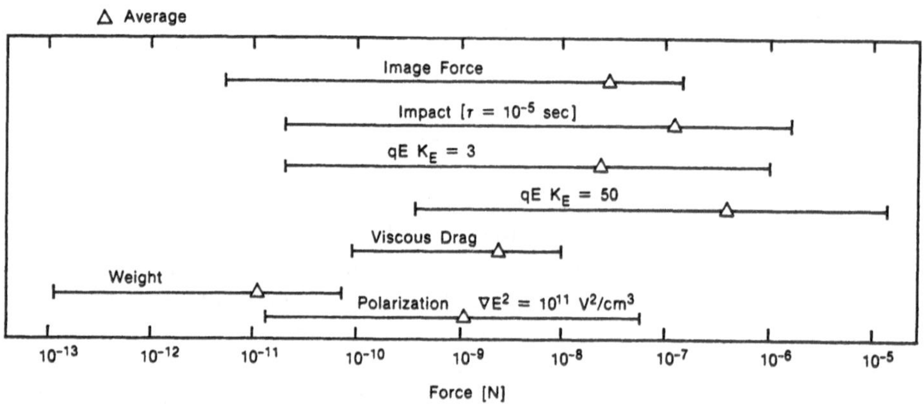

Fig. 5.3 Range of potentially important forces on toner particles.

gravity can be ignored. The other five forces are all the same order of magnitude, making it conceivable that a variety of mechanisms could be causing development. One point of view is [5.2] that "there are such a large number of factors and forces involved in the process that there results wide variations in the fate of individual particles. However, in practical systems these variations are averaged out and consistent and reproducible results can be readily obtained. It is this averaging that makes the process commercial, but it also tends to obscure from view the basic mechanisms of development."

5.1 Development Mechanisms

Let us begin the discussion of cascade development by describing in qualitative terms the suggested mechanisms of development. The crux of the problem can be phrased simply: by what mechanism(s) do toners leave the carrier and end up on the latent image. The objective of such a development theory is to predict the amount of toner (mass per unit area) that ends up on the latent image.

The equation should explicitly contain the important hardware and material parameters (listed in Table 5.1).

5.1.1 Airborne

Sullivan and *Thourson* [5.6] in 1967 were the first to suggest the airborne mechanism. It is envisioned that toner is stripped from a carrier particle at a point remote from the image (by, for example, the inertial forces that occur when a carrier impacts the photoreceptor, an electrode or another carrier bead) and is carried to the image by the electrostatic field of the image (see carrier B in Fig. 5.4) By calculating the volume of space between electric field lines (Fig. 5.5) and assuming a distribution of free toner in space, a prediction

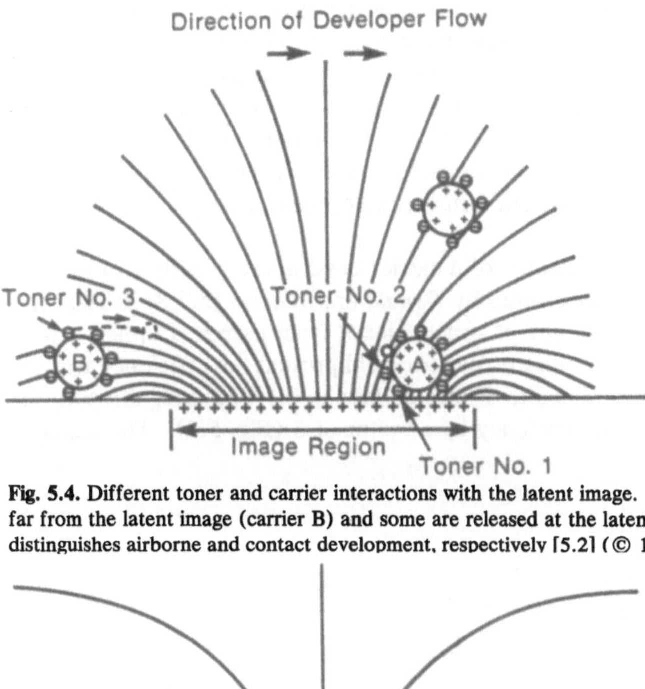

Fig. 5.4. Different toner and carrier interactions with the latent image. Some toners are released far from the latent image (carrier B) and some are released at the latent image (carrier A). This distinguishes airborne and contact development, respectively [5.2] (© 1972 IEEE)

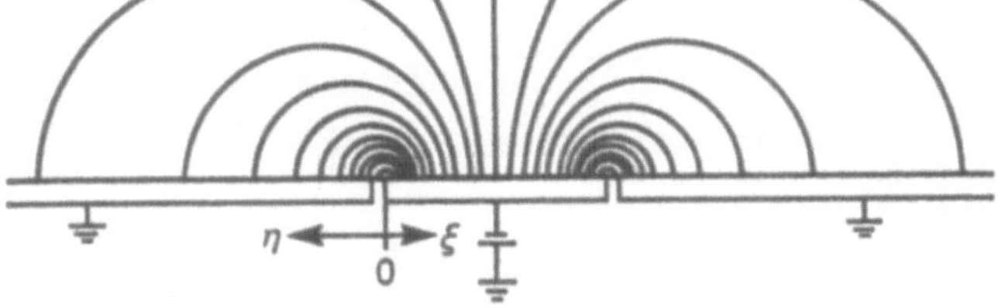

Fig. 5.5. Fringe field configuration around a line and a coordinate system used for the airborne development theory. The charge on the photoreceptor is simulated with a constant voltage [5.6]

of mass per unit area can be made. There were two weaknesses of this model. First, significant approximations were made in defining the conditions under which free toner is generated; therefore, important physics and parameters were missing from the model. Second, it did not predict the observed high development at the edge of lines because the volume of space from which a line edge could collect toner was too small (Fig. 5.4).

5.1.2 Contact

The theoretical analysis of the airborne mechanism and comparison with experiment led to the concept of contact development, in which the source of toner is from carrier beads which contact the photoreceptor on the latent image. *Thourson* [5.2] distinguishes between two types of contacts, ones in which simultaneous contact is made between the toner and both the carrier and the photoreceptor (toner 1 on carrier A, Fig. 5.4.) and ones in which only the carrier contacts the photoreceptor (toner 2 in Fig. 5.4). No follow-up of the potential different behavior of these two types of toner particles appears in the cascade literature. The simultaneous contact was critical in understanding the mechanism of insulative magnetic brush development, as will be discussed in Chap. 6.

The importance of the impact of a carrier bead in the presence of an electric field was clearly demonstrated by *Donald* and *Watson* [5.7,8] in 1970 in a measurement of the efficiency of toner release from carrier dropped a fixed height in an electric field. With no electric field present the efficiency was low and increased with drop height. Fields as low as 0.1 V/μm significantly increased the toner release efficiency by a factor of 5 (Fig. 5.6). The explana-

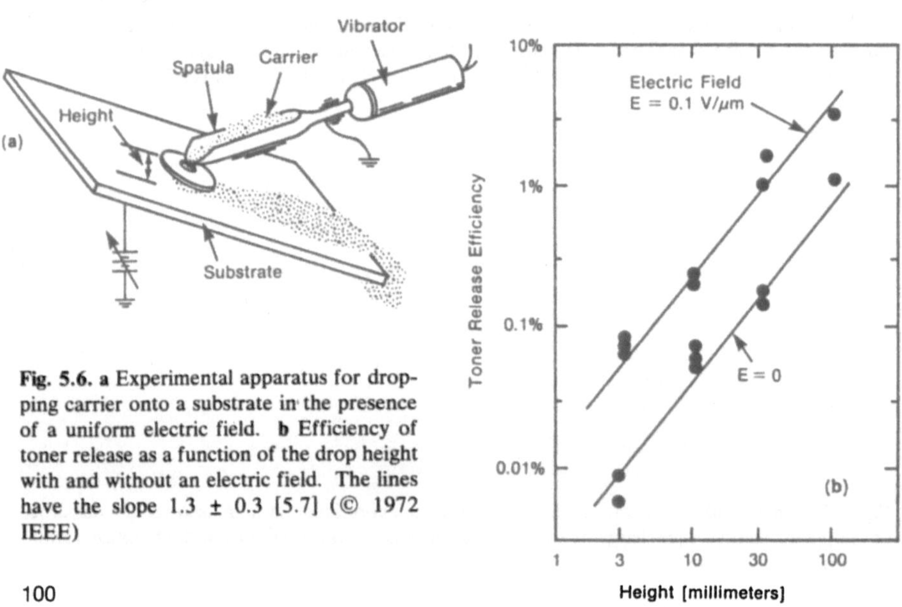

Fig. 5.6. **a** Experimental apparatus for dropping carrier onto a substrate in the presence of a uniform electric field. **b** Efficiency of toner release as a function of the drop height with and without an electric field. The lines have the slope 1.3 ± 0.3 [5.7] (© 1972 IEEE)

tion of this experiment and its application to cascade development led eventually to a quantitative theory of development (see below).

5.1.3 Scavenging

Another important concept was introduced by *Hauser* and *Menchel* [5.14] in 1968. They showed that an effect called scavenging plays an important role in cascade development: toner already on the image can be picked up by an impacting carrier particle, especially if the carrier has been depleted of some of its toner. As one might expect, larger diameter toner particles are more readily scavenged and a continual shift of toner particle size towards smaller size with development time is seen on the developed latent image.

5.1.4 Electrode Source

In the paper by *Bickmore* et al. [5.3] another thought is introduced. Deposition and remobilization of toner on the electrode in the conventional or inverted cascade configuration (Figs. 5.1a) can augment the toner carried directly by the carrier beads to the latent image.

5.2 Experimental Work

The result of an extensive study of the effect of the parameters on development was published by *Bickmore* et al. [5.3] in 1970. They presented data on solid area, line, and background development with and without a counter electrode. Their measured parameter was optical reflection "density" on paper, which they showed is linear in developed mass per unit area. An important complementary study of carrier bead collision rate and velocity was reported by *Stover* in 1974 [5.12].

5.2.1 Solid Area Development

Solid area development was shown to be linearly dependent on "development time," i.e., the length of the counter electrode divided by the photoreceptor velocity. It is also linear in the potential difference V (Fig. 5.7) across the development zone (photoreceptor minus electrode bias potential) *if* the electrode is *not* cleaned between each measurement. With cleaning, higher bias potentials, for the same potential difference, give lower development. Bickmore et al. suggest this occurs because higher biases remove toner from the carrier beads upstream of the image, depleting the mix of toner. Consistent with this observed linear dependence on potential difference, i.e., the electric field, their data indicate development is inversely proportional to the electrode-photoreceptor gap.

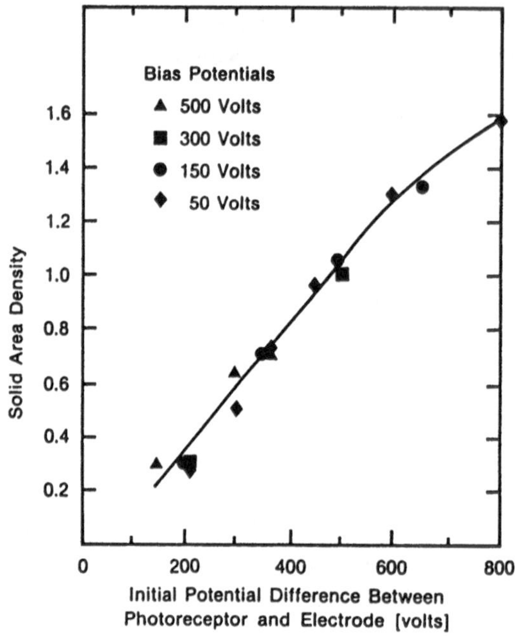

Fig. 5.7. Reflection density as a function of potential difference between photoreceptor and electrode for various bias potentials. A single curve indicates that development depends on the electric field in the development zone [5.3]

Of course, without an electrode, one would not expect solid area development, as shown in Fig. 5.8. However, the surprising result is that, with an electrode, it does not matter whether the system is upside down (with the carrier beads bouncing on the electrode). This is further shown in studies of the dependence on development zone angle (Fig. 5.1) as shown in Fig. 5.9.

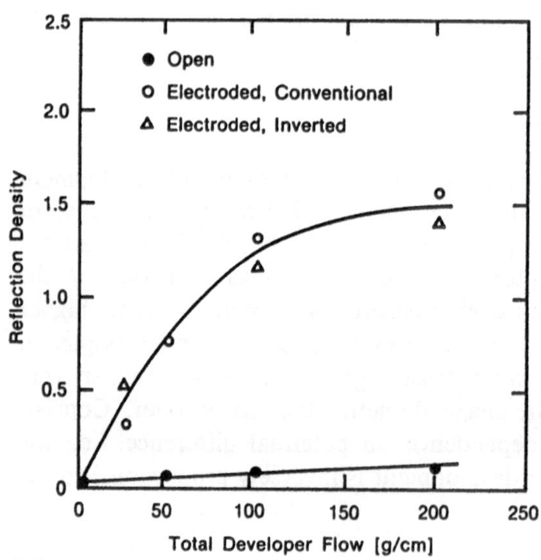

Fig. 5.8. Reflection density in solid areas as a function of total developer flow for open and electroded (both conventional and inverted) development configurations [5.3]

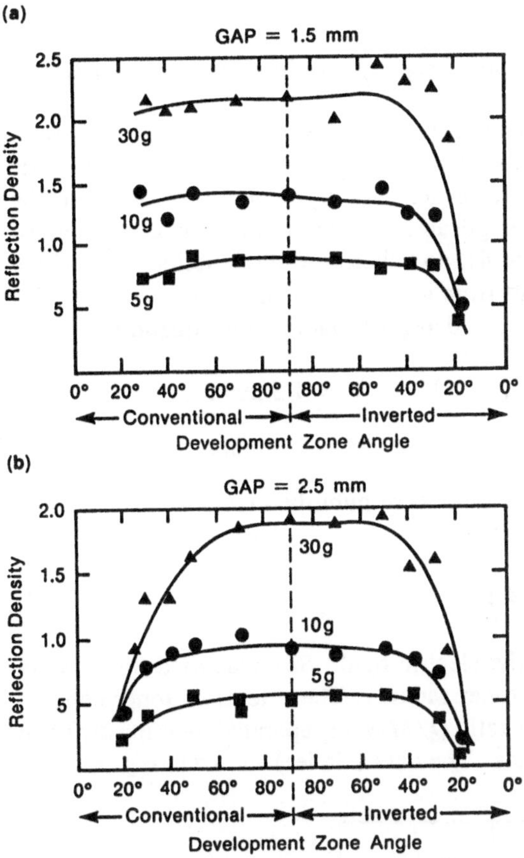

Fig. 5.9. a,b. Reflection density as a function of development zone angle (Fig. 5.1) for various total developer flows (per cm width) in an electroded system. **a** Electrode to photoreceptor gap of 1.5 mm. **b** Spacing of 2.5 mm. Developed density appears to be independent of development zone angle for wide ranges of θ_M about the vertical position [5.3]

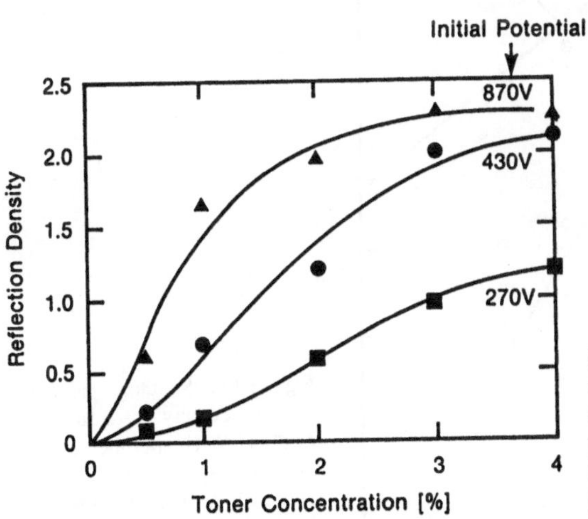

Fig. 5.10. Reflection density of developed solid areas as a function of toner concentration for various initial potential differences between the electrode and the photoconductor [5.3]

Development is independent of zone angle down to angle θ_M, which depends on the electrode-photoreceptor gap. $|\theta_M| = 20°$ for 1.5 mm gap; $|\theta_M| \approx 45°$ for 2.5 mm gap.

The effects of toner concentration C_t are shown in Fig. 5.10. This is an important machine variable because in an actual copy machine control of the optical reflection density of images is achieved by adding fresh toner to the carrier-toner mix. It is also a complicated variable because as C_t is changed the toner charge changes (Chap. 4), and therefore its adhesion is changed. The curves $f(C_t)$ appear to depart from linearity in an S shape.

One can summarize the solid area density D data by the equation

$$D \propto t \frac{V}{L} \theta° f(C_t), \quad |\theta| > 20° \quad L = 1.5 \text{ mm},$$
$$|\theta| > 45° \quad L = 2.5 \text{ mm}. \tag{5.2}$$

Later, solid area experiments of Knapp were published (Fig. 5.11) by *Maitra* et al. [5.9] in which it was shown that

$$D \propto t C_t \exp\left(\frac{-kr}{E} Q/M \right), \tag{5.3}$$

where k is a constant and E is the electric field. Size-classified toners were used to determine the effect of toner radius r. The effects of toner concentration C_t and the charge-to-mass ratio Q/M were separated by using different carrier coatings and blending times. The data (circles) shown in Fig. 5.11 are

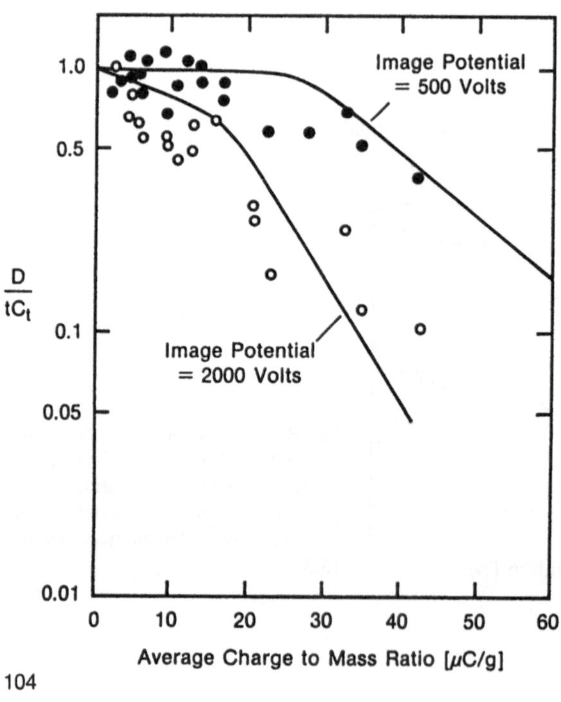

Fig. 5.11. Semilog plots of "optical density" as a function of Q/M. The experimental data are for image potentials of 500 and 2000 V; the solid lines are from the theory of *Maitra* et al. [5.9] (© 1974 IEEE)

development rate (D/t) normalized to toner concentration versus average charge-to-mass ratio for electric field differences of 500 and 1000 V (The lines drawn through the data are based on the theory of *Maitra* et al. which will be discussed below.) A difficulty in this experiment should be clear after reading Chap. 4. While average Q/M was varied, it is not clear that the distribution of Q (or Q/r^2, the independent variable chosen by *Maitra* et al. [5.9]) is unchanged as blending time and carrier coatings are changed. It was not even possible to measure these distributions at the time these measurements were reported. Also note there appears to be a difference in the observed electric field dependence of D between the earlier (5.2) and later experiments (5.3).

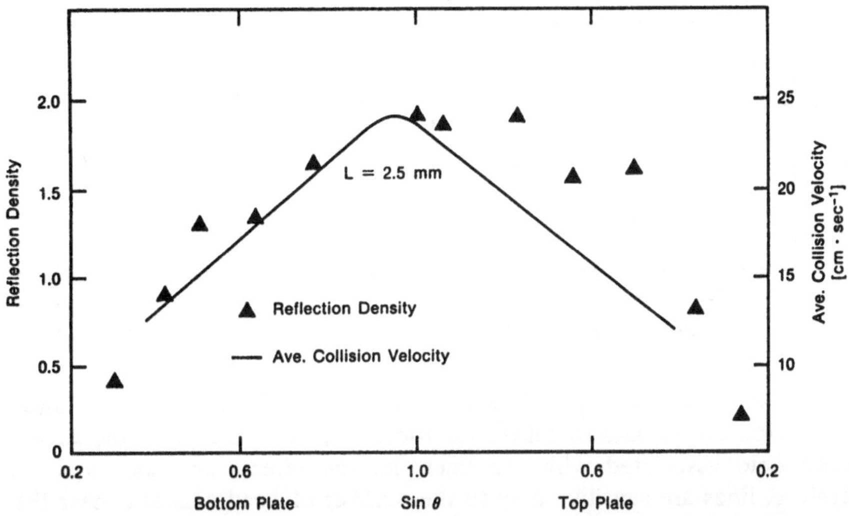

Fig. 5.12. Solid area reflection density as a function of the development zone angle correlates with the average collision velocity in an electroded cascade experiment [5.12] (© 1974 IEEE)

The suspicion that bead velocity effects were important in determining solid area development was qualitatively confirmed by *Stover* [5.12] who published bead velocity data over a range of angles which could be compared to Bickmore et al.'s solid area data. Shown in Fig. 5.12 are solid area density (on the left axis, taken from Fig. 5.9b at 30 g/cm total flow) and average collision velocity (on the right axis) under the same conditions. Both have a peak at sin $\theta = 1.0$ and both fall with lower sin θ by approximately the same amount. Furthermore, in the open cascade configuration, Bickmore et al. showed solid area density peaked at 60° (Fig. 5.13a), the same result Stover obtained for average collision velocity (Fig. 5.13b). Lines, to be discussed below, had a very different behavior. Their density fell continuously as θ increased, qualitatively correlating with collision rates (Figs. 5.13a and b). Stover interprets his data as follows: the release of toner from carrier requires both the inertial

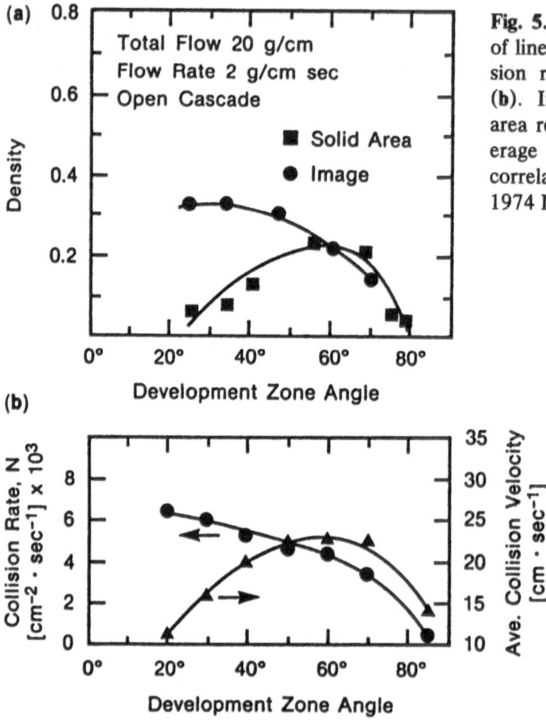

Fig. 5.13. A correlation of optical density of lines and solid areas (**a**) with bead collision rate and average collision velocity (**b**). In an open cascade experiment, solid area reflection density correlates with average collision velocity but line images correlate with collision rate [5.12] (© 1974 IEEE)

force due to a bead bounce and the electric field of the latent image to lower the adhesion barrier and to collect the toner. In the presence of the strong electric field associated with the lines, inertial effects are not required. Therefore, lines are sensitive only to the number of bead collisions near the line image, i.e., the collision rate. The fields associated with solid area images are much weaker. Therefore, inertial forces are important, and correlation with average bead velocity is expected.

5.2.2 Line Development

Less data were presented by Bickmore et al. for line development. The dependence of line development on "time," voltages, and toner concentration are not discussed. This may be due to the fact that line development is a more complicated function. For example, as the development zone angle is varied, the stroke width changes (Fig. 5.14). Evidently, optical reflection density or mass per unit area are not sufficient quantities to characterize line development. Note again that vertical development falls on a continuous curve with normal and inverted cascade configurations. Some copy quality defects can be seen: note the broken nature of the characters at small angle inverted cascade and the toner scatter around the images at small angle conventional cascade.

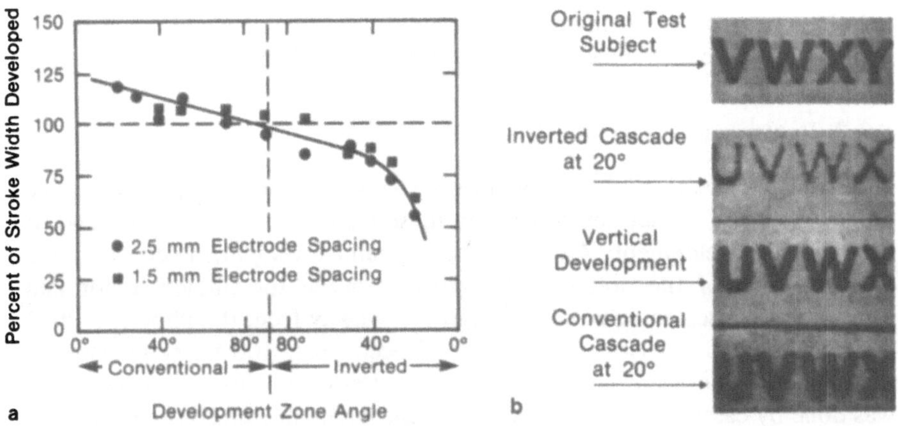

Fig. 5.14. a Developed image stroke width as a function of development zone angle. **b** Line-copy samples for various development zone angles [5.3]

An interesting effect first discussed by *Sullivan* and *Thourson* [5.6] is shown in Fig. 5.15 [5.3] in which reflection density is measured near the edge of a solid area. For large electrode spacing in the inverted cascade configuration a dip in reflection density is observed. This is attributed to the small volume of space from which toner can be collected. This picture assumes an airborne mechanism and that toner follows field lines. A schematic of the field lines is shown in Fig. 5.15c. This (and some of the background data presented

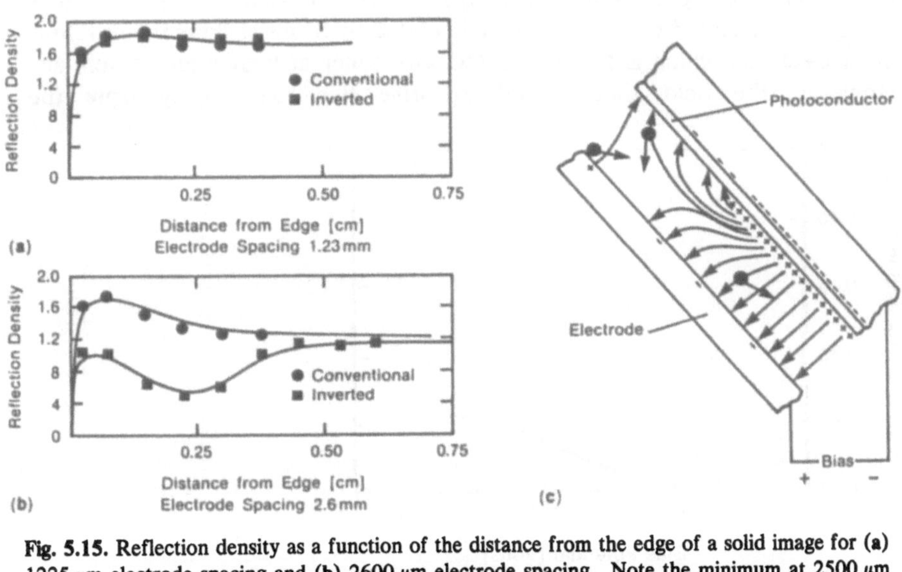

Fig. 5.15. Reflection density as a function of the distance from the edge of a solid image for **(a)** 1225 μm electrode spacing and **(b)** 2600 μm electrode spacing. Note the minimum at 2500 μm from the edge for inverted cascade in **(b)**. **c** Application of the airborne mechanism of development can account for this minimum: carrier particles (shown) colliding with walls (or each other) release toner which tends to follow lines of force [5.3]

below) appears to be strong evidence that at least two mechanisms of development, airborne as well as contact, occur in a cascade development system.

5.2.3 Background Development

A critical requirement of development systems is to keep toner away from nonimage areas. Toner developed into these regions is called background or background development. A curve of background development versus reverse bias (the sign of the potential difference between the photoreceptor and electrode is chosen to drive correct sign toner away from the photoreceptor) is shown in Fig. 5.16. Note that a minimum occurs near 100 V. To understand the minimum one needs to understand several effects. First, the experiment was done by cascading carrier beads over uncharged and unexposed selenium plates. Just as toner and carrier have triboelectric interactions, toner-photoconductor and carrier-photoconductor triboelectric interactions are possible. Indeed, these have been directly observed in other systems (Sect. 4.4 and [5.15,16]). Therefore, it is possible that the developer charges up the photoconductor, creating an electric field that attracts toner. Zero potential difference does not imply zero electric field in the presence of insulators! A finite reverse bias of ≈100 V, observed in Fig. 5.16, is a very typical result for all development systems and may be just the amount of uniform field required to overcome an electric field generated by developer-photoreceptor triboelectric interactions.

The second effect, the rise in background development above 100 V, has several possible explanations. One, suggested by *Bickmore* et al. [5.3], is that it may be associated with a toner cloud created by beads striking the development electrode which is heavily coated with toner at higher bias potentials. However, the field should tend to drive this toner away from the

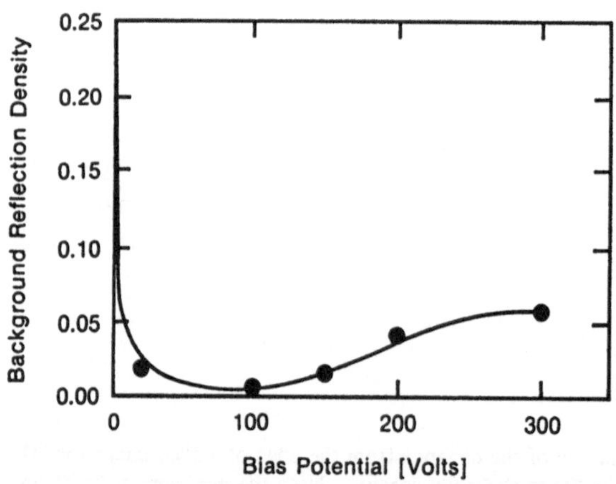

Fig. 5.16. Reflection density in background areas as a function of bias potential. A minimum occurs near 100 V [5.3]

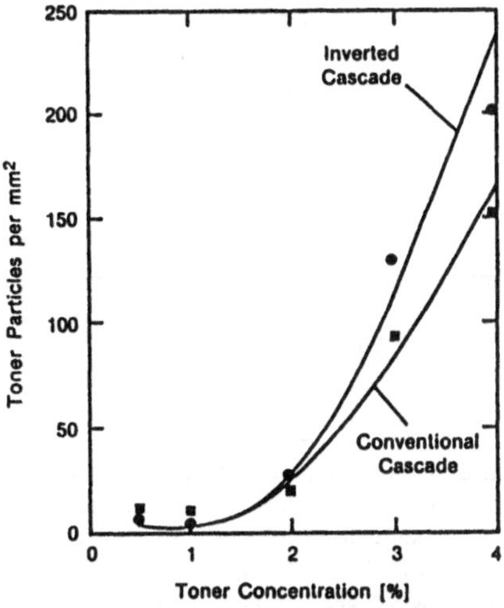

Fig. 5.17. Background development as a function of toner concentration for inverted cascade and conventional cascade configurations [5.3]

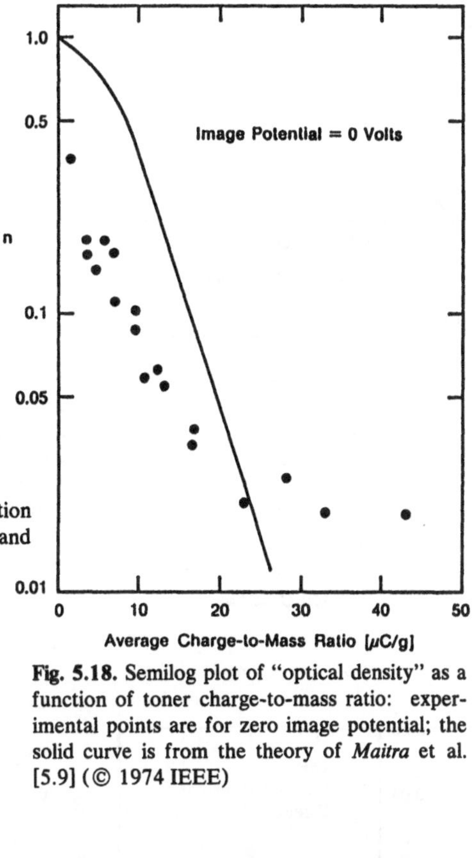

Fig. 5.18. Semilog plot of "optical density" as a function of toner charge-to-mass ratio: experimental points are for zero image potential; the solid curve is from the theory of *Maitra* et al. [5.9] (© 1974 IEEE)

photoconductor. Another possible explanation is the presence of reverse sign toner which "develops" more as the bias is increased. This is consistent with another observation of Bickmore et al. (Fig. 5.17), namely, that background (this time measured by counting particles) rises rapidly with toner concentration. As toner concentration rises, the average charge-to-mass ratio decreases, and the possibility that the distribution includes wrong sign toner increases. Indeed, the strong effect of average Q/M on background was later confirmed in a paper by Maitra et al. (Fig. 5.18). It was also demonstrated by *Stover* and *Schoonover* [5.17] who measured the charge/diameter distribution of developers (Fig. 5.19) and correlated the shape of the distribution with background. High background was correlated with the existence of wrong sign toner (Fig. 5.19b), which could also result from aged developer (Fig. 5.19c).

In terms of the dynamics of the developer flow, Bickmore et al. find, see Fig. 5.20, that background development increases as the total flow increases (and is higher for the open cascade system), and decreases as the development zone angle increases in open cascade (i.e., correlates with collision rate, not

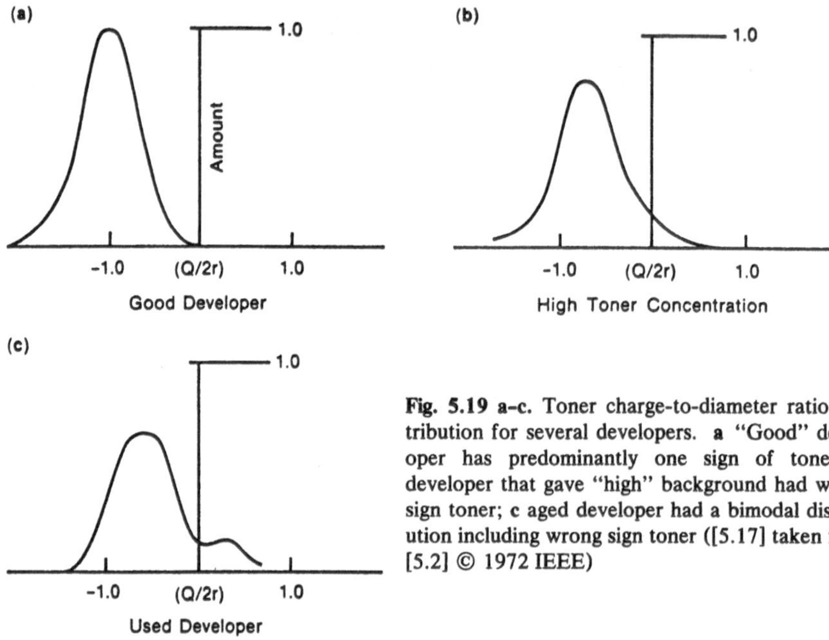

Fig. 5.19 a–c. Toner charge-to-diameter ratio distribution for several developers. **a** "Good" developer has predominantly one sign of toner; **b** developer that gave "high" background had wrong sign toner; **c** aged developer had a bimodal distribution including wrong sign toner ([5.17] taken from [5.2] © 1972 IEEE)

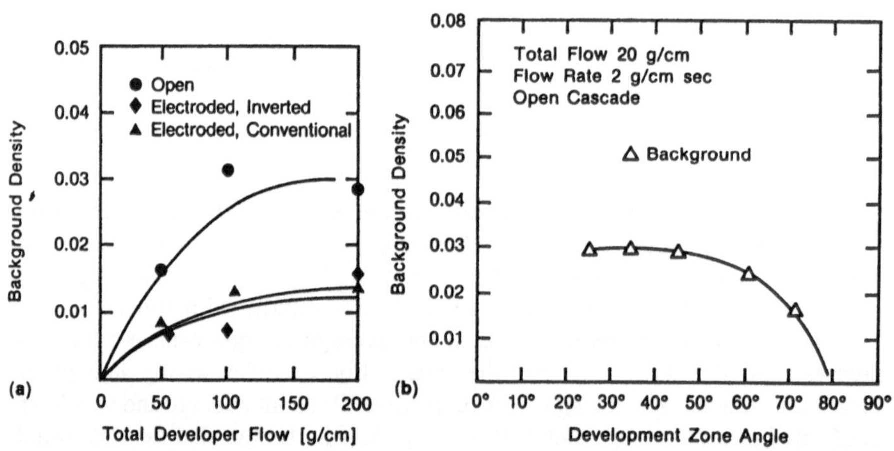

Fig. 5.20. **a** Reflection density in background area as a function of total developer flow for open and electroded configurations. **b** Reflection densities of background as a function of development zone angle in an open development system [5.3]

average collision velocity, see Fig. 5.13). As a function of flow rate, small flow rates produce more background (Fig. 5.21). This is because, according to Bickmore et al., at low flow rates the beads act independently, bouncing high with each impact, while at high flow, the developer moves as a blanket down the photoconductor with much less energetic bounces of the individual beads.

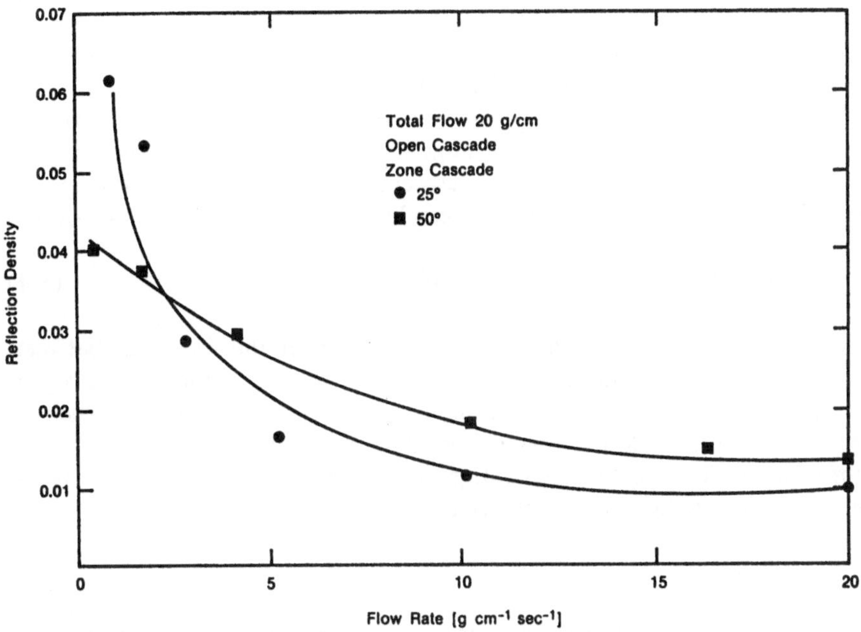

Fig. 5.21. Reflection density of background as a function of developer flow rate for an open cascade system [5.3]

This concept, in which velocity of bead contacts are assumed important, appears to differ from the explanation of Fig. 5.20, in which background correlated with bead collision rate. Perhaps both variables are important in different regions of the parameter space.

5.3 Theory

The data suggest at least two separate mechanisms, airborne and contact development, are contributing to cascade development.

5.3.1 Airborne Development

Airborne development occurs when toner is released in the bulk of the developer by bead-bead, bead-electrode or bead-photoreceptor collisions (far from the latent image) and is collected by the electric field. It appears to contribute significantly only in special circumstances, e.g., low developer flow rates, and manifests itself either indirectly as a drop in the density near the edge of solids or directly as enhanced background. The theory, worked out by *Sullivan* and *Thourson* [5.6], calculates the volume of space between field lines emanating from the electrostatic image (Fig. 5.5). It is assumed that (1) the source of toner is only on the photoreceptor, (2) toner is stripped from the carrier if the Coulomb force exceeds the toner-carrier adhesion force, and (3) this toner-carrier adhesion force is constant for all toner particles.

Following Sullivan and Thourson we define, using Fig. 5.5, $N(\xi, r)d\xi dr$ as the number of particles in the size range between r and $r + dr$ which deposit on the image electrode in the region between ξ and $\xi + d\xi$ per unit length of the electrode assembly. This quantity can be determined from the relationship

$$N(\xi, r)d\xi dr = P_r(\eta, r)N_0(\eta, r)d\eta dr \,, \tag{5.4}$$

where $N_0(\eta, r)d\eta dr$ is the number of toner particles in the size range between r and $r + dr$ in the developer per length of the electrode assembly in the region between η and $\eta + d\eta$, and $P_r(\eta, r)$ is the probability that a toner particle of size r at position η will be released from the carrier. The point η is related to ξ through the electrostatic field configuration so that they are connected by the same electrostatic field line. This relationship is

$$\eta(\xi) = \xi/(1-\xi) \;. \tag{5.5}$$

If the developer is distributed uniformly over the electrode assembly, then $N_0(\eta,r)$ can be written $N_0(\eta,r) = N_0 f(r)$, where N_0 is the number of toner particles per unit area regardless of size and $f(r)$ is the toner particle size distribution function. The function $f(r)$ is normalized such that

$$\int_{r_0}^{\infty} f(r)dr = 1 \,, \tag{5.6}$$

where r_0 is the minimum diameter particle observed in the experiment. Therefore, $f(r)dr$ represents the fraction of toner particles in the developer with diameters between r and $r + dr$.

To account for the observed particle size classification effects, Sullivan and Thourson assume that the charge per particle Q_t increases with particle radius r according to

$$Q_t = k_1 r^2 \,, \tag{5.7}$$

where k_1 is a constant. They argue that this assumption is reasonable, since charging is a surface effect, but acknowledge that there probably is a variation of the magnitude of charge for particles of the same size. Equation (5.7) would then represent the average charge per particle of size r.

Assuming the Coulomb force $Q_t E_{air}$ must overcome the toner adhesion to the carrier, F_t, a critical toner charge Q_c may be defined as

$$Q_c(\xi) \equiv \frac{F_t}{E(\eta(\xi))} \;. \tag{5.8}$$

The toner which deposits at ξ is stripped from the carrier at the point $\eta(\xi)$ of the outer electrode and Q_c is the minimum charge necessary for stripping in the field, $E(\eta(\xi)')$, at that point. The function $\eta(\xi)$ is given by (5.5). Through (5.7) we can define a critical size

$$r_c(\eta(\xi)) \equiv \left(\frac{F_t}{k_1 E(\eta(\xi))} \right)^{1/2} . \tag{5.9}$$

This represents the minimum size particle that will deposit at ξ by airborne development. With these considerations, we can now write the stripping probability as

$$\begin{aligned} P(\eta(\xi), r) &= 0, \qquad r < r_c(\eta(\xi)) \\ &= 1, \qquad r \geq r_c(\eta(\xi)) \; . \end{aligned} \tag{5.10}$$

Using (5.5), the distribution function $N(\xi, r)$ can be written as

$$N(\xi, r) = N_0(1 - \xi)^{-2} P_r(\eta(\xi), r) f(r) \; , \tag{5.11}$$

where P_r is given by (5.10) and $f(r)$ can be obtained by independent size analysis of the toner sample.

From the function $N(\xi, r)$ several experimentally measurable quantities can be calculated. For example, the number of particles per unit area as a function of ξ can be obtained by integrating $N(\xi, r)$ over the range of particle sizes. Thus,

$$\begin{aligned} N(\xi) &= \int_{r_0}^{\infty} N(\xi, r) dr \\ &= N_0(1 - \xi)^{-2} \int_{r_0}^{\infty} P_r(\eta, r) f(r) dr \end{aligned} \tag{5.12}$$

or, using (5.10), this can be written as

$$N(\xi) = N_0(1 - \xi)^{-2} \int_{r_1(\xi)}^{\infty} f(r) dr \; , \tag{5.13}$$

$$\begin{aligned} r_1(\xi) &= r_0 \qquad \text{when } r_c(\eta(\xi)) < r_0 \\ &= r_c(\xi) \qquad \text{when } r_c(\eta(\xi)) \geq r_0 \; . \end{aligned}$$

Other quantities can be calculated from these expressions, such as the average diameter of toner particles deposited at some position ξ and optical density of the image, but we will limit our discussion to the calculation and comparison with experiment of $N(\xi)$. To make comparison with experiments, even in this relatively simple situation, many parameters need to be determined or estimated. The toner particle size distribution can be measured rather precisely with a Coulter Counter. The toner adhesion F_t was taken as a constant, 2×10^{-5} dynes, despite the fact that it should have a significant size and charge dependence if it is dominated by image forces. The constant k_1 was taken to be $10^{-16}\,C/\mu m^2$ corresponding to an average charge-to-mass ratio of $16\,\mu C/g$. While the charge-to-mass ratio is typical, this assumption ignores the distribution of charge for a given size.

With these assumptions, (5.13) can be evaluated and compared with experiment. Figure 5.22 is a plot of the relative number of toner particles per unit area as a function of position on the image. The results are plotted only for one-half the distance across the image. The experimental points include the peak density at the edge of the solid for which this theory does not account (and requires contact development to describe, see below). The minimum and second peak observed experimentally are to be compared with the theory. Qualitative agreement is obtained, although the predicted width appears to be too large.

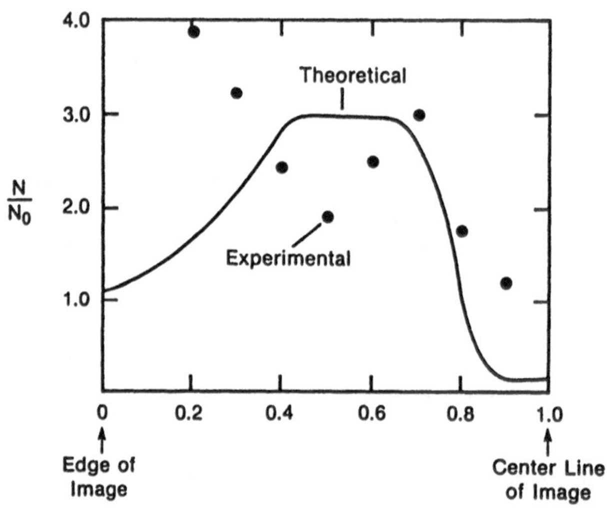

Fig. 5.22. Relative number of particles/unit area as a function of distance from the edge [5.6]

5.3.2 Contact Development

Contact development occurs when the carrier bounces on the latent image. In this case the toner is subject to three forces, its adhesion to the carrier, the Coulomb force due to the latent image, and the inertial force due to the bead bounce.

In Fig. 5.23 we indicate how complicated this problem actually is. There are five distribution functions. First, the bead velocity varies within a mix under a given set of parameters, including zone angle, bead flow, and bead flow rate. Second, for a line image, the electrostatic field varies spatially both

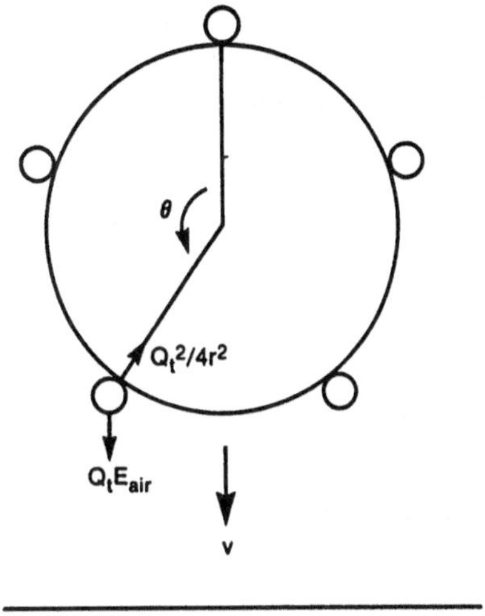

Fig. 5.23. In actual development experiments five parameters are distributed: carrier velocity v, angle θ, electric field E, toner charge Q_t and toner radius r

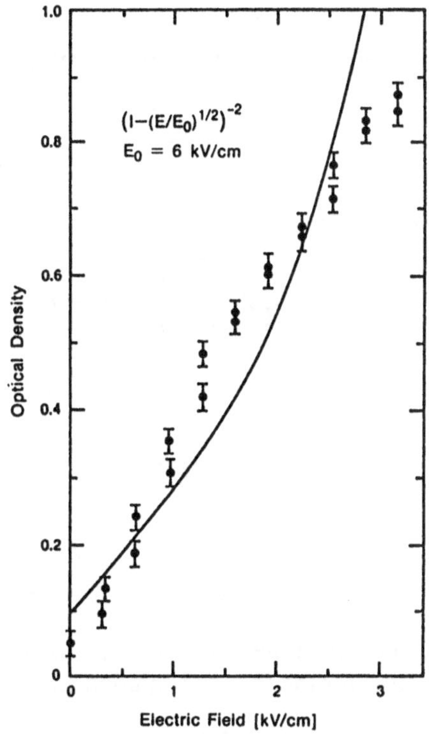

Fig. 5.24. The efficiency of toner release as a function of electric field as predicted by *Donald* and *Watson* with $E_0 = 6$ kV/cm [5.7] (© 1972 IEEE)

along the image and perpendicular to the photoreceptor plane (Sect. 3.2). Third, the angle θ of the toner on the carrier bead at impact can vary from 0 to 360°. Fourth and fifth, the toner radius and charge are distributed.

The first to simplify this problem were *Donald* and *Watson* [5.7,8] who eliminated two distribution functions (the field and velocity) by dropping the carrier beads a fixed height in a uniform electric field. As size-classified toner was not used, three distributions remain. Their experiments and experimental results are shown in Fig. 5.6. Basically, the toner release efficiency (fraction of toner on a carrier released) increased algebraically with drop height. Fields as low as 0.1 V/μm increased the efficiency by a factor of 5. The dependence of η on field is shown in Fig. 5.24.

There are various ways to treat this problem; we follow *Donald*'s and *Watson*'s second approach [5.8]. The basic idea is that in the presence of an electric field, the binding energy of toner to carrier is reduced. When this binding energy equals the kinetic energy during impact, the toner is freed and allowed to develop.

The kinetic energy can be expressed as the potential energy of the bead dropped from a height H_t,

$$\text{kinetic energy} = M_t g H_t \ , \tag{5.14}$$

where M_t is the toner mass and g is the acceleration due to gravity. The toner adhesion $F(x)$ is the image force reduced by the electric field. This is given by

$$F(x) = \frac{Q_t^2}{16\pi\varepsilon_0(r+x)^2} - Q_t E_{air} \tag{5.15}$$

at a distance x from the carrier. This force vanishes at x_m

$$x_m = \sqrt{\frac{Q_t}{16\pi\varepsilon_0 E_{air}}} - r\ . \tag{5.16}$$

The binding energy U is the integral of the adhesion force from $x = 0$ to x_m

$$\begin{aligned} U &= \int_0^{x_m} F(x)dx \\ &= \frac{Q_t^2}{16\pi\varepsilon_0 r}\left(1-\sqrt{\frac{12\varepsilon_0 E_{air}}{(Q_t/M_t)r\rho_t}}\right)^2 , \end{aligned} \tag{5.17}$$

where ρ_t is the toner density. Setting this equal to $M_t H_t$, the kinetic energy, gives for the release condition as a function of electric field

$$H_t = \frac{3Q_t^2}{64\pi^2\varepsilon_0 g\rho_t r^4}\left(1-\sqrt{\frac{12\varepsilon_0 E_{air}}{(Q_t/M_t)r\rho_t}}\right)^2 . \tag{5.18}$$

This is a threshold condition for fixed Q_t, r, and θ. A distribution in any of the parameters (and all three were distributed in Donald's experiments) will smear out the predicted result.

Donald makes two arguments that this theory is a reasonable description of the data. First, he points out that the magnitude of the field effect roughly corresponds to the observation. For a "typical" toner ($r = 6.5\mu$m, $Q/M = 10\mu$C/g) a field of 0.1 V/μm should reduce the drop height for the same efficiency by 60%, using (5.18), which qualitatively agrees with the data shown in Fig. 5.6. Also, from Fig. 5.6, it is reasonable to assume that the toner release should go inversely with the drop height, which suggests

$$\eta \propto H(E)^{-1} \propto \left(1-\sqrt{E_{air}/E_0}\right)^{-2} , \tag{5.19}$$

where E_0 is a constant which is observed to approximately describe the data (Fig. 5.24).

The obvious next step is to apply the knowledge gained in the bead drop experiment to describe actual cascade development experiments. This step was taken first by *Maitra* et al. [5.9] and later by *Herbert* et al. [5.10].

In actual development experiments the various distributions of parameters must be handled differently than above. For example, instead of a fixed carrier velocity, a distribution of velocities exists. The velocity v distribution was taken in both papers from *Stover*'s work [5.12] in which he showed the number of impacts with an incident velocity between v and $v + dv$ has a Maxwellian distribution,

$$h(v)dv = \frac{2v dv}{v_0^2} \exp\left[-(v/v_0)^2\right] , \tag{5.20}$$

where v_0 is a constant characteristic of the flow condition. For the data taken, only one condition was used (presumably zone angle, total flow and flow rate were fixed) and v_0 was taken to be 26.9 cm/s. The r distribution was significantly reduced by using size-classified toner. The electric field was reduced to a single value by using solid area development only. This leaves the θ and charge distributions. Both papers ignore the θ distribution. Both papers replace the charge distribution with a Q/r^2 distribution. Maitra et al. use a log-normal distribution, i.e., if

$$y = \ln Q/r^2 \tag{5.21}$$

the distribution function in the variable y has the form

$$f(y) = \frac{1}{\sigma\sqrt{2\pi}} \exp\left[-(y-\alpha)^2/2\sigma^2\right] , \tag{5.22}$$

where α is the mean value and σ is the standard deviation of y. They choose

$$\sigma = 0.2 + 0.05\,\alpha^2. \tag{5.23}$$

The value α was chosen to describe the average charge-to-mass ratio of each mix. Herbert et al. use a Gaussian distribution with a constant half width. As noted earlier, to obtain different average Q/M 's in the experiment the carrier coating and mixing times were varied. A constant standard deviation or half width is an obvious approximation, probably as good as one can do considering that such distributions were not measurable at the time.

The goal of both papers was to fit the solid area development data of Knapp (Fig. 5.11) which showed that

$$\frac{D}{tC_t} = \exp\left(-\frac{kr}{E_{air}}\,\frac{Q}{M}\right) . \tag{5.24}$$

The basic idea in the theories is that one combines (1) the threshold carrier velocity v_{th} (which depends on toner charge Q_t, radius r, and electric field E_{air}) above which toner will be released from the carrier and developed onto the photoreceptor and (2) the velocity distribution $h(v)$ obtained from *Stover*'s work [5.12]. The two papers use slightly different expressions for the threshold velocity v_{th}. Maitra et al. use

$$v_{th} = C_1 \frac{Q_t}{r^2} - C_2 \left(E_{air} \frac{Q_t}{r^2} \right)^{1/2} , \tag{5.25}$$

where C_1 and C_2 are constants, and Herbert et al. use

$$v_{th}^2 = C_3 Q_t / r^2 \left(1 - \sqrt{\frac{E_{air}}{C_4}} \right)^2 , \tag{5.26}$$

where C_3 and C_4 are other constants.

Both papers obtain the result that development, i.e., density per unit time per unit toner concentration D/tC_t,

$$\frac{D}{tC_t} \propto \exp[-(Q/M)]. \tag{5.27}$$

This result is obtained analytically by Herbert et al. from their velocity and Q/r^2 distribution. The distributions assumed by Maitra et al. do not give this result analytically but they present a simple argument to rationalize this result. Consider the case in which $\sigma \to 0$, i.e., the q/r^2 distribution becomes a delta function. Then v_{th} is a single number and D/tC_t , which is proportional to $h(v)$, is

$$\frac{D}{tC_t} \propto h(v) \propto \exp\left[-(Q/M)^2\right] . \tag{5.28}$$

This is easy to understand: as Q_t increases, v_{th} increases and there are fewer carriers with sufficient velocity to free toner. Now consider the case in which $\sigma \to \infty$, i.e., the Q_t/r^2 distribution is flat. In this case, a shift to higher mean charge-to-mass ratio still leaves a large tail of low Q_t/r^2 values. Hence, for almost any value of Q/M, there is a relatively large number of toner particles that can be developed with a modest value of v. In fact, it can be shown that n now has an algebraic dependence on Q/M. For "realistic" values of σ (≈ 0.2 assumed) an "intermediate" result is expected, such as (5.27).

Comparison of Maitra et al.'s theory with experiment is shown in Fig. 5.11 in which the solid lines are the theory. The general trends appear to be accounted for by the theory. The theory appears somewhat higher than the data at 500 V. The authors point out this may be corrected by adding a small nonelectrostatic term to the adhesion expression. Herbert et al. can also semiquantitatively explain the field dependence.

Obviously questions about solid area cascade development remain. For example, neither theoretical paper addressed the effect of changing bead collision rate or average velocity. Bickmore et al. certainly presented many data in which this distribution was varied by varying flow conditions (Figs. 5.6–12) and Stover characterized the bead behavior over a range of flow conditions. In addition, the observed linearity of M/A on V (Fig. 5.7) is inconsistent with the prediction of both contact development theories and suggests consideration of the effect of charge buildup on the carrier particles (an equilibrium theory, discussed in Chap. 6). However, at about the time these papers appeared, interest turned to the magnetic brush development system and motivation to refine and expand the cascade studies virtually disappeared. The one exception is recent work by *Mukherjee* and co-workers [5.18] on replacing powder cloud development with cascade development in less expensive xeroradiographic applications. Also little theoretical work was done on background development. A few qualitative comments made by *Bickmore* et al. [5.3] concerning contributions from aerosols are reinforced by *Maitra* et al. [5.9]: above 30 μC/g they claim that the major source of background is airborne toner (Fig. 5.18). Maitra et al. tried to fit Knapp's background data to their theory, but a fundamental problem arose. The background did not increase linearly with development "time." Hence, one does not know whether the discrepancy shown in Fig. 5.18 between theory and experiment is because (1) at zero volts, a mechanism other than contact development causes background or (2) at zero volts, the arbitrarily chosen 1s of development time makes the scaling factor incorrect. The data of *Bickmore* et al. [5.3], and *Stover* and *Schoonover* [5.17] clearly suggest that wrong sign toner can contribute to background development. However, no direct characterization of background toner was reported and no theory was proposed which could be used to quantify the phenomenon.

5.4 Summary

As one reviews the cascade development literature, it is apparent that questions remain both theoretically and experimentally. However, in the larger sense, the work was successful. It was clearly demonstrated that two mechanisms of development are operative in solid area development. The most important is contact development in which the electric field assists the inertial force, caused by a bead bounce, to release toner. Under special circumstances such as low flow rates, airborne development can also occur from toner released far from the latent image. There is qualitative evidence that line development is determined by the number of bead contacts, with inertial effects being less important. There is also qualitative evidence that background development is due in part to wrong sign toner.

6. Insulative Magnetic Brush Development

By the late 1960s, electrophotography was established by the products of the Xerox Corporation as the most important copying technology, effectively displacing special-paper copying technologies. Concurrently, the public's perception changed from thinking of copying as a convenience to a necessity. The need for higher speed and higher quality became obvious. In the development system, this required a change from cascade to magnetic brush development. Magnetic brush development was invented at RCA in the late 1950s [6.1,2]. The first copier that used magnetic brush development was the Copytron 1000 manufactured by the Charles Bruning Company. It was introduced in the early 1960s and was based on RCA's Electrofax process, using ZnO coated paper. The first "plain paper" electrophotographic copiers with magnetic brush development systems began to appear in the early 1970s, including the IBM Copier II in 1972 and the Xerox 3100 in 1973. Today it is the most prevalent development system (Table 1.1).

A magnetic brush development system is shown schematically in Figs. 3.6 and 6.1. The carrier beads are made of magnetically soft material so that they can be moved by the magnetic fields. The carrier beads with attached toner are introduced into the vicinity of a roller inside of which are stationary mag-

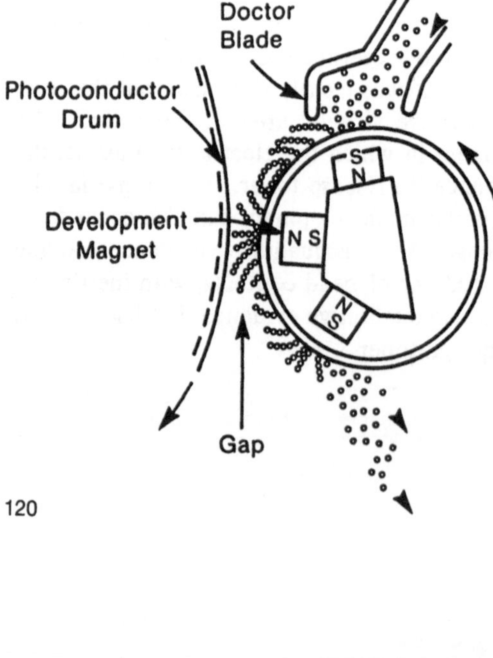

Fig. 6.1. Schematic diagram of a magnetic brush developer system. Toner is mixed and charged by contact with carrier beads. The magnets are stationary and the roller rotates, transporting the magnetically soft carrier beads with attached toner past the electrostatic image on the photoreceptor. The carrier beads are recirculated and fresh toner is added to maintain a constant toner concentration

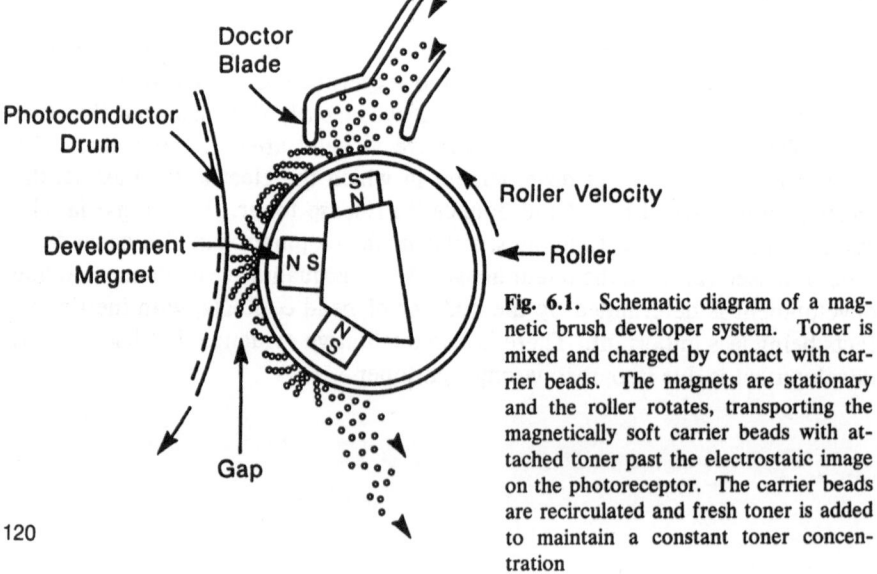

nets. The friction force due to the magnetic fields causes the carrier beads to be carried around the rotating roller. In the gap between the roller and photoreceptor toner moves from the carrier beads to the photoreceptor. Identifying the mechanism(s) by which this toner movement occurs is the primary goal of this chapter.

As discussed in Chap. 3, a complete magnetic brush development system has additional functions, including the sensing of depleted toner and the addition and mixing of fresh toner, which we will not discuss. Neither will we discuss the dynamics of carrier bead flow around the roller. While the qualitative explanation given in the above paragraph appears to describe the physics of this flow correctly, the tuning of the magnets to optimize the flow is done in practice empirically. Quantifying this problem [6.3] involves (1) calculating the magnetic field and forces on the carrier beads, (2) taking into account the presence of a high packing density of magnetically soft carrier beads on the magnetic field, and (3) finally predicting the flow characteristics. The latter is a difficult, unsolved particle flow problem.

The carrier shape turns out to be a crucial property that determines the physics governing toner development. In this section, only spherically shaped particles are considered. It will be shown that for spherically shaped carrier particles, the carrier chain is insulating – hence the term insulative magnetic brush development. A sponge shape (see next chapter) leads to a conductive chain – hence the term conductive magnetic brush development.

The difficulty in understanding this system is similar to the one that exists for cascade development: there are many parameters, some of which cannot be adequately measured. As might be expected, and similarly to the literature on cascade development, early papers [6.4–10] were limited to qualitative discussions while later work dealt increasingly with the underlying physics. Shown in Table 6.1 are the parameters of this development system and the values chosen by the present author [6.11–15] in his studies. The toner charge distribution was not measured because it was not measurable at the time of these papers. The carrier diameter is usually chosen between 100 and 300 μm. Only Minolta uses smaller diameters although published information is not available. A method of handling such an enormous list of parameters is to fix most and to vary those few that one hopes will reveal the underlying physics of the process. *Schein* [6.11–14] chose to vary the photoreceptor and roller velocities and the photoreceptor voltage. Also only solid area latent images were used, in order to "simplify" the electric field. As will be seen below, even under these conditions, specifying the electric field is a difficult task.

As a guide for the reader, we discuss in Sect. 6.1 a qualitative comparison of the proposed mechanisms of development. These mechanisms are described in detail in Sect. 6.3. The driving force for all of the development theories is the electric field of the latent image, discussed in Sect. 6.2. Experimental data are shown and compared with theory in Sect. 6.4. Line and

Table 6.1. Parameters of the magnetic brush development system

Machine variable	Value	Symbol
Photoreceptor velocity	Varied (2.5–50 cm/s)	v_p
Roller velocity	Varied (7.5–75 cm/s)	v_r
Flow rate	10 g/cm s at v_r = 25 cm/s	F
Photoreceptor to roller spacing	1350 μm	L
Photoreceptor potential	Biased Al (Fig. 6.3b)	V
Roller voltage	Grounded	
Toner type		
Radii	Classified and unclassified (Fig. 6.22)	
Concentration	1.4%, 0.8%	C_t
Charge-to-mass ratio	20 μC/g, 30 μC/g	Q/M
Carrier: Radius	50 μm	R
Shape	Spherical	
Coating	2 μm polymer coating	
Magnets: Strength and configuration		
Roller: Number and size, surface	One, 3.75 cm diameter, rough texture	
Mode: Velocity of photoreceptor, with or against the roller	With, against	
Photoreceptor: Type	Aluminum plate, Mylar, a-Se, organic	
Thickness		d_s
Dielectric constant		K_s
Relative humidity	Noted	
Dark decay	None (Al) or corrected for	
Developer transient flow	Checked, none	
Angle with respect to gravity	Perpendicular	
Reservoir	5000 g	

background development are discussed in Sects. 6.5 and 6.6. Ideas for improving this development system, which evolved from the theoretical understanding of the development mechanism, are discussed in Sect. 6.7. A summary is given in Sect. 6.8.

6.1 Qualitative Comparison of Development Mechanisms

Figure 6.2 illustrates the three theories of solid area development that have been proposed to describe the mechanism by which toner leaves the carrier particles and ends up on the photoconductor surface: the field stripping theory, the powder-cloud theory and the equilibrium theory. Four measurements can be used to distinguish among the three theories (see bottom half of Fig. 6.2). Three of these involve measurement of the developed mass per unit area M/A as a function of (1) roller velocity, (2) toner concentration, and (3) voltage V across the photoreceptor-roller gap. The fourth involves measurements of the toner charge-to-mass ratio as a function of V.

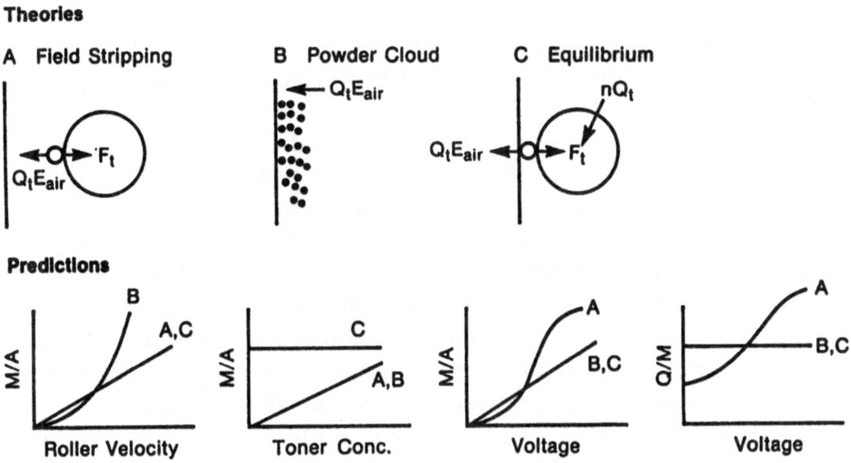

Fig. 6.2. The field stripping theory (theory A), the powder cloud theory (B) and the equilibrium theory (C) are schematically indicated at the top of the figure and their predictions are indicated on the bottom four graphs. Only the powder cloud theory (B) predicts nonlinear mass per unit area versus roller velocity; the equilibrium and field stripping theories can be distinguished by the other three measurements

In the field stripping theory [6.16–19], theory A in Fig. 6.2, the Coulomb force Q_tE_{air} due to the latent image on the photoreceptor overcomes the forces (the electrostatic image force and van der Waals force) that attract the toner to the carrier beads F_t. All particles whose adhesion force is less than Q_tE_{air} are developed from the carrier beads onto the latent image. Because it takes a minimum force to strip toner off the carrier particles, there should be no development at low voltages. Development curves, i.e., mass per unit area versus voltage, are proportional to integrals over the toner adhesion distribution. Because the adhesion distribution is not expected to be rectangular, the development-versus-voltage curve should be nonlinear. Toner particles with lower adhesion or lower charge develop first, so the developed toner charge-to-mass ratio Q/M should increase with applied voltage (see predictions in lower half of Fig. 6.2). The developed mass per unit area should be linear in roller velocity and linear in toner concentration (for constant toner charge-to-mass ratio) because development should increase as the amount of available toner increases.

In the powder-cloud theory [6.20–22], theory B in Fig. 6.2, toner is freed from the carrier by inertial forces during carrier-carrier and carrier-photoreceptor collisions. The electric field associated with the latent image then attracts the free toner to the photoreceptor. If this theory describes the development mechanism, the developed toner mass per unit area should be proportional to the product of the carrier flow and a function of the inertial forces on the carrier beads. The flow of carrier particles is proportional to the roller velocity and the inertial forces on the carrier beads increase with increasing roller velocity. As a result, the developed mass per unit area should exhibit a superlinear dependence on roller velocity, as indicated in the predictions in Fig. 6.2. This prediction distinguishes the powder-cloud theory from the other theories and can be used to test for the presence or absence of this development mechanism. Predicting the outcome of the other measurements requires additional assumptions about the forces exerted on the toner particles. If development of toner depends only on the force exerted by the electric field on the toner, then developed mass per unit area should be linear in the applied voltage and toner concentration. If the amount of toner freed from the carrier depends on the inertial forces alone, the developed toner charge-to-mass ratio should be independent of the electric field. On the other hand, if the release of toner from the carrier depends on both inertial forces and toner charge (for example, via toner adhesion), one expects more complicated behavior.

The equilibrium theory [6.13,14], theory C in Fig. 6.2, assumes that toner continues to come off each of the carrier beads until the Coulomb force of the latent image balances the force of attraction of the toner to the carrier beads, i.e., until a force equilibrium is reached. In this theory the usual forces of attraction of toner to carrier beads (due to image and van der Waals forces) are ignored because it is assumed that development only occurs in three body contact events between carrier, toner and photoreceptor. In this case such forces are cancelled to first order by similar forces between the toner and photoreceptor. The predominant force of attraction between toner and carrier is assumed to be due to the carrier building up a net charge as a result of toner particles developing, i.e., moving from the carrier to the photoreceptor. If n toner particles develop from a carrier bead, a net charge of nQ_t builds up, which attracts the next toner particle considering whether to develop. When this force equals the Coulomb force $Q_t E_{air}$, toner particles cease coming off the carrier bead. Therefore n, the number of toner particles developed per carrier bead, is linear in E_{air}, or the applied voltage. This leads to the prediction that M/A, which is proportional to n, is linear in V. The developed mass per unit area depends linearly on roller velocity because development increases linearly with the number of carrier beads brought into contact with a point on the photoreceptor. This theory predicts that developed mass per unit area is independent of the amount of available toner (toner concentration) for a constant toner charge because development continues until an equilibrium of forces is achieved, independent of how much toner is on a carrier bead. Because toner continues to develop until a force equilibrium is reached, i.e., the "average" toner particle develops, the developed toner charge-to-mass ratio should be independent of electric field or the voltage across the development gap.

For completeness we mention two published efforts to construct computer models of electrophotography. A development model is one of the required inputs. *Paxton* [6.23] simply assumed development was linear in potential difference across the development zone. *Hill* and *Griesmer* [6.24] assumed development continues until the toner charge per unit area neutralizes the photoreceptor charge per unit area.

It is clear from the above qualitative discussion that the electric field drives the development process in all theories. Therefore, we will devote the next section to its specification.

6.2 The Electric Field

Electric fields are determined by charges and dielectric constants. Therefore, an analysis of the electric field during development requires identification of all charges in the system and an understanding of the dielectric properties of the space in which these charges reside.

6.2.1 Charges

The charge and expose process steps create a latent image, consisting of a surface-charged photoreceptor, which is introduced into the development system. This latent image creates an electric field in the development zone which the toner particles experience. We begin this section by calculating the electric field due to a solid area latent image, i.e., a photoreceptor charged uniformly with a charge per unit area of σ_{p}. We establish the equivalence of the electrostatic potential associated with σ_{p} and an applied voltage across the development zone. Then we consider the effect on the electric field of toner developed onto the latent image.

To obtain the electric field due to σ_{p}, we will assume the photoreceptor and magnetic brush roller are parallel planes (Fig. 6.3a) and the developer is specified by an effective dielectric constant K_{E} (see next section). (We will

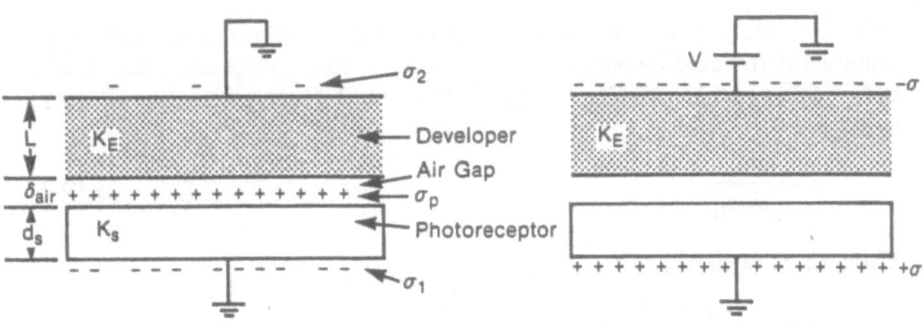

Fig. 6.3a,b. The electric field acting on toner particles in the air gap is calculated by assuming the photoreceptor and roller are parallel planes and the developer is specified by an effective dielectric constant K_{E}. In (**a**) we assume the surface of the photoreceptor is uniformly charged. In (**b**) a potential V is applied across the development zone [6.11]

also assume the charges deposited on the photoreceptor during the charging step remain on its surface. While this is true of a-Se and organic photoreceptors, it is not true for ZnO photoreceptors as shown by several authors [6.7–9].) Charges flow into the ground planes under the photoreceptor and magnetic brush roller depending on their mutual capacitance. If the charge per unit area on the photoreceptor surface is σ_p, and the charge per unit area on the ground planes is σ_1 and σ_2, then

$$\sigma_p = \sigma_1 + \sigma_2 \ . \tag{6.1}$$

Using Gauss's Law, the electric field in the photoreceptor E_p is $\sigma_1/K_s\varepsilon_0$, the electric field in the developer E_D is $\sigma_2/K_E\varepsilon_0$ and the electric field in the air gap just above the photoreceptor in which a toner is trying to decide whether to develop is

$$E_{air} = \sigma_2/\varepsilon_0 \ , \tag{6.2}$$

where K_s is the photoreceptor dielectric constant and K_E is the developer "effective" dielectric constant (Sect. 6.2.2). Since no voltage is applied across the system the sum of the voltage drops must be zero

$$-\frac{\sigma_1}{K_s\varepsilon_0} d_s + \frac{\sigma_2}{\varepsilon_0}\delta_{air} + \frac{\sigma_2}{K_E\varepsilon_0} L = 0 \ , \tag{6.3}$$

where d_s is the photoreceptor thickness, L is the developer thickness and δ_{air} is the small air gap in which the toner resides. Using (6.1–3), the electric field in the air above the photoreceptor is

$$E_{air} = \frac{\sigma_p}{\varepsilon_0}\frac{d_s}{K_s} \Bigg/ \left(\frac{d_s}{K_s} + \frac{L}{K_E} + \delta_{air}\right) \ . \tag{6.4}$$

The electrostatic potential V above a charged photoreceptor is defined as the potential that would be measured with no upper electrode present, i.e., $L = \infty$ in Fig. 6.3a. In this case $E_{air} = 0$,(6.4), and $\sigma_2 = 0$, (6.2), giving $\sigma_1 = \sigma_p$, (6.1) and

$$V = \frac{\sigma_p d_s}{K_s\varepsilon_0} \ . \tag{6.5}$$

Using (6.5) in (6.4) gives

$$E_{air} = V \Bigg/ \left(\frac{d_s}{K_s} + \frac{L}{K_E} + \delta_{air}\right) \ . \tag{6.6}$$

This same result would be obtained if the photoreceptor were uncharged and the counter electrode were raised to a potential V above ground

(Fig. 6.3b). In this case an equal and opposite charge per unit area σ flows onto the two ground planes and the electric fields in the developer, air gap, and photoreceptor are $\sigma/K_E\varepsilon_0$, σ/ε_0, and $\sigma/K_s\varepsilon_0$ respectively. Since the sum of the potential drop must equal the applied potential,

$$\frac{\sigma L}{K_E\varepsilon_0} + \frac{\sigma\delta_{air}}{\varepsilon_0} + \frac{\sigma d_s}{K_s\varepsilon_0} = V, \tag{6.7}$$

the air gap electric field is

$$E_{air} = \frac{\sigma}{\varepsilon_0} = V\Big/\left(\frac{d_s}{K_s} + \frac{L}{K_E} + \delta_{air}\right) , \tag{6.8}$$

the same result as obtained in (6.6). Therefore, the same electric field in the air above the photoreceptor is produced if the photoreceptor is charged to a potential V or a voltage V is applied across the photoreceptor-roller gap.

During development, charged toner moves to the photoreceptor surface. The electrostatics of the charged toner on the photoreceptor outside the development system have been analyzed by several authors with similar results [6.11,19,25,26]. To first order the charge per unit area on the photoreceptor is reduced by the presence of the toner charge per unit area σ_t . Since

$$\sigma_t = \left(\frac{Q}{M}\right)\left(\frac{M}{A}\right) \tag{6.9}$$

the new electric field is obtained by replacing σ_p with $\sigma_p - \sigma_t$. This first order approximation is adequate for most calculations. The toner voltage V_t is defined as

$$V_t = \frac{\sigma_t d_s}{K_s\varepsilon_0} = \frac{Q}{M}\frac{M}{A}\frac{d_s}{K_s}\frac{1}{\varepsilon_0} . \tag{6.10}$$

Using this approximation, the effect of toner development on E_{air} can be obtained by replacing σ_p (or V) with $(\sigma_p - \sigma_t)$ (or $V - V_t$) , so that E_{air} becomes

$$E_{air} = (V - V_t)\Big/\left(\frac{d_s}{K_s} + \frac{L}{K_E} + \sigma_{air}\right) . \tag{6.11}$$

Second order effects are obtained by considering the finite thickness of the toner layer. The developed charged toner, resting on the photoconductor, is assumed to have a constant volume charge density ρ_{tv}, thickness d_t and dielectric constant K_t and to rest on a photoreceptor of thickness d_s and dielectric constant K_s. Using Gauss's Law the charged toner creates an electrostatic potential V_t of

$$V_t = \frac{\rho_{tv} d_t}{\varepsilon_0} \left(\frac{d_s}{K_s} + \frac{d_t}{2K_t} \right) \tag{6.12}$$

when the counter electrode is absent ($L = \infty$ in Fig. 6.3). In terms of measurable parameters, ρ_{tv} and d_t are

$$\rho_{tv} = (Q/M)\, \rho_t p_t \ , \tag{6.13}$$

$$d_t = \frac{M}{A \rho_t p_t} \ , \tag{6.14}$$

where ρ_t is the toner mass density (in g/cm^3) and p_t is the toner volume packing. Using (6.13) and (6.14), Eq. (6.12) can be rewritten

$$V_t = \frac{M}{A} \frac{Q}{M} \frac{1}{\varepsilon_0} \left(\frac{d_s}{K_s} + \frac{d_t}{2K_t} \right) . \tag{6.15}$$

Actually, toner particles are discrete. As the first monolayer of toner is formed, the charge on each toner particle resides at a constant average height d_{t1} above the photoconductor, while the developed mass per unit area is increasing. This gives the voltage due to toner below a monolayer

$$V_t = \frac{M}{A} \frac{Q}{M} \frac{1}{\varepsilon_0} \left(\frac{d_s}{K_s} + \frac{d_{t1}}{2K_t} \right) \quad \text{for } \frac{M}{A} < \text{one monolayer,} \tag{6.16}$$

where d_{t1} is constant. (This value of V_t is actually a spatial average voltage since M/A is also not continuous, being finite where a toner particle has developed on the photoreceptor and zero otherwise.) Above the first monolayer

$$V_t = \frac{Q}{M} \frac{M}{A} \bigg|_1 \frac{1}{\varepsilon_0} \left(\frac{d_s}{K_s} + \frac{d_{t1}}{2K_t} \right) + V_2 \ , \tag{6.17}$$

where

$$V_2 = \frac{Q}{M} \left(\frac{M}{A} - \frac{M}{A} \bigg|_1 \right) \frac{1}{\varepsilon_0} \left(\frac{d_s}{K_s} + \frac{d_{t2}}{K_t} \right) , \tag{6.18}$$

where d_{t2} is the distance to the average toner charge postion in the second layer above the photoreceptor and $M/A|_1$ is M/A at one monolayer. The quantity d_{t2} could be either continuously varying, related to M/A (6.14) giving the usually quoted quadratic dependence on M/A, or a fixed number if the toner in the second monolayer is developed at a uniform height above the photoreceptor.

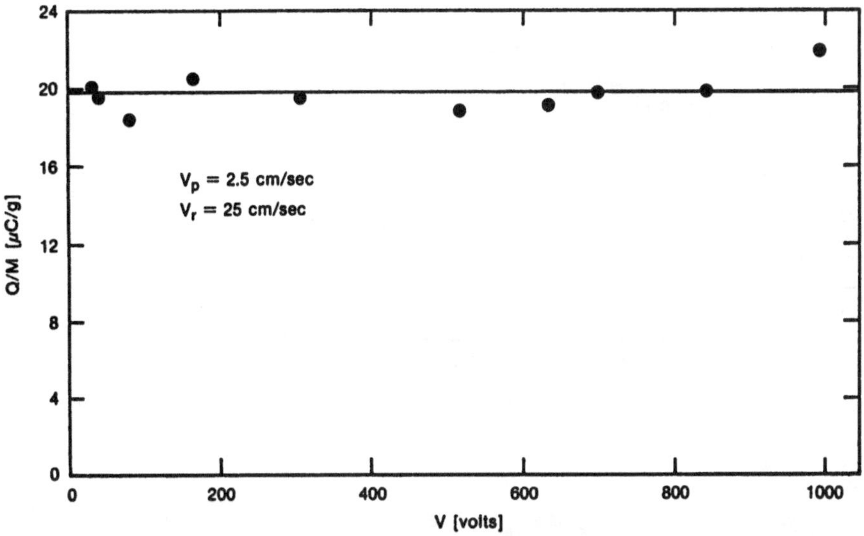

Fig. 6.4. The charge-to-mass ratio Q/M of the developed toner measured by a photoreceptor blowoff measurement is usually observed to be independent of the initial photoreceptor potential. A plate blowoff measurement is done by blowing toner off a photoreceptor with an air jet. The mass loss and charge loss are measured [6.11]

These equations relate the observable electrostatic voltage due to developed toner V_t to the developed mass per unit area M/A. It was assumed for simplicity that Q/M, the toner charge to mass ratio, is a constant as M/A increases, which is usually experimentally observed in magnetic brush development systems (Fig. 6.4). (Generalizing to variable Q/M is straightforward.) The values of d_{t1} and d_{t2} can be obtained experimentally by developing toner onto a metal plate (so that $d_s = 0$, thereby emphasizing the second, smaller term) or by discharging the photoconductor by exposure to light (making it a conductor) and then measuring the remaining potential. In both cases, the result is

$$
\begin{aligned}
&V_t \Big/ \left(\frac{Q}{M}\frac{M}{A}\right) = \frac{d_{t1}}{2K_t}\frac{1}{\varepsilon_0} \,, \\
&\qquad \frac{M}{A} < \text{one monolayer}, \qquad d_s = 0 \,, \\
&(V_t - V_t|_1) \Big/ (Q/M)\left(\frac{M}{A} - \frac{M}{A}\Big|_1\right) = \frac{d_{t2}}{K_t}\frac{1}{\varepsilon_0} \,, \\
&\qquad \frac{M}{A} > \text{one monolayer}, \quad d_s = 0 \,,
\end{aligned}
\tag{6.19}
$$

where $V_t|_1$ is the toner voltage at one monolayer. Experiments to check (6.19) are shown in Fig. 6.5 for size-classified 10 μm toner. In Fig. 6.5a,

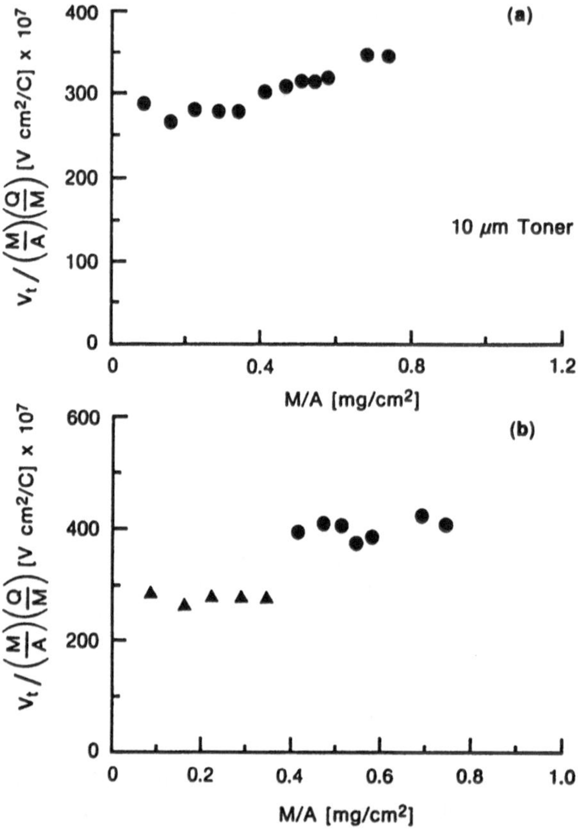

Fig. 6.5a,b. The toner voltage V_t measured after development onto a metal surface. In **(a)** $V_t/(Q/M)(M/A)$ is plotted versus M/A the constant value is the first monolayer. In **(b)** the values for the first monolayer are subtracted (see text); the formation of the second monolayer can be seen

$V_t/(M/A)(Q/M)$ is plotted. A clear change in slope at 0.4 mg/cm^2 is observed, where one would expect the first monolayer to end since

$$\left.\frac{M}{A}\right|_{\text{monolayer}} = \frac{4}{3} r \rho_t p_t \ , \tag{6.20}$$

which equals 0.4 mg/cm^2 for $r = 5\mu$m, $\rho_t = 1$ g/cm^3 and $p_t = 0.6$. In Fig. 6.5b, $V|_1$ and $M/A|_1$, identified from Fig. 6.5a, are subtracted from the data above the first monolayer. It would appear that for classified toner d_{t2} is constant.

It is important to check the magnitude of these effects during development to verify that all charges have been accounted for [6.11]. Figure 6.6 shows ΔV, the change in photoreceptor potential upon development (difference between the electrostatic potential above the photoreceptor after and before

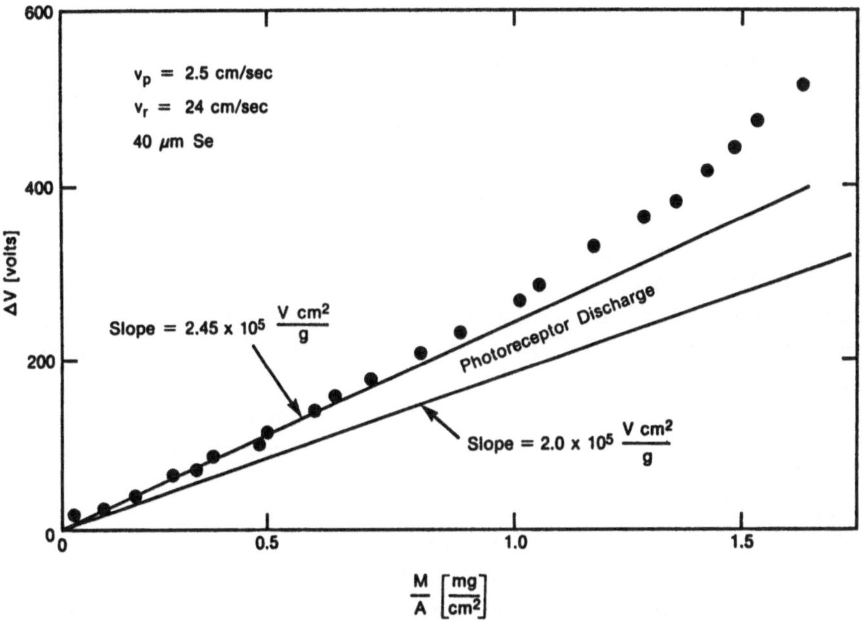

Fig. 6.6 The change in photoconductor potential upon development versus developed mass per unit area. The contribution from the photoreceptor discharge is shown [6.11]

development with appropriate corrections for photoreceptor dark decay). The question being addressed is whether ΔV is due to V_t, the voltage due to toner developed onto the latent image. Qualitatively, the data agree with (6.16–18), with a linear and nonlinear term evident. However, calculation of Q/M from the slope [using $d_t/2K_t = 3.5\mu$m in (6.15)] gives 28.2 μC/g in disagreement with an independent measurement of Q/M by plate blowoff: 19.5 μC/g (Fig. 6.4). Another phenomenon appears to be contributing to ΔV.

The missing phenomenon is photoreceptor charging, the addition of charge to the photoreceptor surface during the development step. Not only does the toner's charge change the photoreceptor charge per unit area during the process of developing, but also the interaction of the carrier and toner with the photoreceptor puts additional charge on the photoreceptor. This is directly observable by removing the toner from the photoreceptor surface (in the dark) and measuring the change in photoreceptor electrostatic potential. After developing and removing toner, it is less than the initial potential. This demonstrates that the net charge on the photoreceptor has been reduced. Subtracting this additional photoreceptor charge (Fig. 6.6) brings the predicted and observed Q/M into virtual agreement [6.11]. (Identical effects are observed with organic photoreceptors [6.14].)

By varying the photoreceptor velocity v_p, even more effects can be observed! It is observed that $\Delta V/(M/A)$, which should be independent of v_p,

(6.16–18), if it were just due to toner, instead increases as v_p decreases. About one-third of the effect is due to an increase of the photoreceptor charging, which is reasonable: the longer a point on the photoreceptor is in the development zone, the more time for charge exchange interactions. The other two-thirds are due to size classification of the toner as determined by Coulter counter analysis of the developed toner. As development "time" increases, the size distribution shifts to smaller particles, which generally have higher Q/M's and therefore higher V_t's. This is direct evidence that scavenging (removal of developed toner from the photoreceptor surface by carrier beads) occurs in a magnetic brush development system and is entirely analogous to similar effects observed by *Hauser* and *Menchel* [6.27] (discussed in Sect. 5.1.3) in cascade development.

The next logical question might be, what is the source of the photoreceptor discharge? It could be the interactions of the photoreceptor with the toner or the carrier particles. Such effects have been discussed by *Cassiers* and *Van Engeland* [6.28]. Direct observations of both effects were made by *Schein* [6.11]. Toner interactions with Se photoreceptor were observed by a double blowoff experiment: toner is blown out of a Faraday cage toward a Se plate. The charge accumulated on the plate is measured. Then the toner is blown off the plate. The amount of charge leaving is always less than the amount arriving and the fraction of charge left behind increases as the toner diameter decreases. Not surprisingly, attempts to understand these results were not successful since these experiments are in the complicated world of triboelectricity (Chap. 4). Carrier-Se interactions are easily observed by running bare carrier against charged Se. Again, large effects are observed; again, the results were not "understood."

Summarizing this section, the electric field in the development zone has been related to the latent image as modified by developed toner. Applying charge conservation, it has been shown that the photoreceptor electrostatic potential is made up of three terms: the initial potential (determined by charging, exposure, and dark decay of the photoconductor), the effect of the charged toner resting on the photoconductor, and the charging of the photoreceptor during the development process. In addition, toner size classification, leading to a higher developed toner charge to mass ratio, occurs as the photoconductor velocity is reduced. The photoreceptor charging could be due to either toner-Se or carrier-Se interactions. Both effects have been independently observed. Neither effect is "understood" in terms of the microscopic physics involved. A corollary to the above results is that, in order to do experiments in which the electric field is "known," one needs to eliminate these effects by, for example, replacing the photoreceptor with a metal plate.

6.2.2 Effective Dielectric Constant

The second quantity needed to calculate the magnitude of the electric field is the dielectric constant of the developer mix. The determination of the dielectric properties of the developer mix logically belongs here. Unfortunately this problem has turned out to be much more difficult than originally envisioned and remains, in the opinion of this author, unsolved.

The first to discuss the dielectric properties of the developer mix was *Schein* [6.12]. He pointed out that the electric field in the developer mix could be characterized by (1) the average electric field, which is determined by the dielectric constant, (2) the peak electric field and (3) the electric field seen by toner particles, which is inhomogeneous but might be described on average by an "effective" dielectric constant. Since toner particles are much smaller than carrier, they could in principle be influenced by peaks in the electric field directly under carrier chains. The electric field distribution hypothesized in 1975 is shown in Fig. 6.7.

The average electric field in a dielectric (dielectric constant K) within a capacitor is simply V/L where V is the potential across the capacitor and L is the spacing between the electrodes. If a narrow air gap exists within the capacitor, then the average electric field in the air gap is

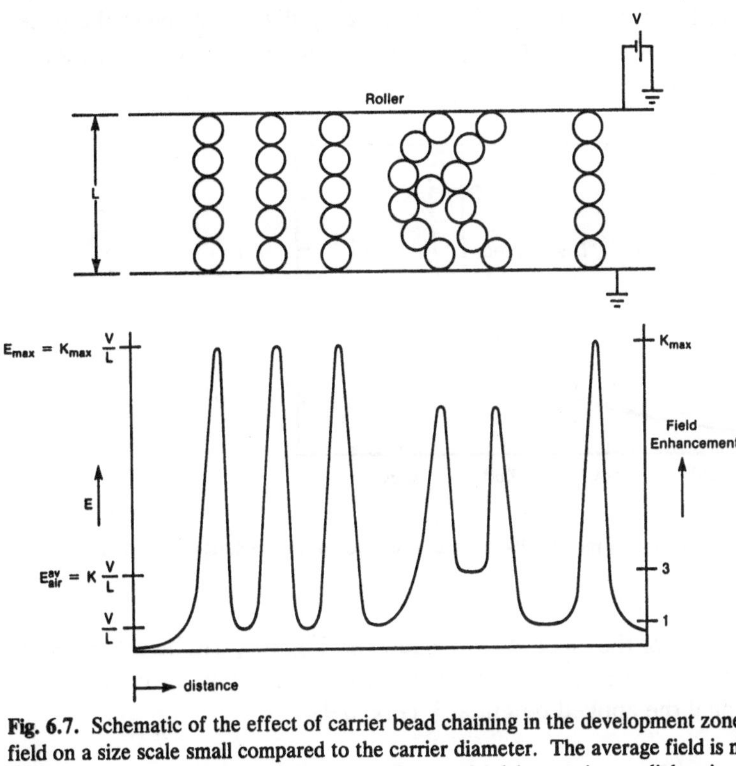

Fig. 6.7. Schematic of the effect of carrier bead chaining in the development zone on the electric field on a size scale small compared to the carrier diameter. The average field is measured by the dielectric constant. The peak fields are characterized by maximum dielectric constant K_{max} [6.12]

$$E_{\text{air}}^{\text{av}} = \frac{V}{L} K \ . \tag{6.21}$$

In a two component development system the "dielectric" is composed of carrier beads, toner and air. Formulas for the overall dielectric constant of two component mixtures have been given by *Maxwell* [6.29], *Garnett* [6.30] and *Lord Rayleigh* [6.31]; for reviews of more recent literature, see [6.32]. Lord Rayleigh's formula for metal spheres with packing p in air, to second order, is

$$K = 1 + \frac{3p}{1 - p - 1.65p^{10/3}} \ . \tag{6.22}$$

With a metal (carrier particles) and a packing fraction p of 50%, $K = 4.75$. Using Rayleigh's procedure, but with the assistance of a computer, *McPhedran* and *McKenzie* [6.33] have calculated K in all orders for several geometries, as shown in Fig. 6.8. At $p = 50\%$, McPhedran and McKenzie obtain $K = 6$ for the disordered geometry. Experiments carried out by *Hays* [6.34] appear to be consistent with these results. He found $K = 5.3$ by placing the beads in a capacitance cell. He used 250 µm diameter carrier particles and found a weak dependence on toner diameter (which presumably slightly changed the packing). The value of K was 5.3 for 8 µm diameter toner; 4.8 for 16 µm toner; 4.7 for 28 µm toner.

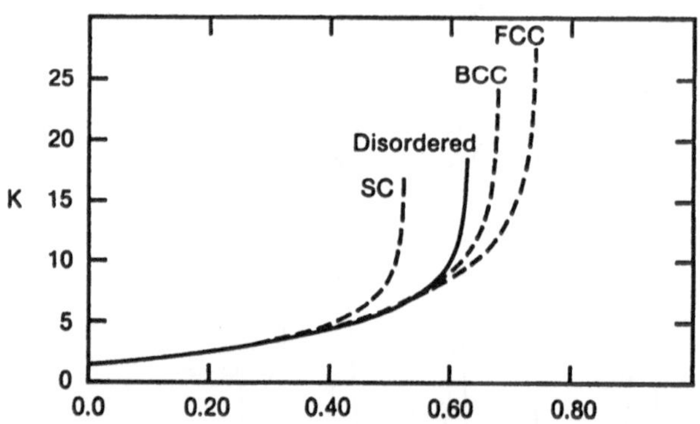

Fig. 6.8. Dielectric constant of metal balls in air as a function of the packing of the metal balls for various lattices [6.33]

The peak value of the electric field, shown in Fig. 6.7, can be estimated as follows. We note that the integral of the electric field through the center of the beads must equal the applied potential V (Fig. 6.9)

$$\int_0^L E \, dy = V \ . \tag{6.23}$$

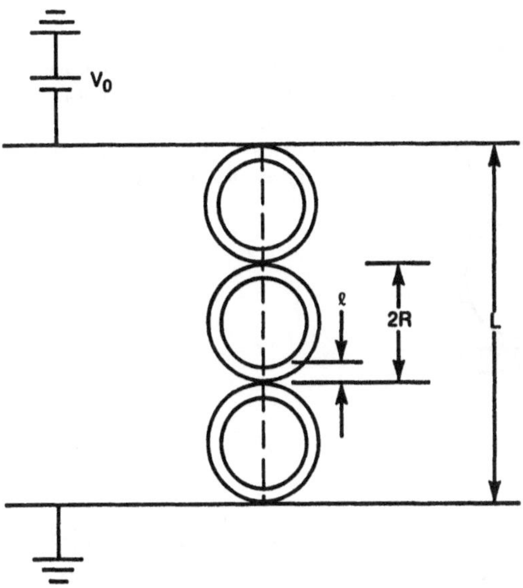

Fig. 6.9. The maximum electric field can be found by integrating the electric field along the centers of the bead chain [6.12]

Since E vanishes inside the metal beads, and is constant in the dielectric coating of thickness ℓ, then for n beads between the two plates

$$E^{\max} 2n\ell = V \, . \tag{6.24}$$

The maximum electric field $E_{\text{air}}^{\max}$ in a small air gap in which toner develops equals $K_{\text{p}} E^{\max}$ where K_{p} is the dielectric constant of a polymer (taken to be 3). We define $K_{\max}$ such that

$$E_{\text{air}}^{\max} = \frac{V}{L} K_{\max} \, . \tag{6.25}$$

Solving for $K_{\max}$, using (6.24 and 25), gives

$$K_{\max} = \frac{LK_{\text{p}}}{2n\ell} = \frac{RK}{\ell} \tag{6.26}$$

since L/n equals a carrier particle diameter $2R$. If toner separates the carrier particles it is easily shown that

$$K_{\max} = \frac{RK_{\text{p}}}{\ell + r} \, , \tag{6.27}$$

where r is a toner particle radius. These peak fields were directly observed by *Hays* [6.34] in 1978 by measuring the field intensification with a 75 μm wire probe imbedded in an electrode (Fig. 6.10). An observed profile is shown in

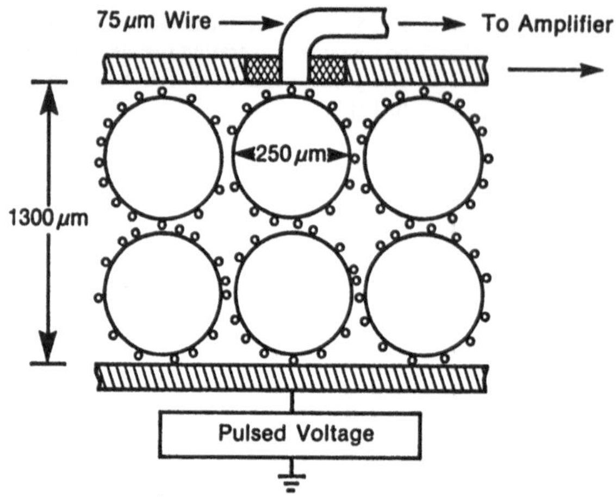

Fig. 6.10. Apparatus for measuring the electric field profile at a developer-electrode interface. A Cober high-voltage pulse unit supplies a 1 ms pulse every 250 ms. The electrode moves at a speed of 2 mm/min [6.34]

Fig. 6.11 and confirms qualitatively the picture given in 1975. Peak electric field enhancements over the value observed in the absence of carrier beads, i.e. K_{max}, of 31 were observed for 8 μm diameter toner (18 for 16 μm and 12 for 28 μm). The sensitivity to toner diameter clearly suggests (6.27) is more appropriate than (6.26). The observed magnitudes are in qualitative agreement with (6.27), which predicts K_{max} = 75, 42, and 26 using $\ell = 1\mu m$ and $2r = 8\mu m$, $16\mu m$ and $28\mu m$, respectively. The approximate factor of 2 discrepancy may be due to the averaging caused by the use of a large (75 μm) diameter probe relative to the size of the carrier beads. Hence it would appear that the electric field has been characterized. However, the question of which electric field the toner senses, remains.

That the toner moves in response to the electric field is clear. The question is whether the toner responds to the average electric field, the peak electric

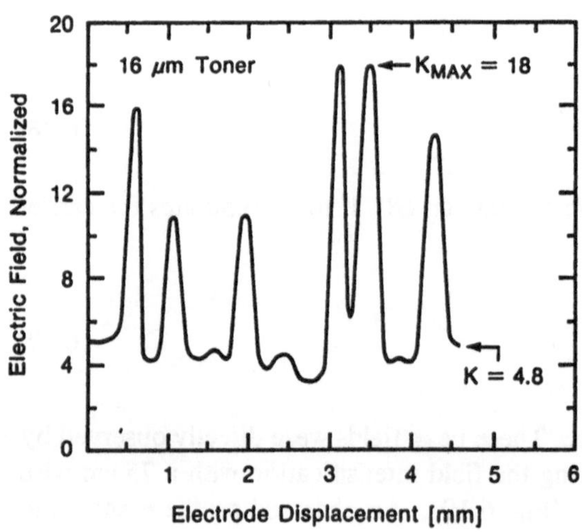

Fig. 6.11. Profile of the electric field, normalized to the empty cell field, as a function of the electrode displacement. The developer is 16 μm toner mixed with 250 μm steel-core beads [6.34]

field, or some special average determined by the physics of the development process. In other words, is K_E, the effective dielectric constant of the mix determined by the electric field experienced by toner, which appears in (6.6)

$$E_{air} = V \Bigg/ \left(\frac{d_s}{K_s} + \frac{L}{K_E} + \delta_{air} \right), \tag{6.6}$$

equal to K, Eq. (6.22), K_{max}, Eq. (6.27) or some other value. In order to answer this question, an experimental determination of K_E is required. Clearly, this experiment must involve an observation of the change in the behavior of toner particles as the electric field is changed.

The first quantitative attempt [6.12] to measure the electric field sensed by toner was made by measuring the change in M/A as the photoreceptor thickness was changed, holding everything else constant (especially the photoreceptor-roller gap L). The idea was that if development is proportional to the electric field, then since

$$\frac{M}{A} \propto E_{air} = V \Bigg/ \left(\frac{d_s}{K_s} + \frac{L}{K_E} \right) \tag{6.28}$$

[a good approximation to (6.6)], by changing the photoreceptor thickness d_s and observing M/A change, one could determine K_E. This idea made two assumptions, both of which turned out to be incorrect. It (1) assumed a small signal situation, i.e., the developed toner mass per unit area is proportional to the *initial* electric field, and (2) ignored the photoreceptor charging (see previous section). It was later learned that development proceeds until the electric field goes to zero. Also, when the photoreceptor is charged, the effect of the counter charge left behind in the developer must be considered. Therefore, we reanalyze the data using our current understanding of development physics, discussed in detail in Sect. 6.3.4.

The correct expression for M/A is required to understand the effect of changing d_s. In Sect. 6.3.4 it is shown (6.57) that developed mass per unit area M/A is related to the applied potential V by (6.57),

$$\frac{M}{A} = V\varepsilon_0 \Bigg/ \frac{Q}{M} \left(\frac{d_s}{K_s} + \frac{\Lambda}{\nu} \right), \tag{6.57}$$

where ν is the speed ratio, and Λ is L/K_E. Using the first-order approximation for V_t (6.10), Eq. (6.57) can be rewritten as

$$\frac{V}{V_t} = 1 + \frac{L/K_E}{d_s/K_s} \frac{1}{\nu}, \tag{6.29}$$

or

$$\frac{V}{V_t + V_m} = 1 + \frac{L/K_E}{d_s/K_s}\frac{1}{\nu}. \tag{6.30}$$

Equation (6.30) was obtained by inserting the voltage due to the charging of the photoreceptor (actually Mylar, V_m) into (6.29). Adding V_m to V_t is the proper method of correcting for Mylar charging since the transfer of charge to the Mylar surface is completely analogous to the effects of the toner charge developing onto the Mylar. This formula turns out to be model independent (Chap. 7). It indicates that as $\nu \to \infty$ or $d_s \to \infty$, the toner voltage approaches the applied voltage V, the neutralization condition. To obtain L/K_E one plots $V/(V_t + V_m)$ versus $(d_s/K_s)^{-1}$.

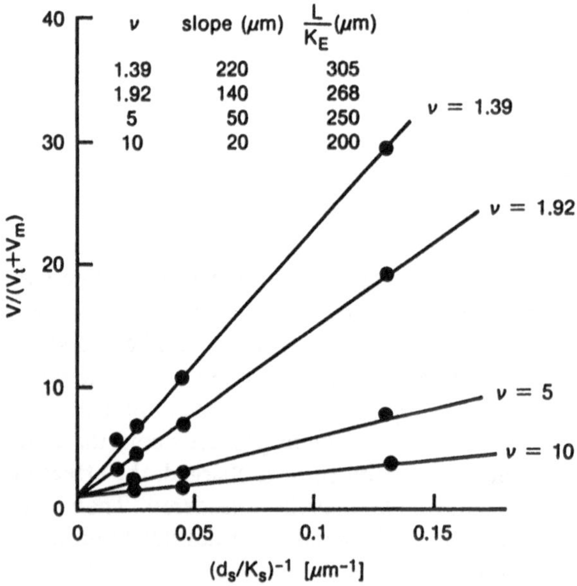

Fig. 6.12. The effective dielectric constant can be obtained by plotting $V/(V_t + V_m)$ versus d_s^{-1}. Here we reanalyze data from [6.14]. Λ is 250 μm, independent of ν, the speed ratio, giving $K_E = 5.3$ (for $L = 1325$ μm)

Figure 6.12 shows data of $V/(V_t + V_m)$ versus $(d_s/K_s)^{-1}$ from the 1975 paper, after reanalysis, and Fig. 6.13 shows data taken on experimental hardware with IBM Series III developer. Each data point in the figure was obtained from the slope of curves in which $V_t + V_m$ were either calculated (1975 paper) or measured directly with an electrostatic voltmeter (IBM Series III data). On each, the speed ratio is a parameter. Within experimental error, L/K_E is independent of ν and is 140 μm for the Series III and 250 μm for the 1975 data. Given the gap L, K_E can be calculated: 8.9 for Series III and 5.3 for the 1975 data.

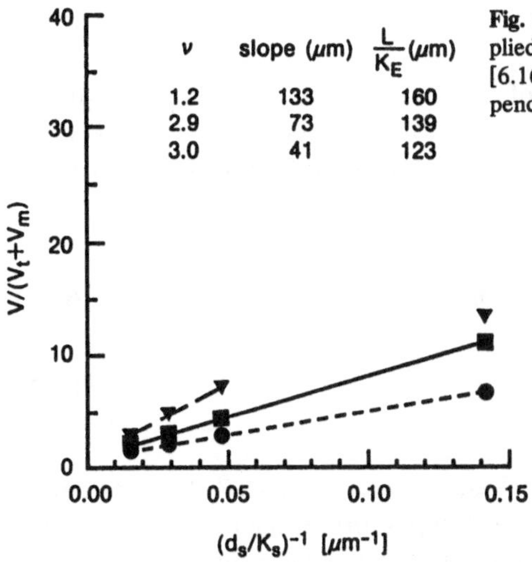

Fig. 6.13. The analysis used in Fig. 6.12 is applied here to the material and hardware used in [6.16]. Here $\Lambda = 140\ \mu m$ and K_E is 8.9, independent of ν

The unsolved problem is accounting for the observed magnitudes of K_E. *Lee* and *Beardsley* [6.35] have asserted that the proper value for K_E is just the average dielectric constant, which they estimate. Using Fig. 6.8 it can be seen that it would be difficult to explain the value obtained for the Series III developer, 8.9, at reasonable packings (40%–50%). Furthermore, given the success of the equilibrium theory in accounting for the data (Sect. 6.4), one would expect the toner to experience electric fields closer to the peak value. However, (6.27) also is not adequate, as the predicted magnitudes are much too large (37.5 and 21 for Series III and the 1975 paper, respectively).

It appears to this author that calculations of the electric field due to the carrier charge must take into account the shape and packing of the carrier beads. The carrier charge is clearly absent where no carrier exists and is distributed across the carrier surface to maintain the carrier beads at an equipotential. However, attempts to take these factors into account have not been successful in predicting the observed magnitude of K_E.

6.3 Theories of Solid Area Development

Qualitative descriptions of three development mechanisms can be found in Sect. 6.1; here, we derive all known theories quantitatively.

6.3.1 Neutralization

The simplest theory assumes that the toner charge per unit area σ_t completely neutralizes the photoreceptor charge per unit area σ_p. The quantity σ_p is determined by the electrostatic potential V to which the photoconductor is charged (6.5):

$$\sigma_p = K_s \varepsilon_0 V/d_s \ , \tag{6.31}$$

where K_s is the dielectric constant of the photoconductor and d_s is its thickness. Because the toner charge per unit area equals its charge per unit mass (Q/M) times the developed mass per unit area (M/A), the predicted toner mass per unit area is

$$\frac{M}{A} = \frac{V\varepsilon_0}{(Q/M)(d_s/K_s)} \ , \quad \text{or} \tag{6.32}$$

$$V_t = V \ , \tag{6.33}$$

where (6.10) is used for the toner voltage V_t.

The most definitive test for neutralization is given by (6.33): after development, the toner voltage should equal the photoreceptor potential. As this is not usually observed (Fig. 6.24), other phenomena must limit development.

6.3.2 Field Stripping

Perhaps the easiest theory to visualize is field stripping, pictured schematically in Fig. 6.2. In this theory, toner is stripped from carrier when the Coulomb force due to the electric field of the latent image overcomes the force of adhesion of toner to carrier. This model has been worked out by *Harpavat* [6.16], *Williams* [6.17,18] and *Kondo* and *Kamiya* [6.19]. We will discuss (and slightly simplify for clarity) Williams's approach first.

In Williams's field stripping theory, the developed toner mass per unit area M/A can be separated into three factors:

$$\frac{M}{A} = \left(\frac{\text{toner mass}}{\text{carrier bead}}\right) \times \left(\frac{\text{number of carrier beads}}{\text{area of image}}\right) \times \tilde{F} \tag{6.34}$$

or, factor for factor,

$$\frac{M}{A} = (C_t M_c)\left(\frac{p\nu}{\pi R^2}\right)\tilde{F} \ , \tag{6.35}$$

where $\tilde{F}$ is the fraction of toner field stripped from one carrier bead, C_t is the toner concentration (ratio of toner mass to carrier mass), M_c is the carrier mass, ν is the speed ratio (ratio of roller to photoreceptor velocities), πR^2 is the projected area of a carrier bead on a photoreceptor, and p is the surface packing of the carrier beads against the photoreceptor (Williams assumed $p = 0.91$). Obviously, the heart of the calculation is determining $\tilde{F}$.

First, we demonstrate that the number of carrier beads that contact a unit area on the photoreceptor is $p\nu/\pi R^2$ with arguments taken from [6.13].

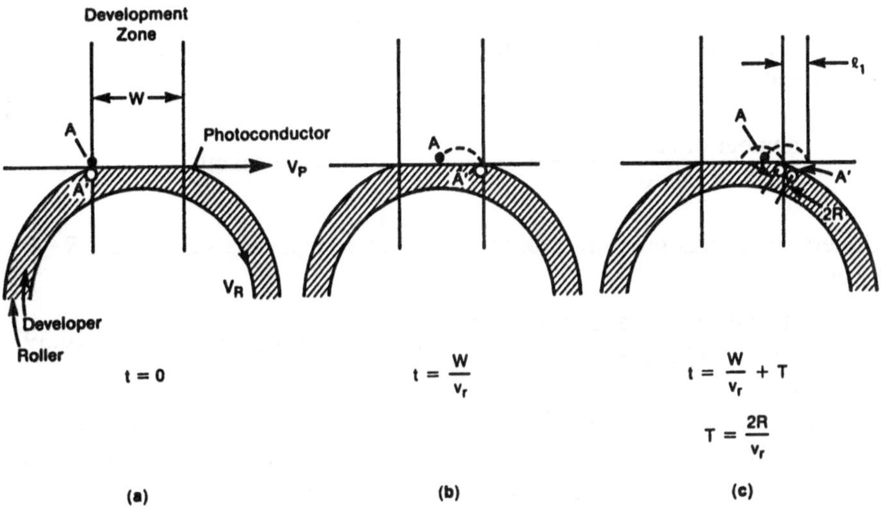

Fig. 6.14a–c. The derivation of the number of carrier beads that contact a unit area on the photoreceptor uses these diagrams [6.13]

Clearly, if $\nu = 1$, (Fig. 6.14), each carrier particle contacts an area on the photoreceptor of $\pi R^2/p$. If $\nu > 1$ (Fig. 6.14), then the roller has a higher velocity than the photoreceptor and each carrier particle contacts additional photoreceptor area. Consider, for example, a carrier bead at point A' on the roller (in Fig. 6.14a) opposite point A on a photoconductor, both points just about to enter the development zone (of width W). At time t (Fig. 6.14b), when the carrier at A' is just leaving the development zone, point A is still within the zone. The carrier at A' has contacted the circled region on the photoconductor in Fig. 6.14b. The next carrier behind the one at A' also contacts a circled region on the photoconductor, displaced backwards by a distance ℓ_1 (Fig. 6.14c); ℓ_1 is equal to the photoconductor velocity ν_p multiplied by the time it takes the second carrier (of radius R) to reach the end of the development zone, $2R/\nu_\mathrm{r}$ (assuming carrier beads are in contact and cubic packing for simplicity)

$$\ell_1 = 2R\,\frac{\nu_\mathrm{p}}{\nu_\mathrm{r}} \quad . \tag{6.36}$$

Therefore, each carrier bead contacts the circled regions on the photoconductor (Fig. 6.14c) which are displaced by a distance ℓ_1. This is equivalent to each carrier particle contacting a length ℓ_1 of photoreceptor or a photoreceptor area of $\ell_1 2R$:

$$\frac{\text{photoreceptor area contacted}}{\text{no. of carrier beads}} = \ell_1 2R = 4R^2\,\frac{\nu_\mathrm{p}}{\nu_\mathrm{r}} = 4R^2/\nu \tag{6.37}$$

using (6.33), where ν is the ratio of the roller velocity to the photoreceptor velocity. Inverting this, the number of carrier beads which contact a unit area on the photoreceptor is

$$\frac{\text{number of carrier beads}}{\text{area of photoreceptor}} = \frac{\nu}{4R^2}\,, \qquad \text{cubic packing}\ . \tag{6.38}$$

Generalizing from cubic to arbitrary packing p (replacing $4R^2$ with $\pi R^2/p$) gives

$$\frac{\text{number of carrier beads}}{\text{area of photoreceptor}} = \frac{\nu p}{\pi R^2}\,, \qquad \text{arbitrary packing}\ . \tag{6.39}$$

Similar arguments can be used for the "against" mode, i.e., the photoreceptor and roller move in opposite directions, to show that (6.37–39) remain valid: the ν represents the absolute value of the speed ratio.

To obtain $\tilde{F}$, Williams considered the forces on the toner particles. He assumed that every toner particle has a single charge Q_t and radius r. This leads to a single value for the toner-carrier adhesion force F_t

$$F_t = \frac{Q_t^2}{4\pi\varepsilon_0 4r^2} \tag{6.40}$$

(ignoring any van der Waals contributions). The problem then becomes one of calculating the variation of the Coulomb force across the surface of a carrier bead (Fig. 6.15). Whenever the Coulomb force exceeds F_t, toner develops. To calculate this field, Williams assumes an isolated metal carrier bead in a

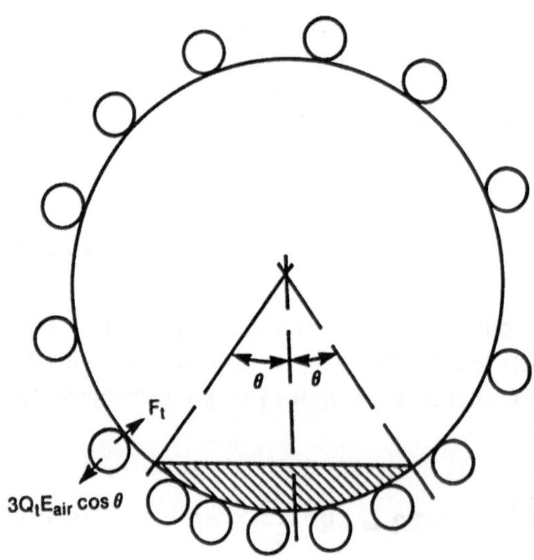

Fig. 6.15. The fraction of toner removed from a carrier is defined as $F = (1/2)(1 - \cos\theta)$, where θ depends on toner charge density and the external electric field E of the latent image [6.18]. (Copyright © 1984 by John Wiley and Sons, Inc.; reprinted by permission)

uniform electric field E_{air}. He shows that near the surface of the carrier bead the radial electric field is $3E_{air}\cos\theta$, where θ is the angle shown in Fig. 6.15. Therefore, developments occur up to an angle θ given by

$$F_t = 3Q_tE_{air}\cos\theta \ , \tag{6.41}$$

where it is assumed that if $F_t > 3Q_tE_{air}$, $\theta = 0$ and $\cos\theta = 1$. The fraction of toner developed from a carrier bead, using solid geometry to relate F and θ, is

$$\tilde{F} = \frac{1}{2}(1-\cos\theta) = \frac{1}{2}\left(1-\frac{F_t}{3Q_tE_{air}}\right) . \tag{6.42}$$

Combining this with (6.35 and 40) gives

$$\frac{M}{A} = \frac{2}{3}C_t\rho_cRp\nu\left(1-\frac{1}{48\pi\varepsilon_0}\frac{Q_t}{r^2E_{air}}\right) \tag{6.43}$$

$$= 0 \qquad \text{if } F_t > 3Q_tE \ .$$

This has a strong threshold behavior in M/A versus E_{air} resulting directly from the assumption of a single value for Q_t and r.

This threshold phenomenon was softened by two modifications to the theory. *Schein* [6.13] in 1975 pointed out that, as toner develops, the carrier particle to which the toner was attached builds up a net charge, increasing the toner adhesive force to the carrier. Williams added this term to obtain a total adhesive force F_t^{tot},

$$F_t^{tot} = F_t + \frac{n_0\tilde{F}Q_t^2}{\varepsilon_0R^2} , \tag{6.44}$$

where n_0 is the initial number of toner particles on a carrier and is related to the toner concentration C_t,

$$n_0 = \left(\frac{R}{r}\right)^3\frac{\rho_c}{\rho_t}C_t \ . \tag{6.45}$$

Equation (6.42) is now resolved, replacing F_t with F_t^{tot}, and a new expression for $\tilde{F}$ is found

$$\tilde{F} = \frac{1}{2}\left(\frac{3Q_tE_{air}}{F_t}-1\right)\Big/\left(\frac{3Q_tE_{air}}{F_t}+\frac{8\pi\varepsilon_0C_tR\rho_c}{r\rho_t}\right) . \tag{6.46}$$

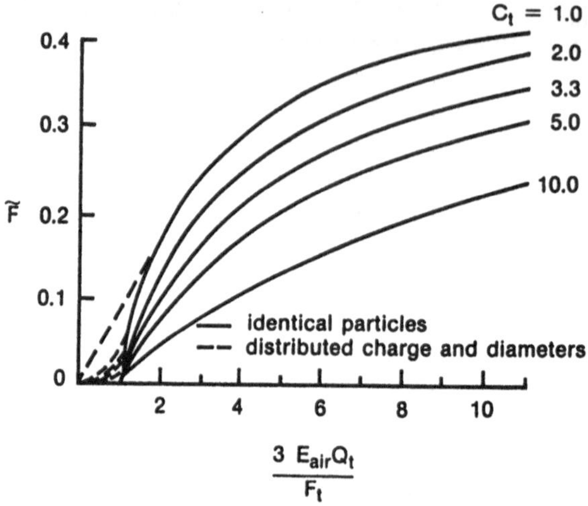

Fig. 6.16. The behavior of F with nondimensional electric field [6.18]. (Copyright © 1984 by John Wiley and Sons, Inc.; reprinted by permission)

This equation is plotted in Fig. 6.16 (solid lines) for various toner concentrations as a function of $3Q_tE_{air}/F_t$. The strong threshold behavior is still evident. Williams then made a second modification by assuming a distribution of charges and diameters (the dotted lines in Fig. 6.16). Although he does not specify the distributions used or how they were added to the theory, this clearly softened the threshold behavior.

Note that all the qualitative predictions made in Sect. 6.1 for a field stripping model are realized. The nonlinear dependence of mass per unit area on voltage is self-evident in Fig. 6.16. It is somewhat softened by the inclusion of distributions of Q_t and r. The linear dependence on speed ratio is explicitly shown in (6.43). The predicted dependence of developed toner charge on applied field is not dealt with explicitly but it is intuitively obvious that higher fields will develop higher charged toner. Finally, the strong dependence on toner concentration, i.e., available toner, is explicitly shown in Fig. 6.16 and (6.43).

Harpavat [6.16] takes a different approach to a field stripping theory. Following unpublished work by Stover, he treats the development process as though it were a rate process. The rate of toner development $d(M/A)/dt$ equals the rate at which toner is deposited less the rate at which toner is removed, i.e., scavenged. The deposition rate contains the essential physics. He assumes toner is field stripped from carrier from a depth h above the photoreceptor surface. The depth h is assumed constant, independent of roller velocity, photoreceptor velocities, or electric fields; values of 300 μm were obtained by fitting data. The deposition rate is therefore the fraction of toner field stripped, $\tilde{F}$, times the total amount of toner flowing above the photoreceptor to a depth h, or $\tilde{F}FC_th/LW$, where F is the developer flow rate (g/cm s), C_t is the toner concentration, L is the photoreceptor to roller gap, and W

is the contact length of the brush against the photoreceptor. Clearly $\tilde{F}$ is a function of the electric field E, toner charge Q_t and radius r. Since the toner with the largest adhesion force that can be field stripped with a field E_{air} is

$$Q_t E_{air} = \frac{1}{4\pi\varepsilon_0} \frac{Q_t^2}{4R^2} \ , \tag{6.47}$$

$$\tilde{F} = \int_0^{E_{air}} f(x)dx \ , \tag{6.48}$$

where

$$x = \frac{1}{4\pi\varepsilon_0} \frac{Q_t}{4R^2} \tag{6.49}$$

and $f(x)$ is the distribution function of x. The scavenging rate was assumed to be equal to $k(FC_t/L)(M/A)$, where k is a constant, giving finally

$$\frac{d}{dt}\left(\frac{M}{A}\right) = \tilde{F}\frac{FC_t h}{LW} - \frac{kFC_t}{L}\left(\frac{M}{A}\right) \ . \tag{6.50}$$

This differential equation has two adjustable constants h and k, and a distribution function of Q_t/R^2.

Knowing that M/A is generally observed to be linear in V, Harpavat chose the only Q_t/R^2 distribution that would reproduce this experimental result, a rectangular distribution. This distribution had the additional advantage that it allowed an analytic solution to the equations. He does state, however, that other distributions were studied using numerical integration techniques and the effect of the width of the distribution was "significant," which is intuitively obvious from Williams's work.

Given this rectangular distribution, Harpavat displays graphically the effect of various parameter variations which again reproduce the qualitative results derived earlier for a field stripping model: (1) M/A increases linearly with ν (for large values of ν, saturation effects are predicted due to thick toner layers modifying the electric field) and (2) M/A increases with C_t for constant Q/M. He does not discuss Q/M dependence on V. As stated above, he predicts linear dependence of M/A on V due to his choice of a rectangular distribution function for Q_t/r^2.

6.3.3 Powder Cloud

It has been suggested by workers at Agfa-Gevaert [6.20,21] that powder cloud development can also contribute to magnetic brush development. Although the amount of information on this theory is limited, it appears that their concept of powder cloud development is an extension of *Sullivan* and *Thourson*'s [6.36] work on cascade development (Sect. 5.3.1). A cloud of toner is produced in the development region, primarily by inertial forces on the toner particles, which is then collected by the electric field associated with the latent image. As in the field stripping, this development process is regarded as a rate process with development increasing exponentially with "time," i.e., the time a point on the photoreceptor is in the development region.

Two observations tend to confirm the existence of power cloud development at least in some regime of the magnetic brush development operating space. First, as the distance between the photoreceptor and roller is increased, development decreases, as expected. It then becomes constant above 6 mm (Fig. 6.17) when contact between the carrier beads and the photoreceptor no longer occurs. If a powder cloud exists, it would contribute to development under these conditions. Second, in this region of large gaps, the "touch-free" region [6.20,21], the observed current to the photoreceptor ground plane exhibits an exponential time dependence, as one would expect for a dynamic process, i.e., either powder cloud or field stripping.

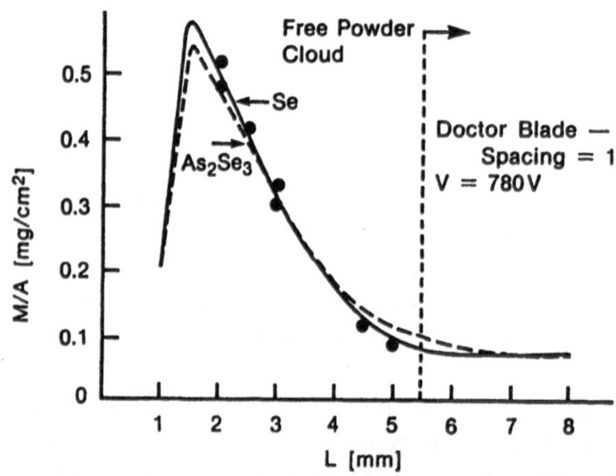

Fig. 6.17. The developed mass per unit area on the photoconductor depends strongly on the photoconductor–magnetic roller spacing [6.21]

Inspection of Fig. 6.17 indicates that the powder cloud component, at least with the toner and hardware used by the workers at Agfa-Gavaert, is at least five times smaller than the development at the peak of the curve, the more usual operating point. Therefore, it would seem to be at most a small contributor to normal development. However, it is possible that it contributes to image defects (Sect. 6.6).

6.3.4 Equilibrium

The equilibrium theory resulted directly from attempts to understand surprising regularities in data taken on a highly controlled magnetic brush development system (Sect. 6.4). For example, (1) it is intuitively obvious that field stripping theories predict threshold behavior for development versus voltage: there should be sufficient toner-carrier adhesion so that no development occurs for low electric fields. Yet no evidence for threshold behavior was observed under a wide variety of conditions. A rationale was required to understand the absence of a threshold voltage. This was accomplished by assuming that toner develops only in three-body-contact events in which the toner simultaneously contacts both a carrier and the photoreceptor. This allowed the toner-carrier adhesion force to be cancelled to first order by a toner-photoreceptor adhesion force. Later microscopic experiments by *Hays* [6.34] support this assumption. (2) Over a wide range of conditions, development was always observed to be linear in the applied potential. Again, that is not expected intuitively based on field stripping theories. One expects development to be proportional to the integral of the toner adhesion distribution up to a value determined by the applied electric field. Only in the very special case of a rectangular distribution of Q/r^2, as assumed by Harpavat, could such data be explained. The solution was to assume toner development was *not* a dynamic process but represented an equilibrium condition determined by the applied field and the buildup of charge on carrier particles as toner was removed. Such a theory was a departure from previous theories because it ignored the dynamics of the development process. Instead, it focused on determining a condition at which development ceased.

There are now three different derivations of the equilibrium model in the literature, all of which give identical results. Because this theory appears to correctly describe insulating magnetic brush development, we will show the equivalence of all three derivations.

The original picture, given by *Schein* [6.13,14], focuses attention on an individual toner particle about to be developed in a three-body-contact event, as shown in Fig. 6.18. The developed mass per unit area equals the number of toner particles n which come off each carrier bead times the toner mass M_t divided by the area that each carrier contacts on the photoreceptor, $\pi R^2/p\nu$ [see (6.39)], giving

$$\frac{M}{A} = nM_t \frac{p\nu}{\pi R^2} \quad . \tag{6.51}$$

Clearly, the heart of the calculation is determining n. It can be determined by considering the forces on the nth toner particle. In Fig. 6.18, two toner particles have already come off the carrier, producing a net charge in the carrier. The forces on the nth toner particle include the Coulomb force $Q_t E_{air}$,

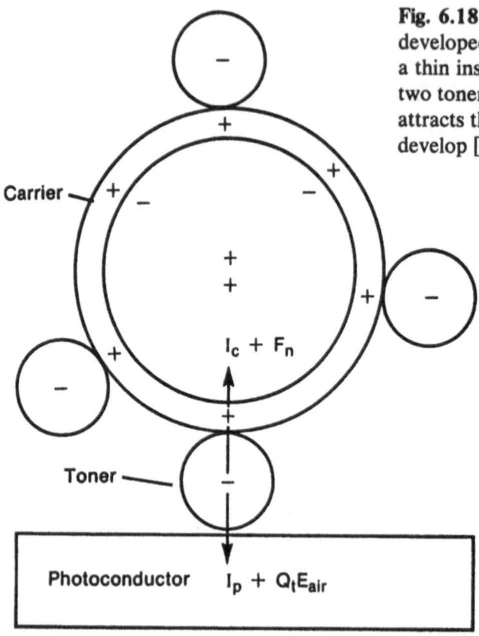

Fig. 6.18. The forces on a toner particle about to be developed. The metal carrier bead is assumed to have a thin insulating coating. This carrier has already lost two toner particles and has built up a net charge which attracts the next toner particles considering whether to develop [6.13].

which is directed toward the photoreceptor [E_{air} is given in (6.6)]. There are short-range image forces toward the photoreceptor (I_p) and carrier (I_c). The carrier has a net charge $(n-1)Q_t$ or a charge per unit area of approximately $(n-1)Q_t p/\pi R^2$. The force of attraction of the nth toner particle to the carrier is its charge Q_t times the electric field due to the charged carrier:

$$F_n = Q_t\left(p\,\frac{(n-1)}{\pi R^2}\,\frac{Q_t}{\varepsilon_0}\right). \tag{6.52}$$

Taking into account the capacitive splitting of the electric field of the carrier bead between the two ground planes gives

$$F_n = Q_t\left(\frac{p(n-1)Q_t}{\pi R^2 \varepsilon_0}\right)\left[\Lambda\Big/\left(\frac{d_s}{K_s}+\frac{L}{K_E}+\delta_{air}\right)\right], \tag{6.53}$$

where we have assumed, to be slightly more general, that the carrier charge is a distance Λ from the roller ground plane. (For insulating magnetic brush development, the carrier charge remains on the carrier beads adjacent to the photoreceptor so that $\Lambda = L/K_E$.) Equating these forces gives

$$\frac{Q_t^2(n-1)p}{\pi R^2 \varepsilon_0}\,\Lambda\Big/\left(\frac{d_s}{K_s}+\frac{L}{K_E}+\delta_{air}\right)+I_c$$

$$= Q_t V \Big/ \left(\frac{d_s}{K_s} + \frac{L}{K_E} + \delta_{air} \right) + I_p \ , \tag{6.54}$$

where V is the voltage due to charge on the surface of the photoreceptor plus any bias voltages on the counter electrode (6.8). Solving for n gives

$$n = \frac{\pi\varepsilon_0}{Q_t/R^2 p} \frac{V}{\Lambda} + 1 + \frac{\pi\varepsilon_0}{p\Lambda} \frac{R^2}{Q_t^2} \left(\frac{d_s}{K_s} + \frac{L}{K_E} + \delta_{air} \right) (I_p - I_c) \ , \tag{6.55}$$

which predicts a term linear in voltage and two terms independent of voltage. Focusing attention only on the voltage-dependent term (the other terms are considered in Sect. 6.6) and using (6.51) gives

$$\frac{M}{A} = \frac{\varepsilon_0 V}{Q_t/M_t} \frac{\nu}{\Lambda} = \frac{\varepsilon_0 V}{Q/M} \frac{\nu}{\Lambda} \ , \tag{6.56}$$

where we have assumed that the proper average of Q_t/M_t is approximated by the average Q/M (measured in a blowoff experiment).

As toner particles develop, the buildup of toner charge on the photoreceptor decreases the electric field experienced by the next toner particle. This can be taken into account by replacing V in (6.56) by $V - V_t$, see (6.11), where V_t is the voltage due to the developed toner (6.10). Making this substitution in (6.56) gives the final result

$$\frac{M}{A} = V\varepsilon_0 \Big/ \frac{Q}{M} \left(\frac{d_s}{K_s} + \frac{\Lambda}{\nu} \right) . \tag{6.57}$$

Note that one can associate with each term specific charges. The Λ/ν term is due to the carrier charge, Λ being the average dielectric distance of the carrier charge from the roller. The d_s/K_s term is due to the buildup of toner charge on the photoreceptor a dielectric distance d_s/K_s from the ground plane.

A second version of this theory, which has been published by Rasmussen (as quoted by *Williams* [6.18]) and *Folkins* [6.37] assumes that development continues until the electric field in the air gap between the separated toner and carrier charge vanishes. The electrostatic problem of finding the electric field in the air gap can be solved by separately calculating the electric field due to the applied voltage, the toner charge, and the carrier charge (Fig. 6.19). If σ is the charge per unit area in the ground planes induced by the applied bias V, then

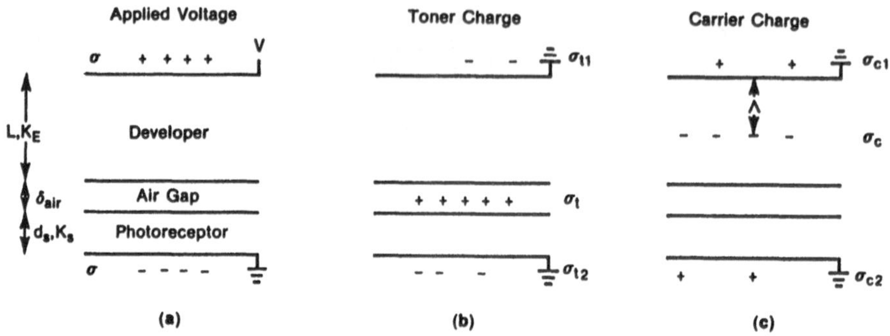

Fig. 6.19a-c. Another method of expressing the equilibrium condition is to assume development occurs until the electric field between the toner and carrier charge (in the air gap) vanishes. The electrostatic problem of finding the total field in the air gap can be solved by separately calculating the field due to the applied voltage, toner charge and carrier charge

$$\sigma = V\varepsilon_0 \Big/ \left(\frac{L}{K_E} + \frac{d_s}{K_s} + \delta_{air} \right) \tag{6.58}$$

and the electric field in the air gap E_{air} due to V is

$$E_{air} = \frac{\sigma}{\varepsilon_0} = V \Big/ \left(\frac{L}{K_E} + \frac{d_s}{K_s} + \delta_{air} \right) . \tag{6.59}$$

If σ_t is the developed toner charge per unit area (which induces charge σ_{t2} in the lower ground plane and σ_{t1} in the upper ground plane), it is easily shown, using $\sigma_t = \sigma_{t1} + \sigma_{t2}$, that the electric field in the air gap due to the toner charge E^t_{air} is

$$E^t_{air} = \frac{\sigma_{t1}}{\varepsilon_0} = \frac{\sigma_t}{\varepsilon_0} \frac{d_s}{K_s} \Big/ \left(\frac{L}{K_E} + \frac{d_s}{K_s} + \delta_{air} \right) . \tag{6.60}$$

If σ_c is the carrier charge per unit area at a dielectric distance Λ from the upper ground plane which induces σ_{c2} in the lower ground plane and σ_{c1} in the upper ground, it is easily shown that the electric field in the air gap due to the carrier charge E^c_{air} is

$$E^c_{air} = \frac{\sigma_{c2}}{\varepsilon_0} = \frac{\sigma_c}{\varepsilon_0} \Lambda \Big/ \left(\frac{L}{K_E} + \frac{d_s}{K_s} + \delta_{air} \right) . \tag{6.61}$$

Adding the electric fields in the air gap (6.59–61) and setting them equal to zero (the equilibrium condition) gives

$$-V + \frac{\sigma_t}{\varepsilon_0} \frac{d_s}{K_s} + \frac{\sigma_c}{\varepsilon_0} \Lambda = 0 . \tag{6.62}$$

We now need the relationship between σ_c and σ_t. Since the charge is the same (the charge left behind in the carrier beads by toner development equals in magnitude the charge of the developed toner)

$$\frac{\sigma_t}{\sigma_c} = \frac{\text{area of carrier bead}}{\text{area of photoreceptor}}$$

$$= \left(\frac{\text{no. of carrier beads}}{\text{area of photoreceptor}}\right)\left(\frac{\text{area of carrier bead}}{\text{no. of carrier beads}}\right) \quad (6.63)$$

$$= \left(\frac{\nu p}{\pi R^2}\right)\left(\frac{\pi R^2}{p}\right) = \nu$$

using (6.39). Using $\sigma_t = (M/A)(Q/M)$, we obtain

$$\frac{M}{A} = V\varepsilon_0 \Big/ \frac{Q}{M}\left(\frac{d_s}{K_s} + \frac{\Lambda}{\nu}\right), \quad (6.64)$$

the same result obtained previously (6.57)

A third derivation, suggested by *Lee* and *Beardsley* [6.35], focuses attention on the charges. If the electric field vanishes when development ceases, then the total charge in a Gaussian pillbox with its top in the air gap and its bottom in the photoreceptor ground plane must vanish. Application of this condition, using Fig. 6.19, gives $\sigma + \sigma_{t2} = \sigma_t + \sigma_{c2}$ or

$$\frac{\sigma}{\varepsilon_0} = \frac{\sigma_{t1}}{\varepsilon_0} + \frac{\sigma_{c2}}{\varepsilon_0} \quad (6.65)$$

which is just (6.62) using (6.60 and 61).

Note that the qualitative predictions for the equilibrium theory made in Sect. 6.2 can be verified by inspecting (6.57). It is predicted that M/A is linear in V. It also should be linear in ν for $\Lambda/\nu >> d_s/K_s$, and independent of C_t for constant Q/M. The latter condition also can be expressed as $(M/A)(Q/M)$ or (Q/A) is independent of C_t. While Q/M of the developed toner is not explicitly dealt with, it is implied by the theory that toner develops until the equilibrium condition occurs, i.e., the average toner develops. Therefore, the developed toner's Q/M should be independent of V.

Multiple magnetic brush rollers are often used to increase the amount of development on the photoreceptor. Normally, carrier beads are passed from one roller to the next using magnetic forces. Under such conditions, the space charge due to the carrier charge accumulates, decreasing the effectiveness of subsequent rollers. Two extreme conditions can be easily understood. If the carrier space charge remains on the carrier beads adjacent to the photoreceptor, no development will occur at the second roller. If the carrier space charge moves all the way to the roller surface, the development on the second roller will double M/A (ignoring the field reduction due to the toner developed by the first roller). Rasmussen (as discussed by *Williams* [6.18]) showed that for his hardware, additional rollers could develop approximately 45% of the amount developed by the previous roller.

6.3.5 Depletion

Depletion describes the condition in which all of the toner is depleted from the carrier beads. Such a condition is considered by *Vahtra* [6.38] in his discussion of line development. For solid area latent images, assuming that toner can be developed from only the first layer of carrier beads, an expression for M/A is easily derived:

$$\frac{M}{A} = C_t M_c \left(\frac{p\nu}{\pi R^2} \right) N \ , \qquad (6.66)$$

where C_t is the toner concentration, M_c is the carrier mass, $\pi R^2/p\nu$ is the area on the photoreceptor touched by each carrier bead, see (6.39), and N is the number of rollers. Expressing M_c in terms of the carrier radius R and the carrier density ρ_c, and writing ν as v_r/v_p where v_r is the roller velocity and v_p is the photoreceptor velocity gives

$$\frac{M}{A} = \left(\frac{4}{3} C_t R \rho_c p \frac{N}{v_p} \right) v_r \ . \qquad (6.67)$$

Note that M/A is predicted to be independent of applied voltage and linear in speed ratio. Such depletion effects can be seen in the data shown in Fig. 6.20, which were taken by the author in a two-roll magnetic brush de-

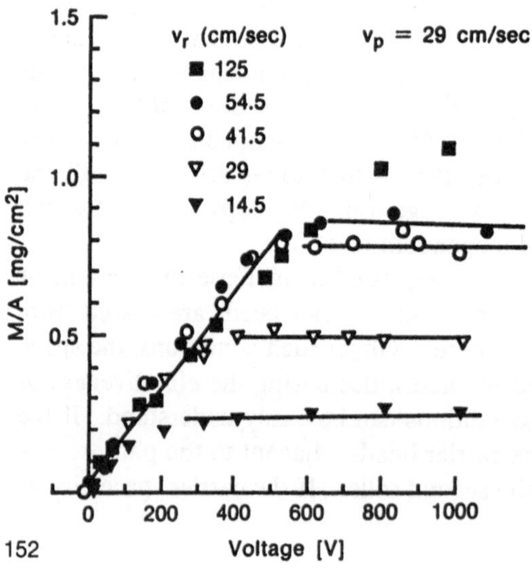

Fig. 6.20. M/A versus V for a very efficient development system and a thick "photoreceptor" Both depletion and neutralization are observed

velopment system using sponge carrier (see next chapter) and an unusually thick "photoreceptor," 75 μm Mylar. As the roller velocity is increased, M/A becomes independent of V at progressively higher voltages. If the value of M/A, which is independent of voltage, is plotted against v_r, the roller velocity, the slope is 1.7×10^{-5} g s/cm^3, to be compared with the theoretical value, from (6.67), of 1.8×10^{-5} g s/cm^3 (assuming $p = 0.4$, $\rho_c = 3.6$ g/cm^3, $C_t = 2.3\%$, $R = 60$ μm, $N = 2$, and $v_p = 29$ cm/s). Note also in this experiment that in the voltage range where $M/A \propto V$, M/A is independent of speed ratio, an indication that the neutralization condition may have been reached. Neutralization predicts (6.32)

$$\frac{M}{A} = V\varepsilon_0 \Big/ \frac{Q}{M}\frac{d_s}{K_s} \ ,$$

which equals 0.35 mg/cm^2 for $V = 200$ V, $Q/M = 20\,\mu$C/g and $d_s/K_s = 25\,\mu$m (75 μm Mylar was used) as compared to the experimental value of 0.32 mg/cm^2. Figure 6.20 shows exciting data that clearly exhibit two different development mechanisms!

6.3.6 "Complete" Theory

It is probably obvious to the reader that considerable simplifications were made in the theories discussed so far. A natural next step is to attempt to build a more realistic theory including the variations in toner diameter and charge, carrier rotation, toner movement during development, etc. *Benda* and *Wnek* [6.39] attempted such an approach. As one might guess, the result was considerably more complex than previous theories. There are 22 parameters in Benda and Wnek's theory, most of which are not directly measurable. The authors state that regression analysis is not possible, so one must assume some arbitrariness was exercised in the choice of values for the parameters used to fit data.

Benda and Wnek begin with a differential equation approach to development. The spatial derivative of the toner number densities on the photoreceptor n_p and on the carrier layer closest to the photoreceptor n_B are

$$v_r \frac{dn_B}{dx} = -D + S + G \ , \tag{6.68}$$

$$v_p \frac{dn_p}{dx} = D - S \ , \tag{6.69}$$

where v_r and v_p are the roller and photoreceptor velocities, x is the position in the nip, D is the deposition rate, S is the scavenging rate and G is a toner migration rate. Here D is taken proportional to n_B, S is taken proportional to

n_p, and the toner migration G is taken as

$$G = f_G^+ n_{B0} - f_G^- n_B \ , \tag{6.70}$$

where the f's are constants and n_{B0} is the initial value of n_B.

The deposition rate is divided into two parts, one originating from the initial contact of the carrier bead with the photoreceptor and one resulting from rotation of carrier beads which exposes new toner as it moves through the development nip. The initial contact part of the deposition rate depends on the balance of forces seen by the toner particle. Benda and Wnek consider image forces, Coulomb forces, induced dipole forces, and nonelectrostatic forces. The electric field is worked out including the toner charge, the buildup of carrier charge as toner develops, bulk space charge, charge due to toner migration during development, and charge due to the conductivity of the brush (of importance for the conductive magnetic brush discussed in the next chapter).

It is of interest to examine their expression for the electric field under the simplifying conditions under which the equilibrium theory was derived. If we ignore the electric field due to the developed toner, bulk carrier charge and toner migration, their Eq. (3) becomes (in our notation)

$$E \propto V + \frac{\sigma_c}{\varepsilon_0} \frac{L}{K_E} \ , \tag{6.71}$$

where V is the development potential and σ_c is the carrier charge per unit area, which they write [their Eq. (6)] as

$$\sigma_c = -\nu \frac{M}{A} \frac{Q}{M} \ . \tag{6.72}$$

Setting $E = 0$ (the equilibrium condition) gives

$$\frac{M}{A} = \frac{V \varepsilon_0 \nu}{(Q/M)(L/K_E)} \ , \tag{6.73}$$

the equilibrium equation (6.57) derived earlier (with $d_s = 0$). Unfortunately, the authors do not discuss how close to the $E = 0$ condition they come in their attempts to fit data.

The carrier rotation contribution to development requires a carrier motion model. They use four constants to build this model, including brush slippage relative to the roller, relative shear between roller and photoreceptor, plus some rotation at synchronous motion, $\nu = 1$. Further, toner size and charge are distributed variables (assumed Gaussian) and the rates are averaged over these distribution functions.

The scavenging rate is calculated in a manner similar to the deposition rate: the force balance away from the photoreceptor is considered. Toner migration rates are derived using the rotational model since macroscopic toner migration requires toner transfer between two carriers and then carrier rotation.

The result of this work is a model which requires computer solution. The published results compare the theory to experiments of Gutman in which M/A and Q/M were measured versus V for a few developer mixes varying in their "conductivity." The "conductivity" is empirically correlated to a breakdown potential measured in a development fixture with a metal plate substituted for the photoreceptor (adding more constants). It was found experimentally and also predicted that higher "conductivity" mixes produced higher M/A's at a given voltage. (However, the mixes with higher conductivity also had lower Q/M, which one would expect to produce large M/A's). They also measured the size classification during development of toner from sponge carrier at 25 and 400 V. Larger average sizes are observed and predicted at the lower voltages. To obtain this result, "it was necessary to restrict carrier rotation severely. It may be that some of the toner is trapped in crevices in the carrier and is unavailable for development."

6.4 Solid Area Development Experiments

As discussed in the introduction to this chapter, experiments on a magnetic brush development system are difficult because of the many parameters that must be controlled (Table 6.1). When development data is taken from a robot that emulates the complete copying process, the parameters needing control are manyfold increased. It is therefore not surprising that relatively few development studies have appeared in the literature [6.4–10,13,14,19] and many of them show experimental data with considerable error bars.

The need for experimental data adequate to distinguish between the models generated two studies by *Schein* [6.13,14], one on an experimental fixture and one on commercial hardware taken from an IBM 6670 printer. Data from other sources [6.4–10,19] are in agreement with Schein's data, within experimental error. As will be seen, the data are consistent with the equilibrium model.

We first address the question, is M/A linear in V (as predicted by the equilibrium model) or are there intercepts and an S-shape to the curve (as predicted by the field stripping model)? Such curves were published for a wide range of roller (7.5–75 cm/s) and photoreceptor (25–50 cm/s) velocities at two toner concentrations (1.4% and 0.8%). Typical results are shown in Fig. 6.21: M/A is linear in V within experimental error. To make absolutely sure that this linearity did not result from a rectangular adhesion distribution and a field stripping mechanism (assumed in [6.16]), the adhesion distribution was changed by narrowing the size distribution of the toner (Fig. 6.22). The result (Fig. 6.23) remains, $M/A \propto V$ with no evidence for the nonlinearities predicted by a field stripping theory.

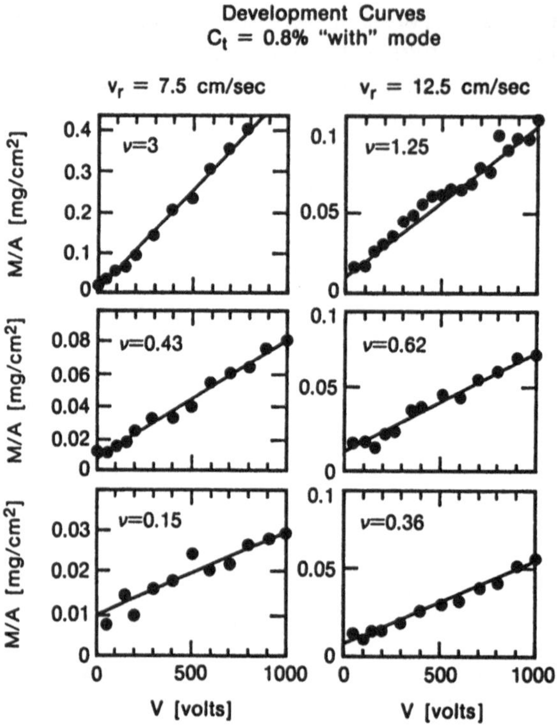

Fig. 6.21. Developed mass per unit area M/A versus the initial potential V for different values of v_r and v_p. The vertical axes differ for each graph. Note the observed linearity independent of v_r or v_p [6.13]

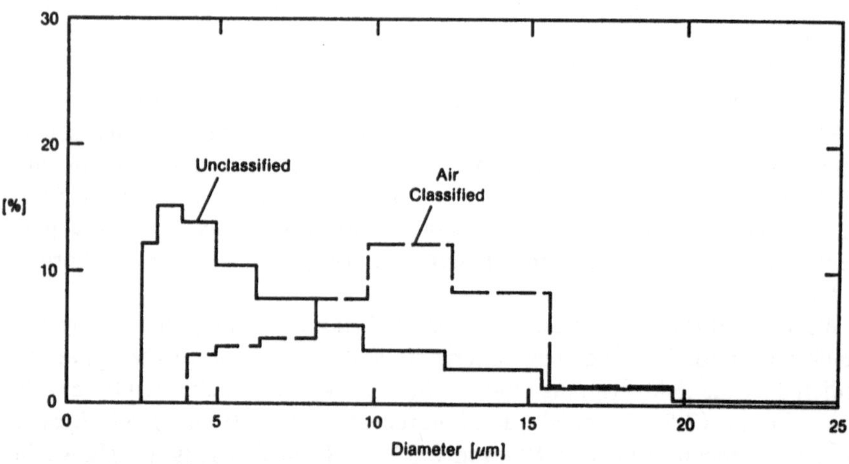

Fig. 6.22. The size distribution of unclassified and classified toner used in the experiments shown in Fig. 6.23 [6.13].

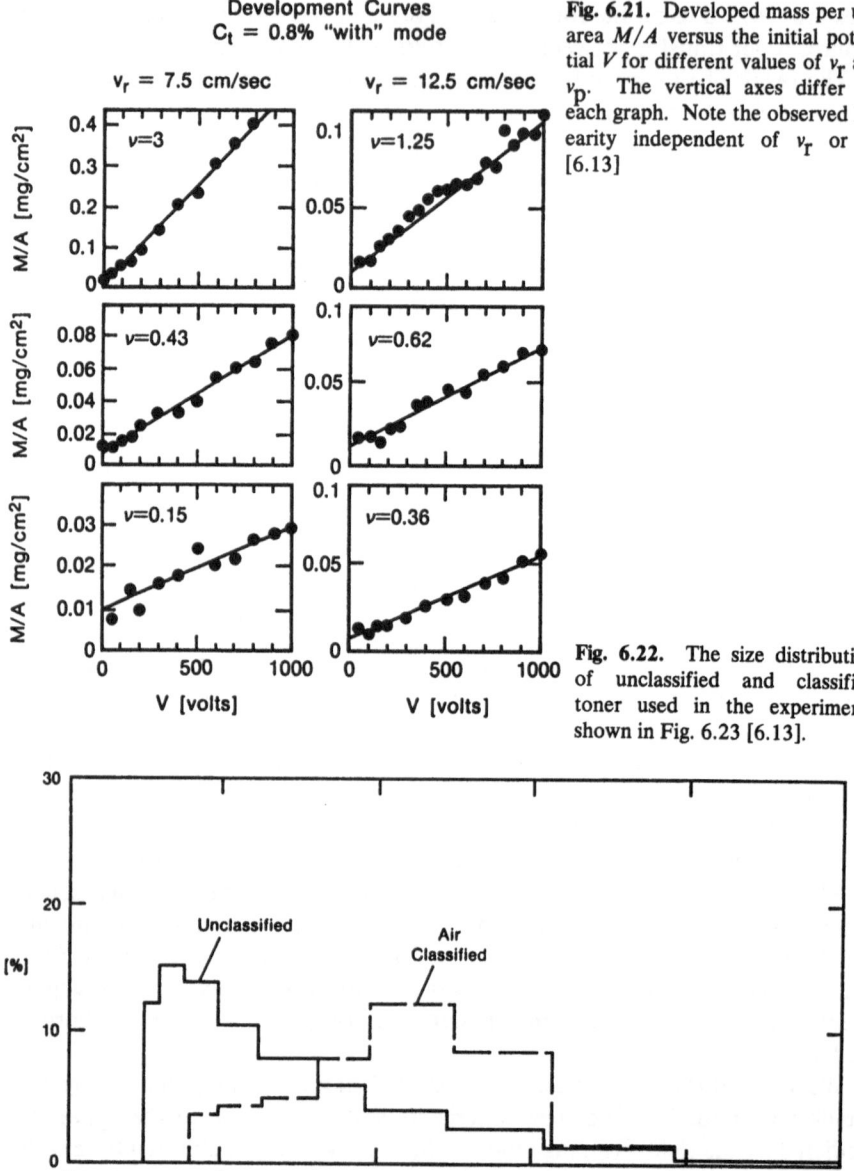

We next address the question whether M/A is linear in roller velocity (as predicted by the equilibrium theory as long as $\Lambda/v >> d_s/K_s$) or nonlinear in roller velocity (as predicted by the powder-cloud theory). Development (M/A) versus roller velocity is shown in Fig. 6.24. This experiment requires that the developer packing in the gap be maintained constant as the roller ve-

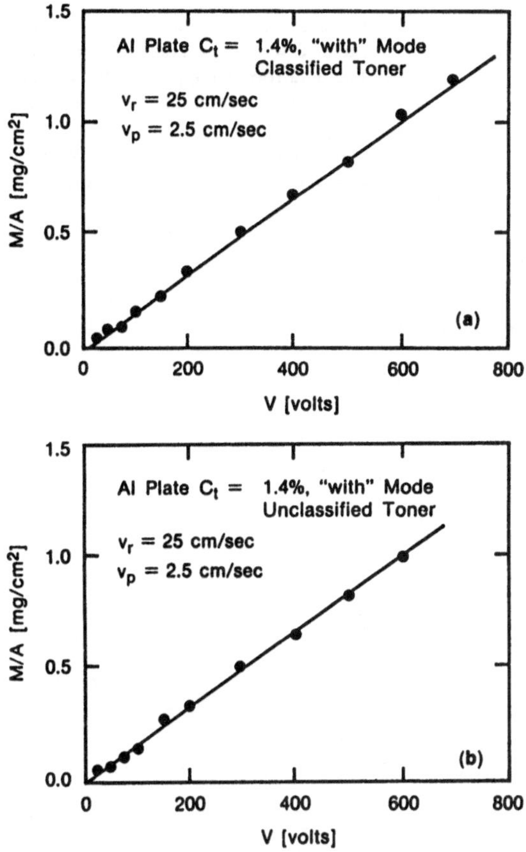

Fig. 6.23. **a** Development versus the initial potential for *classified toner* at 1.4% toner concentration. **b** Development versus the initial potential under the same conditions except *unclassified toner* was used. Both curves are linear [6.13]

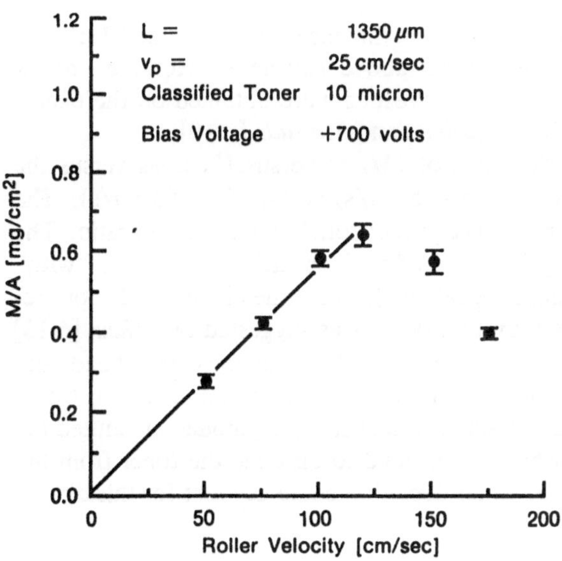

Fig. 6.24. Developed mass per unit area versus roller velocity. Linearity is observed over the range of roller velocities in which developer flow is linear in roller velocity. Even at the highest M/A's, the toner voltage is only about 100 V [using (6.10) with $Q/M = 20\ \mu C/g$], about 14% of the applied voltage [6.14]

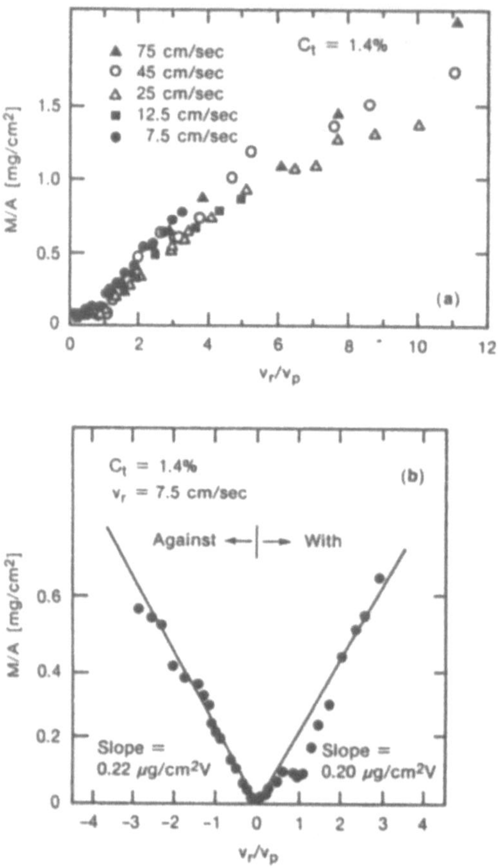

Fig. 6.25. **a** Development per unit voltage at $C = 1.4\%$ versus the speed ratio. The data form an approximate universal curve for an order of magnitude variation in ν_r and ν_p. **b** Development per unit voltage versus the speed ratio at $C = 1.4\%$ and $\nu_r = 7.5$ cm/s [6.13]

locity is increased. Hardware limitations in the experiments on the IBM 6670 prevented this condition from being met above 100 cm/s. However, below 100 cm/s, linearity is observed. Similar results were obtained on the experimental hardware [6.13], and by *Nakajima* and *Matsuda* [6.40].

In Fig. 6.25a are shown the slope of (M/A) versus V curves versus the speed ratio for a wide range of ν_r (7.5–75 cm/s) and ν_p (2.5–50 cm/s). The data appear to follow a universal curve proportional to ν, the speed ratio. The region near $\nu = 1$ is blown up in Fig. 6.25b which includes both the "with" (positive ν) and "against" modes (negative ν). Note the dip at $\nu = 1$, not predicted explicitly by the equilibrium theory. It was suggested by *Schein* [6.13] that at $\nu = 1$ there is not enough relative motion between the carrier beads and the photoreceptor to remove all of the toner the equilibrium condition would predict. *Williams* [6.18], in discussing his field stripping model, attributed the dip to reduced shear forces which are needed to dislodge the toner from the carrier. *Vahtra* [6.38] suggested toner depletion could account for this dip.

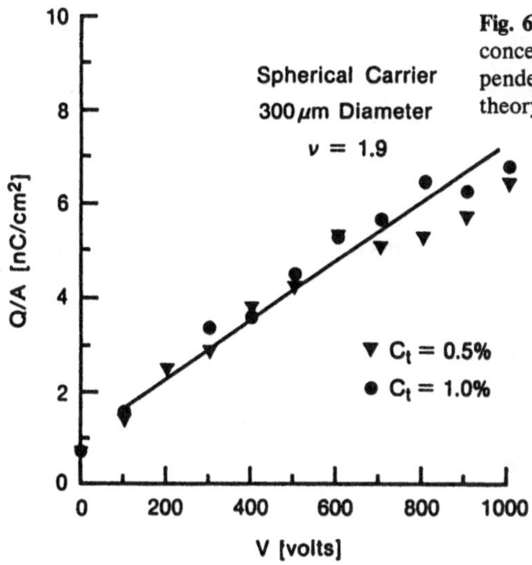

Fig. 6.26. Q/A versus V for two different toner concentrations. The data indicate Q/A is independent of C_t, consistent with the equilibrium theory

The prediction that the developed toner's Q/M is independent of V (equilibrium model) and not monotonically increasing with V (field stripping model) is verified in Fig. 6.4. Similar data were obtained on the IBM 6670 development hardware.

Finally, M/A is shown to be constant independent of toner concentration at constant Q/M (as predicted by the equilibrium model, but not the field stripping model) by plotting Q/A, i.e., M/A times Q/M, versus V for two different toner concentrations (Fig. 6.26).

In summary, two extensive solid area experimental studies have been carried out on the insulative magnetic brush development system. Measurements were made showing M/A is linear in V, M/A is linear in ν for small ν (except near $\nu = 1$), Q/M is independent of V, and M/A is independent of toner concentration at constant Q/M. The data are consistent with the equilibrium theory and are inconsistent with the field stripping and powder-cloud theories.

6.5 Line Development

Despite the fact that most of the scientific studies have been done on solid area latent images, from the point of view of use, line development is at least 10 times more important. But it is a considerably more difficult problem (Sect. 3.2). As pointed out by *Jen* and *Lubinsky* [6.41], the electric field depends not only on the surface charges on the photoreceptor at the spatial point of interest, but also on other points as well. It is thus necessary to track physical states of all the points along a line profile simultaneously through the length of the development zone. Also, history effects, i.e., how the toner in the magnetic brush evolves through the development zone and across a line profile, are

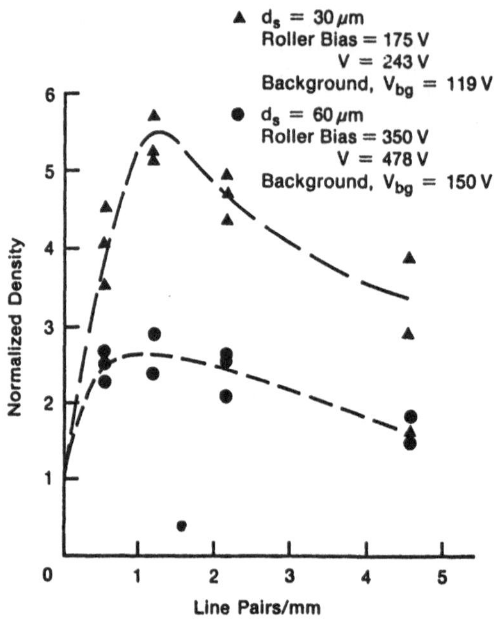

Fig. 6.27. Normalized developed density for an insulative developer measured as a function of line frequency for a normal photoreceptor thickness (60 μm) and a thin photoreceptor (30 μm). Developed density is low for solid areas and rises to a maximum between 0.5 and 1.5 line pairs/mm for both photoconductor thicknesses [6.22]

considerably more complicated. It is known that the mass per unit area developed onto lines is 1.5–2 times more than is obtained for solids and is relatively uniform for most line widths and lengths. An experiment by *Scharfe* [6.22] for a few line widths is shown in Fig. 6.27. The ratio obviously depends on photoreceptor thickness, which is expected based on simple considerations of the electric field associated with lines and solids. Beyond this work, there are almost no systematic experimental studies of line development.

From a theoretical perspective, two attempts have been made to understand the behavior of lines. *Vahtra* [6.38] examined extreme conditions in which line development depleted the carrier of toner. Some depletion, i.e., history effects, can be observed with test patterns that compare a single line at the bottom of a page with and without a page full of closely spaced lines. Vahtra shows that such a test pattern run in a cascade development system produces darker lines for the single line on the page. Vahtra assumes the latent image removes all available toner from the carrier, independent of the nature of the latent image. Naturally, such a model predicts that toner development onto lines depends strongly on line width and spacing between lines, results not generally observed.

Jen and *Lubinsky* [6.41] model lines in the spirit of rate equations, following earlier work of *Benda* and *Wnek* [6.39] and *Harpavat* [6.16]. They succeeded in predicting the density across a line for lines both parallel and perpendicular to the roller velocity direction as a function of the conductivity of the brush. Their basic equations are

$$\frac{d}{dt}\left(\frac{M}{A}\right) = DcE - R_s\left(\frac{M}{A}\right) , \tag{6.74}$$

$$\frac{d\sigma_c}{dt} = gE + \frac{Q}{M} \frac{d}{dt}\left(\frac{M}{A}\right) , \tag{6.75}$$

where D is the deposition rate parameter, c is the toner mass/area at the brush tip, E is the electric field in the brush (which is expressed in terms of spread functions to take into account the effect of neighboring charge), R_s is a scavenging rate constant, σ_c is charge per unit area on the brush surface, g is the developer conductivity (zero for the insulating magnetic brush), and Q/M is the toner charge-to-mass ratio. The last term describes the buildup of charge identified in the equilibrium model as the primary effect limiting development. The functions M/A, c, σ_c, and E all are functions of the two directions in the photoreceptor plane and the differences between with and against modes are explicitly dealt with.

Predictions for parallel and perpendicular lines in the with and against modes for input densities of 1.0 and 0.4 are shown in Fig. 6.28. Predicted asymmetry for perpendicular lines is evident. The evolution of these asymmetries with brush conductivity is also studied. However, no data to support the predicted asymmetries are presented.

Experimental data were given for peak optical density of 280 μm wide parallel and perpendicular lines as a function of $V - V_{bias}$ and toner concen-

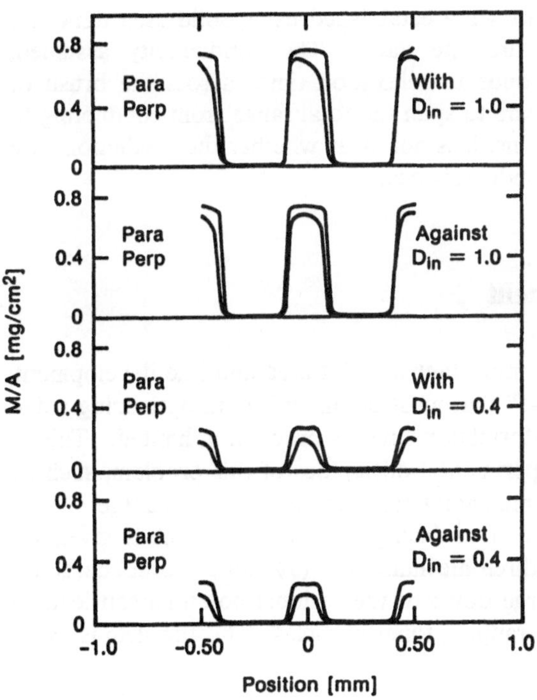

Fig. 6.28. Predicted M/A for lines parallel (Para) and perpendicular (Perp) to the developer flow direction for with and against modes and two densities [6.41]

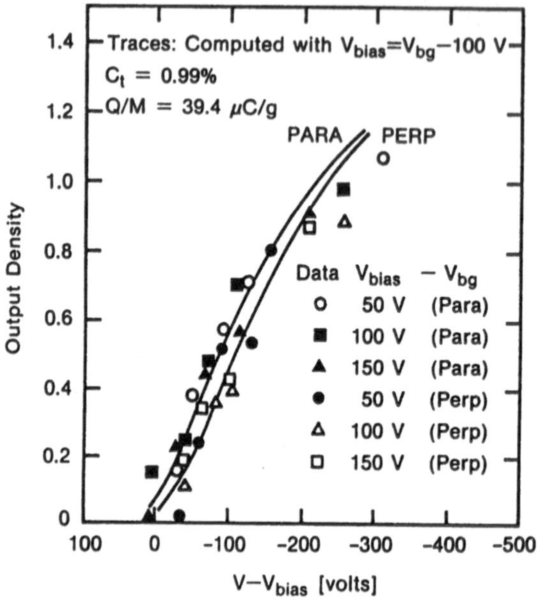

Fig. 6.29. Peak line densities as a function of image minus bias potential with bias minus background potential as a parameter [6.41]

tration. One set of curves is shown in Fig. 6.29. Here $V-V_{bias}$ appears to be a controlling variable, while $V_{bias}-V_{bg}$ variations from 50 to 150 V cannot be differentiated in the data. The term V_{bias} is the potential on the magnetic brush roller; V_{bg} is the potential of the discharged photoreceptor. The predicted curves do appear to account for the data. According to Jen and Lubinsky, the model parameters were established using solid area data and then used unchanged to fit the line data. The conductivity assumed, $10^{-6}-10^{-8}$ $(\Omega m)^{-1}$, corresponds to relaxation times across the brush of $10^{-5}-10^{-3}$ s. As this appears to span the total range from conducting to insulating magnetic brush systems, it is not clear whether the predictions are for the insulating case or the conducting case.

6.6 Background Development

Background development is as important as solid area and line development because it determines the "noise" that must be minimized in optimizing a development system. Yet the information on this problem is limited. This is certainly due in part to the experimental difficulties of this problem, such as collecting and measuring very small amounts of toner mass and charge.

A useful qualitative discussion of background development is given by *Gundlach* [6.4]. He discusses three mechanisms: (1) Nonuniform charge on the photoreceptor after discharge down to the residual potential can lead to small electric fields capable of trapping toner particles. This problem is exacerbated in DAD systems (Sect. 2.1.2): nonuniform dark decay or dielectric breakdown can cause small fringe electric fields.

(2) Toner will develop onto a completely uncharged photoreceptor, indicating that some background may be due to nonelectrostatic forces. Presumably mechanical or contact forces are sufficient to cause toner to leave the carrier. (However, as we discussed in Sect. 6.2.1, the developer can charge the photoreceptor, creating small fringe electric fields).

(3) Wrong-sign toner will be driven, in an electroded development system, into the background regions.

Gundlach points out that two effects tend to minimize background. One is scavenging, observed in solid area cascade and magnetic brush experiments: carrier beads depleted of toner will tend to pick up stray toner from the photoreceptor. The second involves wrong-sign toner: with an electrostatic transfer system, wrong-sign toner tends to be rejected.

Wrong-sign toner could be the result of a wide distribution of toner charges in developer mix. A possible method of reducing the amount of wrong-sign toner, and therefore decreasing background, is to raise the average Q/M (by changing the toner or carrier materials or lowering C_t, see Chap. 4). This is, in fact, a standard operating procedure. It is also a move in exactly the wrong direction to obtain blacker lines and solids. The equilibrium theory (and other theories) predicts that raising Q/M will lower M/A. Hence, a trade-off exists between making the blacks blacker and the whites whiter. No sets of experimental data to support the wrong-sign mechanism of background development have been published.

Williams [6.18] estimated the inertial forces on toner during a carrier-carrier collision. For 125 μm radius carriers colliding at a relative velocity of 175 cm/s, toner experiences an acceleration of 10^9 cm/s^2, which is sufficient to detach toner from carrier. If a carrier impacts a plane of a polymer material, i.e., an organic photoreceptor, the acceleration on toner is still enormous, 2.4×10^8 cm/s^2. Williams's estimate suggests that a powder cloud is to be expected in the development zone which could contribute to background development. No experimental data to support this picture were published by Williams, but the Agfa-Gavaert [6.20,21] data support this possibility (Fig. 6.17).

Schein's equilibrium theory predicts [6.13] a source of background at $V = 0$. Note in (6.55) that n, the number of toner particles developed per carrier particle, is finite at $V = 0$, depending upon the difference between the toner-carrier and toner-photoreceptor adhesion force. If $n = 1$ at $V = 0$, (6.51) predicts

$$\frac{M}{A} = \frac{M_t v p}{\pi R^2} \quad . \tag{6.76}$$

This mechanism can be easily verified by placing a reverse bias across the de-

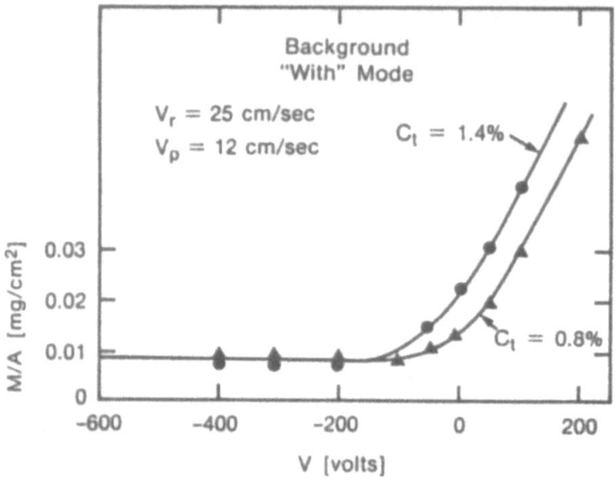

Fig. 6.30. Background development at $v_r/v_p = 2$ versus bias. Background development is independent of reverse bias and independent of toner concentration in the range 0.8%–1.4% within experimental error [6.13]

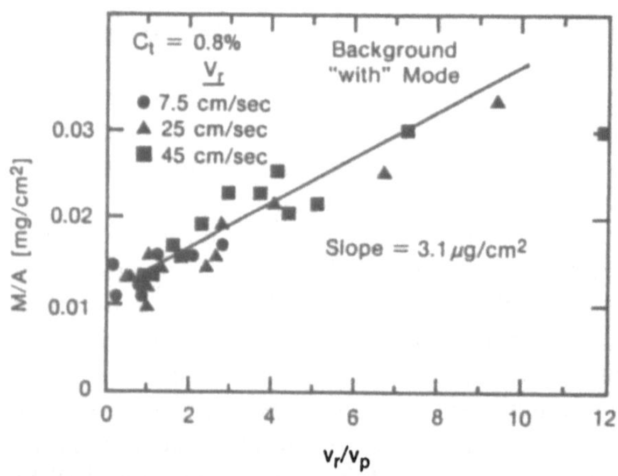

Fig. 6.31. Background development at $C_t = 0.8\%$ versus the speed ratio. Background development is linear in the speed ratio [6.13]

velopment zone; M/A should decrease to zero as the Coulomb force shifts the force balance to decrease the number of toner particles developed per carrier. Such experiments are shown in Figs. 6.30 and 31. In Fig. 6.30 is shown M/A versus bias. Note that the curve is continuous through $V = 0$ and decreases until $V = -150$V is reached. In Fig. 6.31 M/A at $V = 0$ is plotted versus speed ratio. As predicted, it is linear in ν, independent of C_t (the slope is 3.1 μg/cm² at $C_t = 0.8\%$, shown, and 3.5 μg/cm² at $C_t = 1.4\%$, not shown) and close to the predicted value of 4.6 μg/cm² (using $p = 0.4$). However, not predicted but observed is a background development of 0.01 mg/cm² independent of reverse bias and independent of C_t. The source of this background development was not identified. In actual machines that use reverse bias to suppress background, this may be the primary source of background toner.

De Lorenzo and *Garsin* [6.42] have pursued the microfield argument that small electric fields in the photoreceptors caused by charge nonuniformity, photoreceptor defects or imperfect cleaning could attract toner and cause background development. They predict background mass per unit area once the potential of the defect is known by using solid area response data. The main assumption is that microfields will attract toner if the net force due to the electric field, including the photoreceptor residual potential and the bias potential, points toward the photoreceptor. Experimental data to support this model of background were published by *Goldmann* [6.43]. He produced line latent images with small contrast potentials but large and varying charge potentials (for convenience, the roller was grounded). He estimated theoretically that for his experimental setup, the electric field associated with a contrast potential was 13 times greater than the electric field determined by the difference of the charge and bias potentials. His data, shown in Fig. 6.32, show development versus charge potential for several different contrast potentials V_{cr}. The important point to notice is that development of these lines goes to zero when the charge potential is 10 times the contrast potential, in reasonable agreement with the prediction. Goldmann therefore demonstrated that if microfields exist, they develop only when the net force due to the electric field points toward the photoreceptor (De Lorenzo and Garsin's point) and can be suppressed by sufficient electric field generated by the potential difference between the photoreceptor and roller.

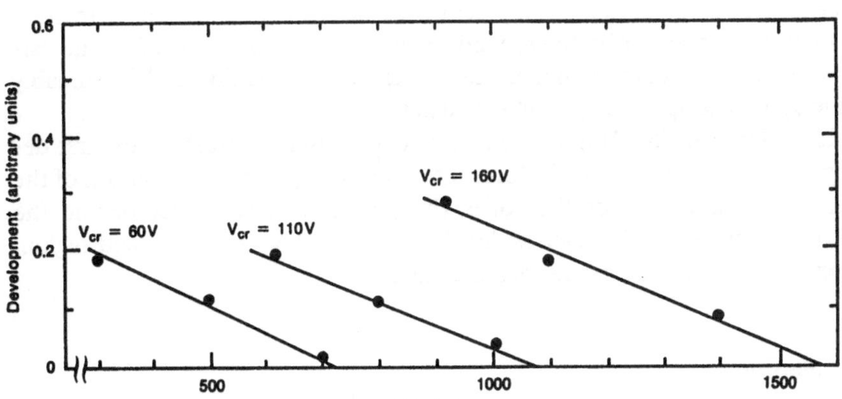

Fig. 6.32. Development as a function of the surface potential V (bias voltage is zero, toner concentration approximately 2.4% and roller speed, 0.75 m/s) with varying contrast potential V_{cr} [6.43].

6.7 Improvements

The data strongly suggest that the amount of toner developed onto the photoreceptor is limited by the buildup of net charge on carrier beads adjacent to the photoreceptor, as postulated by the equilibrium theory. It is natural to ask whether means can be devised to increase development by eliminating these charged carriers. The most successful idea so far was provided by Kasper and May of Kodak and is discussed in the next chapter. Other ideas are discussed here.

Teshigawara et al. [6.44] pointed out that it may be possible to mix the carrier beads adjacent to the photoreceptor with carrier beads in other layers in the development zone by using low permeability carrier, so decreasing the magnetic force between carrier beads. They decreased the permeability by half (to 26 from normally 50 emu/g at 700 Oe) and observed increased output density for a given voltage. The density increase was 50%. However, part of this increase was due to other effects. They decreased the development gap by 20%, which should be responsible for at least an increase in development of 20%; presumably the toner Q/M which effects development linearly was also not identical for the two systems, but this was not specified.

Miskinis and *Jadwin* [6.45] suggested the opposite approach. They suggested making the carrier beads hard magnetically, as characterized by a coercivity of at least 300 G. The diameter of these carrier beads was chosen to be 53–62 μm, smaller than the usual 100–300 μm, and the beads were placed on a roller within which the magnetic poles rotate. They claim "high quality images were obtained from the standpoint of development completion and uniformity" at photoreceptor speeds up to 75 cm/s.

Hays [6.46] has suggested another alternative: a magnetic-field-free development region. He suggests running a belt between two rollers within which are standard magnets. The carrier beads are picked up by the first roller, carried through the development zone on the belt, and then returned to the developer reservoir by the second roller. Between the two rollers the carrier beads are in an approximately zero magnetic field region and presumably will mix, increasing development efficiency.

Lubinsky et al. [6.47] describe a two-roller magnetic brush. The first developer has a portion of the photoreceptor belt wrapped about a portion of the external circumference of the roller (by placing another roller behind the photoreceptor belt). This causes development in a low magnetic field region, presumably causing the carrier beads to mix.

6.8 Summary

Insulative magnetic brush development is both the most prevalent (Table 1.1) and the most studied development technology. While much progress has been made in understanding the mechanisms of development and the electric fields existing in this development system, significant questions remain unanswered.

Electric fields are determined by charges and dielectric constants. The sources of charges have been identified. They include the charges on the photoreceptor surface (determined by charging, exposure and dark decay), charged toner developed onto the photoreceptor, and the charging of the photoreceptor during the development process caused by either toner or carrier triboelectric interactions with the photoreceptor surface. The average and peak electric fields inside the development system have been estimated, and reasonable agreement exists with experiment. However, identification of the electric field experienced by toner (as expressed by the effective dielectric constant) remains an unsolved problem. It has been suggested that the peak and the average field are appropriate. The limited data available indicate the correct answer may be somewhere between these two values.

An extensive list of development theories have been applied to this development system, including neutralization, field stripping, powder cloud, equilibrium, depletion and one attempt at a "complete" theory. This body of work is by itself a valuable source of theoretical ideas on electrophotographic development physics.

Two extensive solid area experimental studies have been carried out on the insulative magnetic brush development system. The data are consistent with the equilibrium theory and inconsistent with the other theories (field stripping, powder cloud, neutralization, depletion).

Line copy development remains puzzling. It is not clear how to generalize the equilibrium theory to a situation with a sparce latent image, yet the constant ratio of line development to solid area development suggests a similar development mechanism is occurring. The one attempt to understand this problem does not identify the physical processes controlling line development.

Background development is even less well understood. Here the qualitative ideas from the literature have been reviewed. Virtually no sets of data exist to compare against these ideas. Only one background development experiment has been reported. One background development mechanism identified was a natural consequence of the equilibrium theory, i.e. n, the number of toner particles which come off each carrier, may be finite at zero voltage due to the force balance. However, below -150 V another source of toner was identified experimentally. It is probably the real source of background in copiers and printers (since reverse bias is usually used). Its physics has not been identified.

The realization that the equilibrium theory describes insulative magnetic brush development has led to the suggestion that improvements in the system can be achieved by removing the charged carrier beads from the vicinity of the photoreceptor. The most successful idea, providing a short circuit to the roller by using "pointed" shaped carrier beads, is discussed in the next chapter. Other ideas to create dynamic mixing of the carrier beads in the development have been patented.

7. Conductive Magnetic Brush Development

In 1975, Eastman Kodak entered the high speed copier business with the introduction of the first of their Ektaprint series of copiers. These copiers were immediately recognized by both electrophotographers and the public as significantly improved products.

What the public could see was dramatically improved copy quality. Figure 3.1c shows lines and solids produced by these copiers. Both darker lines and improved solid areas are evident, as compared with results from other copiers (Fig. 3.1b). This improved copy quality was the direct result of the introduction of a new development system, which is called conductive magnetic brush development [7.1,2].

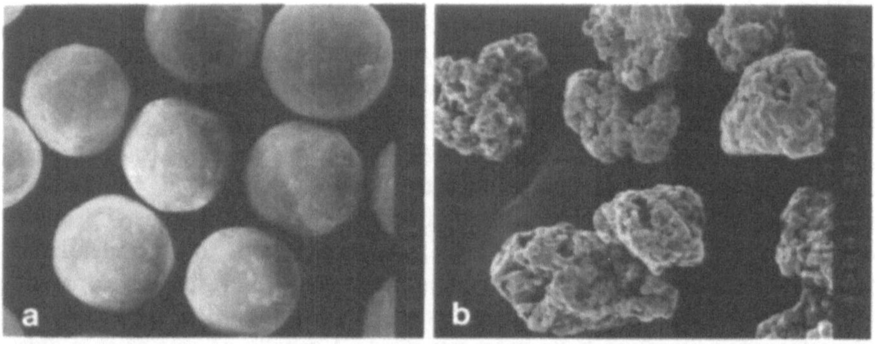

Fig. 7.1 a,b. SEM micrographs of spherical carrier (a) and sponge carrier (b)

The one major difference between this new development system and the insulative magnetic brush development system was the shape of the carrier beads. Insulative systems use spherical beads; the conductive system uses rough-shaped particles (Fig. 7.1). These rough-shaped particles are called sponge carrier (or sponge iron or sponge steel depending on whether the core is porous or solid, respectively) because the surface, viewed under a scanning electron microscope, appears to have a sponge-like appearance. What the rough shape provided was conductivity paths down bead chains by eliminating the toner between carrier beads. That conductivity should improve magnetic brush development should be qualitatively obvious after reading the previous chapter: development is limited for insulative magnetic brush by the buildup of charge on carrier beads adjacent to the photoreceptor. If a conductive path is provided which short-circuits this charge to ground, increased development should result.

Quantitatively, the prediction for a totally conductive system was easily obtained. If the distance from the ground plane to the carrier charge (Λ) vanishes, then the equation of development for the equilibrium theory derived in the previous chapter (6.57) predicts that the developed mass per unit area M/A is

$$\frac{M}{A} = \frac{\varepsilon_0 V}{(d_s/K_s)(Q/M)} \quad , \tag{7.1}$$

i.e., the neutralization condition. However, experimental data indicated that while conductive magnetic brush development produced higher M/A 's than insulative magnetic brush, it did not usually produce toner development as high as predicted by the neutralization condition. This was demonstrated by data taken by Gutman (and published in [7.3]) (Fig. 7.2), in which he observed M/A versus V for a series of "conductive" developer mixes. The mixes differed in their breakdown potential (listed on the figure), the potential at which significantly enhanced conductivity was observed. The neutralization limit has been added to the figure. Clearly all of the mixes fall somewhere between the insulative and neutralization limits. Gutman's data suggests that predicting M/A for conductive magnetic brush development is not a trivial extension of previous theories. This chapter will be devoted to a discussion of the search for a theory of conductive magnetic brush development. Unlike other chapters, this chapter will be developed historically, to give the reader a feeling for

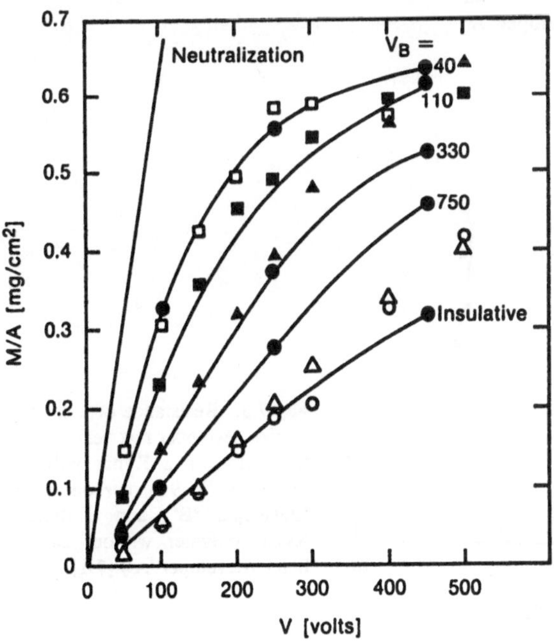

Fig. 7.2. Developed M/A for developers with different breakdown potentials V_B. We have added the neutralization limit for Q/M = 16 μC/g and d_s/K_s = 9.5 μm (after Gutman in [7.3] © 1981 IEEE)

how ideas actually unfold in this field. We will discuss in Sect. 7.1 the initial theoretical ideas. Then, in Sect. 7.2, the extensive experimental work done at IBM will be presented. Comparison of these results with available theories indicated inconsistencies and suggested a new theory, discussed in Sect. 7.3. Comparison with experimental data is given in Sect. 7.4. Discussions of line and background development are given in Sects. 7.5 and 7.6.

7.1 Initial Theoretical Ideas

The Eastman Kodak discussions of the behavior of sponge carrier are contained in one patent [7.1] and a paper [7.2] presented at the 1977 IEEE-IAS annual meeting which describes the complete Ektaprint 150 copier. The discussions are primarily qualitative and are focused on the conditions necessary to obtain a conductivity breakdown. In [7.1] measurements are reported (Fig. 7.3) of resistance of the developer mix versus electric field. At fields of 5 and 25 V/mm, resistance changes of 3 orders of magnitude are reported for sponge carrier partially coated with a Kynar polymer loaded with 9% carbon black (developer 'B') and a thinner Kynar polymer without carbon black (developer 'A'), respectively. They go on to show that the breakdown field depends on toner concentration and gap (between the roller and photoreceptor).

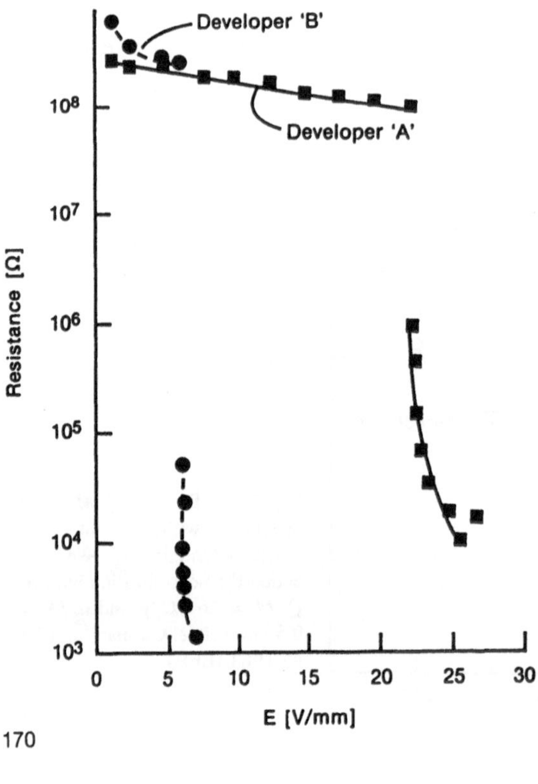

Fig. 7.3. Resistance of a sponge carrier developer mix versus electric field for a Kynar polymer loaded with 9% carbon black (developer 'B') and a thinner Kynar polymer without carbon black (developer 'A') [7.1]

It is postulated (by Kasper and May) that above this breakdown field, the electrode is established at the surface of the carrier beads adjacent to the photoreceptor, thus creating the "strongest theoretical field for development." This was also assumed by *Nakajima* and *Matsuda* [7.4] in their theoretical and experimental study of the effects of developer flow rate on M/A. Later *Jewett* [7.2], also of Eastman Kodak, modified this by stating that the electrode approaches the photoreceptor, consistent with the developer's conductivity. In other words, Jewett introduced the first partially conducting model.

A problem is raised by this discussion, as was pointed out by *Benda* and *Wnek* [7.3]. The discussion suggests that there are two regimes of operation, a high resistance regime and a low resistance regime. In the high resistance regime in which the developer is insulating, the slope of M/A versus V should be small. In the low resistance regime above breakdown in which the developer is conducting, the slope of M/A versus V should abruptly change to a much larger value. However, there is no evidence for an increase in slope of M/A versus V at high V in the data given by Gutman (Fig. 7.2) or Jewett (Fig. 7.4). Two possible explanations are (1) the change in slope occurs at very low voltages, much lower than the observed breakdown voltages reported by Gutman or given in the Kasper and May patent, or (2) the resistance measurements do not measure conductivity properties that correlate with development behavior. That the latter is not unreasonable can be argued as follows. Resistivity measurements of composite powders could be measuring the current flow in a single chain that becomes conducting. Magnetic brush development over large areas would be sensitive to the average conductivity of many chains, a quantity not necessarily measured in the conductivity experiment.

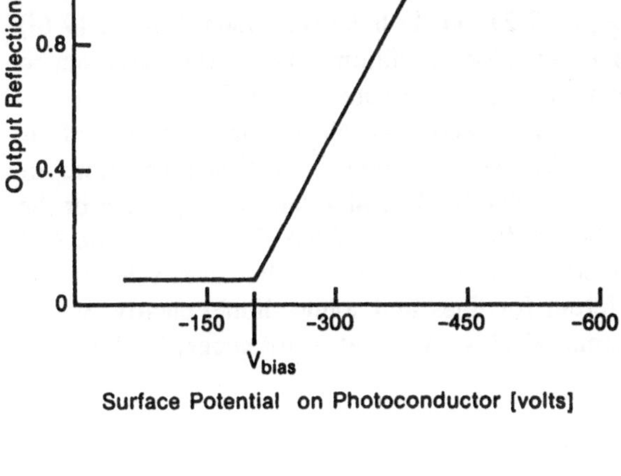

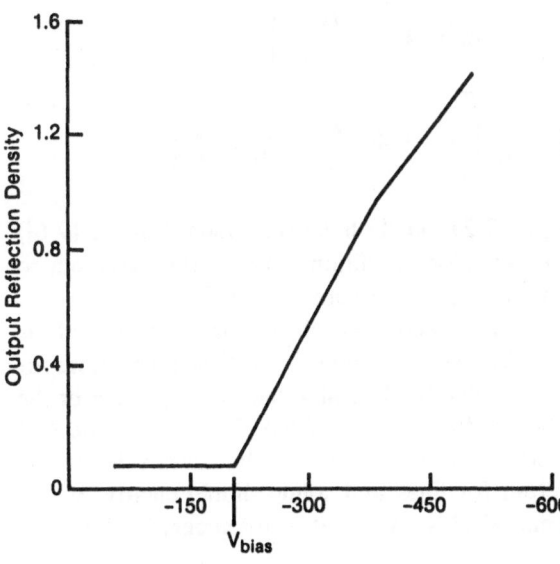

Fig. 7.4. Development (measured as reflection density on paper) versus the photoreceptor voltage. The developer bias potential is approximately −200 V. Therefore zero potential difference across the development gap occurs when the photoreceptor is charged to −200 V [7.1]

Despite the above problem, it is clear that somehow conductivity is most likely associated with the enhanced toner development of sponge carrier and a parameterization of the available development equation is possible. The equlibrium model predicts

$$\frac{M}{A} = \varepsilon_0 V \Big/ \frac{Q}{M} \left(\frac{d_s}{K_s} + \frac{\Lambda}{\nu} \right) , \tag{7.2}$$

where Λ is the dielectric distance from the carrier charge to the roller ground plane. Gutman's data could certainly be fit by making Λ a variable and associating it with the movement of the ground plane towards the photoreceptor as conductivity increases. *Benda* and *Wnek* [7.3] did just this but more formally by including in their system of differential equations a term for changed bead charge due to the brush conductivity. They assume the conductivity is ohmic, although data given by Kodak (Fig. 7.3) suggests this may not be a good approximation. *Folkins* [7.5] handled the problem somewhat differently. In the spirit of the equilibrium model, he assumed development is determined only by the electric fields. In the "with" mode (i.e., the photoreceptor moves in the same direction as the roller), he assumed at first the development system was insulating and therefore described by (7.2). Then as the carrier particles are compacted together going into the nip, breakdown occurs, moving the electrode towards the photoreceptor with a time constant given by $\rho\varepsilon_0$, where ρ is the ohmic resistivity of the developer (standard RC circuit charging). Since the time available for electrode movement is the nip width W divided by the roller velocity v_r, he obtains the result (we have slightly simplified his results):

$$\frac{M}{A} = \left(\frac{M}{A} \right)_{\mathrm{IMB}} \exp\left(- \frac{W}{v_r \varepsilon_0 \rho} \right) + \left(\frac{M}{A} \right)_{\mathrm{CMB}} \left[1 - \exp\left(- \frac{W}{v_r \varepsilon_0 \rho} \right) \right] , \tag{7.3}$$

where he assumes $(M/A)_{\mathrm{IMB}}$ is (7.2) and Folkins (and later *Scharfe* [7.6]) assume $(M/A)_{\mathrm{CMB}}$ is the neutralization condition. Again, this turns out to allow a single variable parameter ρ to describe data such as Gutman's.

However, *Hays* [7.7], at the same meeting at which Folkins presented his ideas, hints at a problem with (7.3). The time used in (7.3) is appropriate for solid areas, the nip width (1 cm) divided by the roller velocity. For lines or the edges of solids the time to establish the position of the electrode can be 100 times less, approximately the size of a line divided by the roller velocity. A time constant 100 times different for lines and solids should clearly be observable in significantly different M/A 's. Yet that is not generally observed.

In fact, one of the primary advantages of conductive magnetic brush development is equal solid area and line densities.

Characterization of the conductivity of developer, begun by *Kasper* and *May* [7.1], was continued by *Hays* [7.7] and *Hoshino* [7.8]. Hays showed the conductivity of sponge carrier depended on toner concentration, magnetic field strength, developer thickness, and developer agitation in a special electroded cell. Hoshino studied the conductivity of spherical carrier and found similar magnitudes as found for sponge carrier, and also found the conductivity was a strong function of magnetic field and depended exponentially on the square root of the electric field, suggesting Schottky emission.

These papers raise another problem with the partially conducting model suggested by (7.3). The time constant for charge flow is determined by the resistivity of the chain, i.e., the resistivity of many carrier contacts. Resistivities of powders are notorious for being irreproducible and sensitive to environmental conditions such as pressure, humidity, etc. If sponge carrier operates in a mode suggested by (7.3), why is it reproducible across a page and from page to page?

Given these problems with a partially conducting model, one might be motivated to reconsider the model in which the bead chain is infinitely conducting (originally suggested by Kasper and May). Three objections occur. The first two have already been discussed. First, an infinitely conducting chain with $\Lambda = 0$ in the equilibrium equation predicts neutralization which is not observed. Second, a change in slope of M/A versus V should occur at the breakdown voltage, which has not been observed. The third objection concerns air breakdown. If the full potential (500 V) is across the air gap between carrier beads adjacent to the photoreceptor and the photoreceptor surface, enormous fields will be generated which should be sufficient somewhere along the carrier surface to generate air breakdown which would discharge the latent image. This is also not observed.

Summarizing the thinking up to 1985, infinitely conductive brushes are not consistent with M/A being less than neutralization, with no change in slope of M/A versus V, or the lack of air breakdown. Partially conducting brushes cannot explain line/solid ratios near 1 or the observed reproducibility in M/A.

An unpublished idea of the present author for circumventing these difficulties is that the enhanced development of sponge carrier is due entirely to an enhanced effective dielectric constant (enhanced K_E). The average chain is therefore regarded as insulating. In this concept, the conductivity measurements are regarded as due to spurious shorts down occasional chains. Most of the chains remain totally insulating so that the bead charge, on the average, remains on the carrier beads adjacent to the photoreceptor. The effective dielectric constant, K_E however, is enhanced by the rough morphology of the sponge carrier forcing toner particles out of the contact region between carriers. Then Λ becomes [associating K_E with K_{max}, (6.27)]

$$\Lambda = \frac{L}{K_E} = \frac{L}{RK_p}(\ell + r) \rightarrow \frac{L\ell}{RK_p} , \tag{7.4}$$

where the toner radius is taken to be zero since it no longer contributes to the dielectric distance between the ground planes and ℓ is very small since thin coatings (0.1 μm) are typically used on sponge carrier. This immediately explains why neutralization is not observed ($\Lambda \neq 0$, but is smaller than in an insulative magnetic brush system), why air breakdown is not observed (the electrode remains on the roller), why line/solid ratios approach 1, and why M/A is reproducible. A problem with the idea was pointed out by *Lee* and *Beardsley* [7.9]: (7.4) does not appear to successfully describe K_E observed for spherical carrier. If the equation for K_E is not valid, conclusions drawn based on this equation will not be valid.

Another idea, suggested by *Lee* and *Beardsley* [7.9], was that the observed conductivity may not be electrical in origin, but may be partially due to toner motion down the bead chains during development. Their view was that such toner motion could contribute to observed currents and, in moving charge down the chain, would effectively move the bead charge up the chain, thereby decreasing Λ. To obtain such toner motion, they had to postulate: (1) toner jumping between carrier particles across air gaps (which they observed directly under slow speed conditions) and (2) that the rough-shaped sponge carrier particles rotate throughout the chain during the time the carrier particles are in the development region. Further, they must also assume the observed conductivity is due to spurious shorts down occasional chains since an electric field in the brush is required to drive the toner motion.

It should be clear at this point that a thorough experimental study of the behavior of sponge carrier would be useful. Some of the questions that need to be addressed are: (1) Where is the counter electrode? (2) Does the measured conductivity involve some or all the chains? (3) Is sponge carrier in an insulating ("enhanced K_E", Lee's picture), partially conducting (Folkins, Benda and Wnek, Jewett) or infinitely conducting (Kasper and May) regime?

7.2 Experimental Data and Discussions

In 1987, *Schein* et al. [7.10] published a detailed experimental study of sponge carrier in an actual magnetic brush development system. The magnetic brush development system used in these experiments was experimental hardware designed so that most variables were easily changed. The variables in this experiment were voltage across the ground planes (0–1000 V), "photoreceptor" thickness (actually aluminized Mylar was used) and speed ratio (the roller velocity was held at 54.5 cm/s and the photoreceptor velocity was varied in the "with" mode). The gap between the photoreceptor and roller was nominally 0.312 cm. However, in one experiment this was increased in steps from 0.125

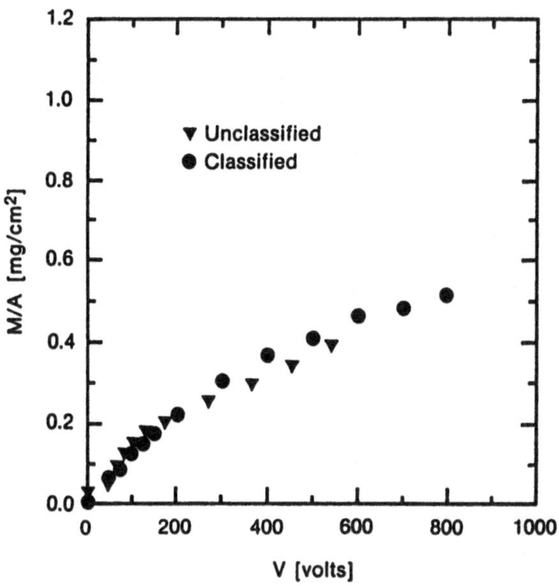

Fig. 7.5. Experimental data showing the dependence of the developed mass per unit area (M/A) on applied voltage V for classified and unclassified toner [7.10]

to 0.375 cm and the flow was changed (by increasing the doctor blade gap) to maintain constant packing. The carrier was nominally 200 μm diameter sponge carrier purchased from Hoeganaes (AST 60/100) and coated with a polymer to a thickness of 0.6 μm (1.5 μm and 0 were also used). The Mylar dielectric thicknesses were determined by capacitance measurements. For the 1, 3, 5, and 10 mil (nominal) Mylar, d_S/K_S was determined to be 7.11, 21.34, 35.56, and 65.65 μm, respectively.

Figure 7.5 shows M/A versus V for both unclassified and classified (10 ± 3 μm) toners. Note that some of the nonlinearity is reduced when size-classified toner is used. Obviously, size classification is occurring during development, a result observed previously [7.11]. How much of the remaining nonlinearity is due to further size classification cannot be determined. These data are in agreement with *Gutman*'s results [7.3]. Any valid theory must predict that M/A is approximately linear in V.

Figure 7.6 shows the developed toner's charge-to-mass ratio Q/M versus V for unclassified and classified toners. Note that for unclassified toner Q/M increases with voltage while for classified toner Q/M is almost independent of V.

Figure 7.7 shows the electrostatic voltages above the developed Mylar versus speed ratio with an applied potential of 400 V. The total electrostatic voltage is mostly due to the toner (Fig. 7.8). Note that this variable is linear in ν for small ν and then saturates near 400 V, the applied voltage.

These three results (Figs.7.5–7) are to first order consistent with an equilibrium type theory, such as proposed by *Folkins* [7.5] or the enhanced K_E picture.

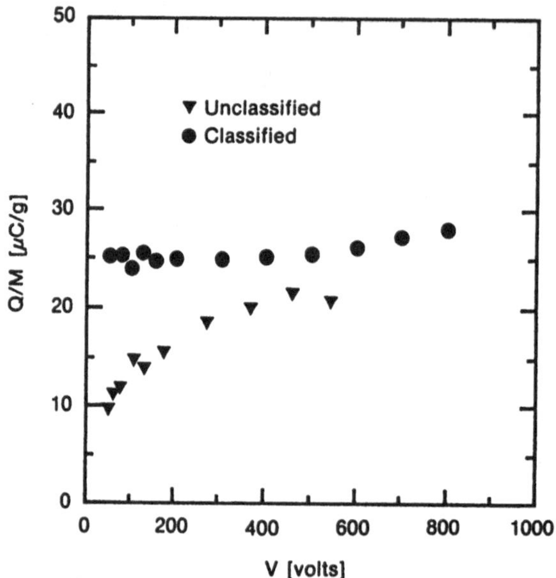

Fig. 7.6. Experimental data showing the dependence of the developed toner's charge-to-mass ratio on applied voltage [7.10]

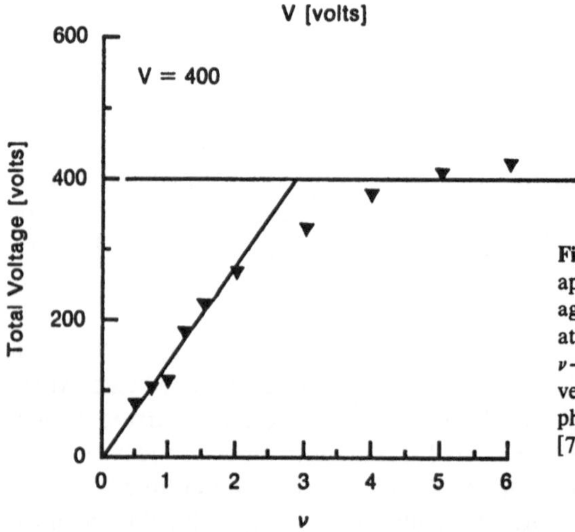

Fig. 7.7. The total voltage, which is approximately equal to the toner voltage (Fig. 7.8), is linear in speed ratio ν at low ν and is independent of ν as $\nu \to \infty$. For these experiments the roller velocity was held constant and the photoconductor velocity was varied [7.10]

In Fig. 7.8, the electrostatic voltage above the developed toner V_{total} is separated into its components by removing the toner and measuring the remaining charge on the Mylar surface. The curve labeled toner voltage is calculated based on (6.10). If in removing the toner no charge is added or subtracted from the Mylar then the sum of the Mylar voltage and calculated toner voltage should equal the original measured voltage V_{total}. This can be seen to occur. Note that some small amount of Mylar charging does occur during these experiments.

A measurement of Λ, the dielectric distance of the carrier charge from the ground plane in the equilibrium theory, is achieved by measuring V_t (i.e., M/A and Q/M) of the developed toner as a function of the Mylar thickness,

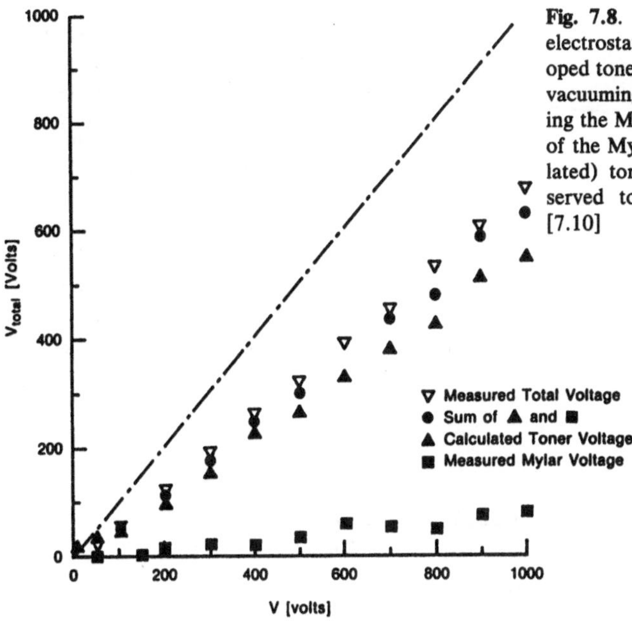

Fig. 7.8. The contributions to the electrostatic voltage above the developed toner layer can be separated by vacuuming off the toner and measuring the Mylar voltage. Note the sum of the Mylar voltage and the (calculated) toner voltage equals the observed total electrostatic potential [7.10]

(6.29). This is shown in Fig. 7.9 with the speed ratio as a parameter. Correction has been made for Mylar charging. Note that, as predicted, straight lines are a reasonable fit to the data and the intercepts are near 1. The value of Λ is obtained by multiplying the slope by the speed ratio (6.26). This is given on the figure. There Λ appears to be (17 $\pm$ 2) μm and independent of speed ratio within experimental error.

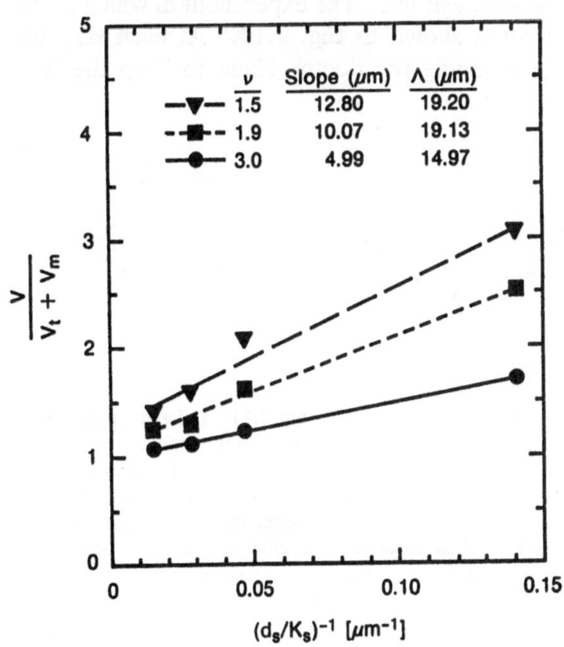

Fig. 7.9. The dielectric distance to the carrier charge in the equilibrium model, Λ, can be obtained by plotting $V/(V_m + V_t)$ versus the inverse of the photoconductor thickness. $\Lambda =$ 17 μm, independent of speed ratio [7.10]

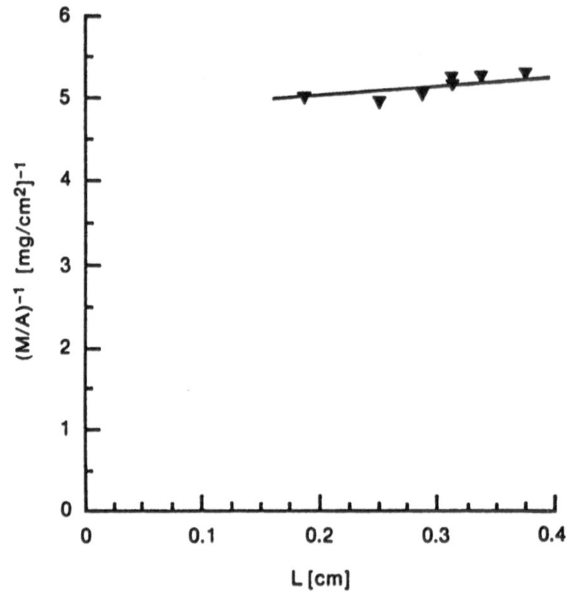

Fig. 7.10. The dependence of Λ on L, the gap, can be determined by plotting $(M/A)^{-1}$ versus L. For this experiment the polymer coat thickness was 0.6 μm, $V =$ 400 V and $\nu = 2$ [7.10]

To test where the carrier charge is, Schein looked for the dependence of Λ on L. If (7.2) of the equilibrium equation is inverted we obtain

$$\left(\frac{M}{A}\right)^{-1} = \frac{Q/M}{V\varepsilon_0}\left(\frac{\Lambda}{\nu} + \frac{d_s}{K_s}\right) . \tag{7.5}$$

If $\Lambda \propto L$, then $(M/A)^{-1}$ should be linear in L. The experiment in which L was varied from 0.125 and 0.375 cm is shown in Fig. 7.10. At each step the packing was held constant by changing the doctor blade to keep the flow

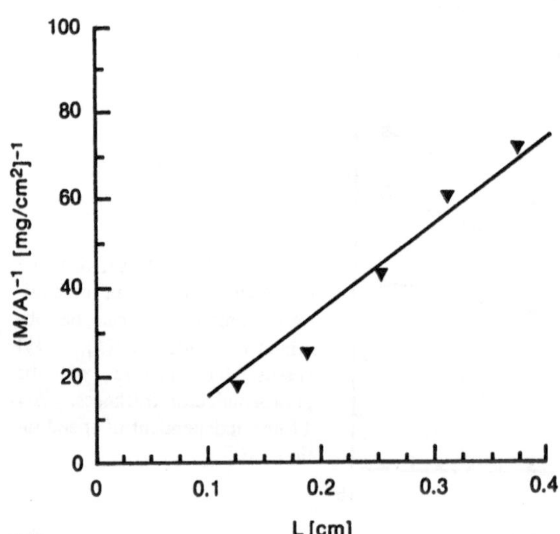

Fig. 7.11. The experiment giving the results shown in Fig. 7.10 was repeated for a carrier with 1.5 μm coat thickness. This experiment is consistent with the predictions of the equilibrium theory [7.10]

constant. This should tend to keep the electric field constant, but the magnetic fields are decreasing at the photoreceptor surface. In fact, $(M/A)^{-1}$ is observed to be independent of L, suggesting the position of the carrier charge is independent of L. Because of the technical difficulties in carrying out this experiment, it was redone with the coat thickness increased to 1.5 μm. In that case (Fig. 7.11), $(M/A)^{-1} \propto L$, as predicted by the equilibrium theory with $\Lambda \propto L$.

This experiment was repeated with sponge carrier with no coating applied. This should drive the system into a totally conductive region; indeed conductivity measurements indicated a short circuit across the developer mix. (To obtain reasonable toner charge-to-mass ratios, the type of toner was changed.) Again $(M/A)^{-1}$ was independent of L (not shown). In addition, M/A versus V was basically unchanged (Fig. 7.12a) from the nominal coat weight material (Fig. 7.5). One interpretation of these data might be that $\Lambda \to 0$ (due to conductivity in the chains). That this is not a satisfactory explanation is dramatically shown in Fig. 7.12b. With $\Lambda = 0$, V_t is predicted to equal V [the neutralization condition, see (7.2)]; this is not observed.

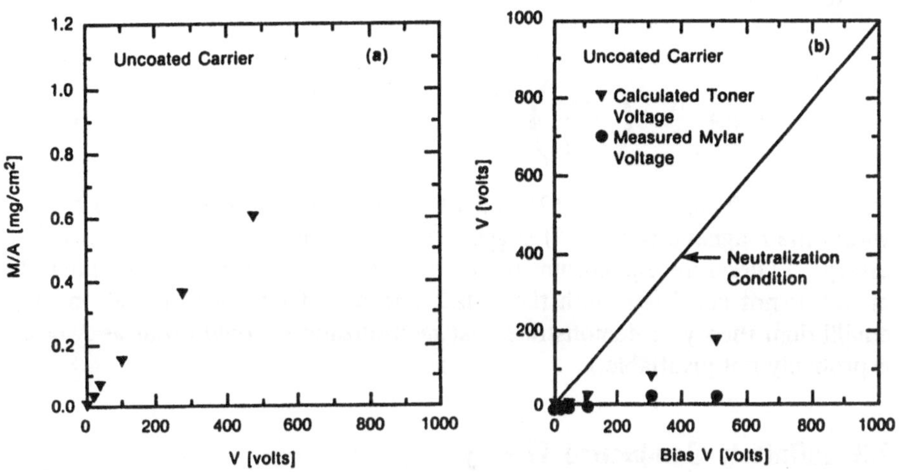

Fig. 7.12. a,b.. Experiments with sponge carrier with no coating. **a** M/A versus V. **b** The measured toner voltage versus V [7.10]

Finally, using the 0.6% coat weight material, the toner concentration C_t was varied from 2.5% to 1% (Fig. 7.13). To correct for Q/M variations, V_t which is proportional to $Q/A = [(M/A)(Q/M)]$, is plotted versus V. Note that Q/A depends on C_t, another piece of data inconsistent with the equilibrium theory (7.2).

These data indicate that M/A is (1) approximately linear in V, (2) independent of carrier coat thickness below 0.6 μm, (3) independent of photoreceptor to developer roll gap, (4) linearly dependent on toner concentration (for constant Q/M), and (5) approaches neutralization by the equation

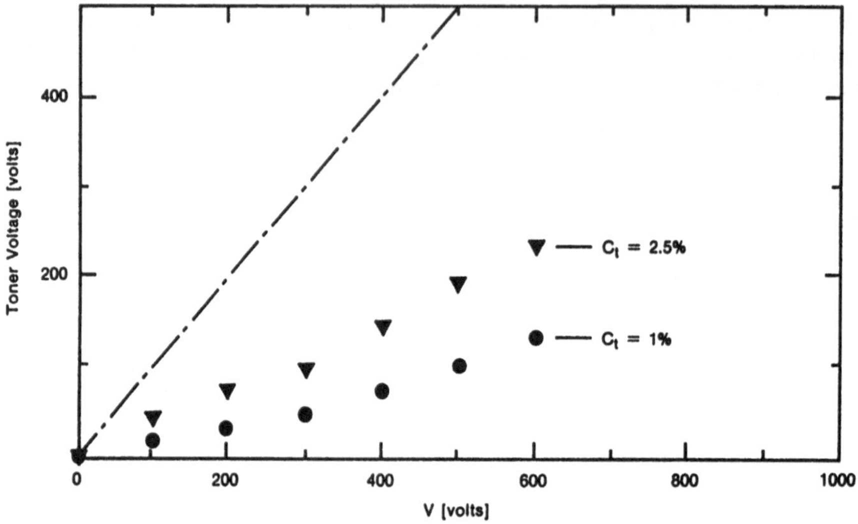

Fig. 7.13. The dependence of V_t on toner concentration. As toner concentration is reduced, V_t decreases [7.10]

$$\frac{V}{V_t} = 1 + \frac{\Lambda}{\nu}\left(\frac{d_s}{K_s}\right)^{-1}, \tag{7.6}$$

where $\Lambda = 17\,\mu$m. Points (2) and (3) suggest that no electric field exists within the magnetic brush; (4) suggests a toner-supply-limited model is necessary. (Therefore equilibrium theory, which is not a toner-supply-limited model, is not consistent with these data and use of equations based on the equilibrium theory to demonstrate that neutralization should occur at $\Lambda = 0$ is probably not justifiable.)

7.3 Infinitely Conductive Theory

A new infinitely conductive theory proposed by *Schein* et al. [7.10] has two assumptions: (1) the roller potential moves down to the last carrier bead adjacent to the photoconductor (as discussed by *Kasper* and *May* [7.1]) and (2) toner is field stripped from the first layer of carrier beads up to angle θ (Fig. 7.14) which is determined by the local electric field (similar to *Williams*'s theory of development [7.12] although the electric field will be treated differently).

The predicted mass per unit area (M/A) at synchronous motion $\nu = 1$ ("with" mode) is then

$$\frac{M}{A} = \frac{n_0 M_t p}{\pi R^2}\tilde{F}, \qquad \nu = 1\ , \tag{7.7}$$

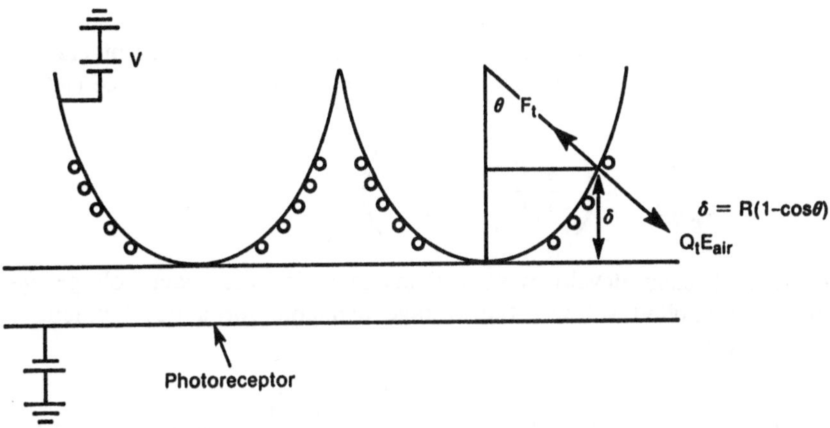

Fig. 7.14. Schematic of the new infinitely conductive, field-stripping theory. The last carrier adjacent to the photoconductor is at the roller potential; toner is field stripped up to an angle θ which depends on the applied voltage [7.10]

where n_0 is the total number of toner particles on a carrier bead, $\tilde{F}$ is the fraction of available toner stripped off by the electric field, M_t is the mass of a toner particle, πR^2 is the projected area under a carrier particle, and p is the carrier surface packing.

From solid geometry

$$\tilde{F} = \frac{1}{2} \; (1 - \cos\theta) \; , \tag{7.8}$$

where θ is the angle between the vertical and the last toner particle stripped off. At the angle θ the force of attraction of toner to carrier F_t just equals the Coulomb force Q_tE_{air}, which can be shown [7.10] to be well approximated by

$$F_t = Q_tE_{air} \approx \frac{Q_tV}{\delta + d_s/K_s} \; , \tag{7.9}$$

where δ is the air gap from the top of the photoconductor to the carrier surface (Fig. 7.14). Since δ depends on θ as

$$\delta = R(1 - \cos\theta) \; , \tag{7.10}$$

$$\tilde{F} = \frac{1}{2}\frac{\delta}{R} = \frac{1}{2}\left(\frac{Q_tV}{F_tR} - \frac{d_s/K_s}{R}\right) \tag{7.11}$$

and therefore

$$\frac{M}{A} = \frac{n_0M_tp}{2\pi R^2}\left(\frac{Q_tV}{F_tR} - \frac{d_s/K_s}{R}\right) , \qquad \nu = 1 \; . \tag{7.12}$$

To generalize this to speed ratios other than 1, it is a reasonable approximation to assume more toner is developed proportional to the number of carriers which contact a point on the photoconductor, which is linear in ν, see (6.39), giving

$$\frac{M}{A}=\frac{n_0 M_t p}{2\pi R^2}\,\nu\left(\frac{Q_t V}{F_t R}-\frac{d_s/K_s}{R}\right), \qquad \text{for all } \nu\,. \tag{7.13}$$

Finally, for efficient development systems in which the toner voltage approaches the applied voltage, the voltage available for a development is $(V - V_t)$ giving

$$\frac{M}{A}=\frac{n_0 M_t p}{2\pi R^2}\,\nu\left(\frac{Q_t (V-V_t)}{F_t R}-\frac{d_s/K_s}{R}\right) \qquad \text{for large } V_t\,. \tag{7.14}$$

Since the toner concentration C_t is

$$C_t = n_0 r^3 \rho_t / R^3 \rho_c\,, \tag{7.15}$$

where r is the toner radius, ρ_t (ρ_c) is the toner (carrier) density, and assuming F_t can be described by the image force

$$F_t = \frac{1}{4\pi\varepsilon_0}\,\frac{Q_t^2}{4r^2} \tag{7.16}$$

(this ignores a possible van der Waals force of attraction; *Lee* and *Ayala* [7.13] have shown that the force of attraction can be larger and different than assumed above) we obtain

$$\frac{M}{A}=8\varepsilon_0\,\frac{C_t \nu p}{Q/M}\,\frac{\rho_c}{\rho_t}\,\frac{(V-V_t)}{r}-\frac{2}{3}\,C_t\rho_c\,\nu p\,\frac{d_s}{K_s}\,, \tag{7.17}$$

where we have assumed that the proper average of Q_t/M_t is approximated by the average value of Q/M measured in a blowoff experiment. It is predicted that M/A is linear in V with a voltage intercept (where $M/A = 0$) of

$$V_{\text{intercept}}\left(\frac{M}{A}=0\right)=\frac{\rho_t r}{12\varepsilon_0}\,\frac{Q}{M}\,\frac{d_s}{K_s}\,, \tag{7.18}$$

which is small (17 V for $Q/M = 15\mu C/g$, 75 μm Mylar, 5 μm radius toner). Ignoring the intercept and using

$$V_t = \frac{M}{A}\,\frac{d_s}{K_s}\,\frac{Q}{M}\,\frac{1}{\varepsilon_0} \tag{7.19}$$

we finally obtain

$$\frac{M}{A} = \frac{C_t \nu V p}{Q/M} \frac{\rho_c}{\rho_t} \frac{8\varepsilon_0}{r} \Bigg/ \left(1 + C_t \nu p \frac{\rho_c}{\rho_t} \frac{8}{r} \frac{d_s}{K_s}\right) \tag{7.20}$$

and

$$\frac{V}{V_t} = 1 + \frac{\Lambda'}{\nu} \left(\frac{d_s}{K_s}\right)^{-1} , \tag{7.21}$$

where Λ' is now

$$\Lambda' = \frac{1}{8} \frac{r}{C_t} p \frac{\rho_t}{\rho_c} . \tag{7.22}$$

To first order (far from neutralization) (7.20) can be simplified to

$$\frac{M}{A} = \frac{C_t \nu V p}{Q/M} \frac{\rho_c}{\rho_t} \frac{8\varepsilon_0}{r} . \tag{7.23}$$

Several simplifying assumptions were made in deriving this theory. First, the electric field was assumed to be given by

$$E_{air} = V \Bigg/ \left(\delta + \frac{d_s}{K_s}\right) . \tag{7.24}$$

Finite element analysis (see [7.10]) indicates that for many close packed spheres above a ground plane this is a close approximation (within 10% up to 45°) after which the field rapidly drops to zero. Second, a spherical shape was assumed; one could replace R in the equation

$$\delta = R(1 - \cos\theta) \tag{7.25}$$

with a smaller R derived from a statistical analysis of the shape of the particles. Third, the image force probably underestimates the actual force of attraction of toner to carrier particles [7.13].

7.4 Comparison with Experiment

This theory appears to describe all of the experiments. It is predicted that M/A is approximately linear in V (7.20), which is observed (Fig. 7.5). A small intercept is predicted (7.18). An intercept is apparent in Fig. 7.5, although it is possible that a finite voltage is required to make the bead chain conductive.

The nonlinearity observed at high voltage may be due to the falloff in field at large θ (see [7.10]).

The theory does not explicitly predict the dependence of Q/M on V. However, one would expect for a wide distribution of Q/M 's that as V increases, toner particles with larger carrier adhesion, and therefore larger Q/M 's, would develop at higher V, as observed for unclassified toner in Fig. 7.6.

The theory predicts (7.20) that M/A is linear in ν at low ν and becomes independent of ν at large ν, which is observed in Fig. 7.7. It also predicts M/A is independent of the gap between the photoconductor and roller and the carrier coat thickness, which is observed in Figs. 7.10 and 12. It further predicts M/A is linear in toner concentration for constant Q/M, which is observed in Fig. 7.13.

The theory predicts V/V_t is linear in $(d_s/K_s)^{-1}$ with an intercept of 1 (observed in Fig. 7.9) and

$$\Lambda' = \frac{1}{8} \; \frac{r}{C_t} p \; \frac{\rho_t}{\rho_c} \; , \tag{7.26}$$

which theoretically equals 21 μm (for $r = 5\,\mu$m, $C_t = 2.5\%$, $\rho_t/\rho_c = 1/3$, $p = 0.4$) as compared with the observed value of 17 μm. The slight discrepancy may be due to several factors, including the assumed force of adhesion of toner particles to carrier (if the force is increased, the predicted Λ' would increase), the guess of $p = 0.4$, or the approximation in using a spherical geometry for the shape of a carrier bead.

Air breakdown, and its effect on the latent image, was expressed as a concern for theories which postulate infinitely conducting chains. While air breakdown was not observed at the voltages used to obtain the above data, it

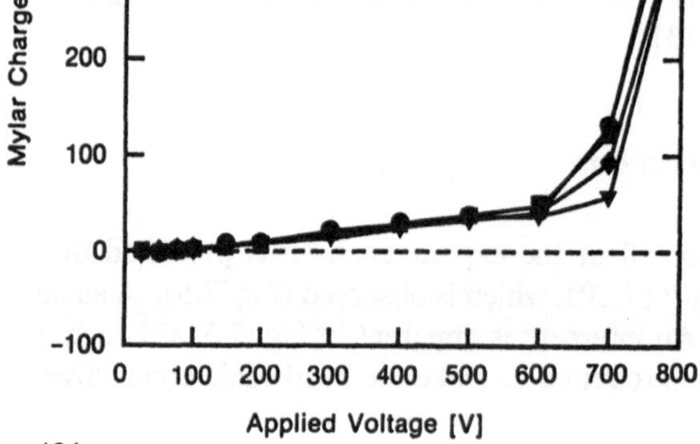

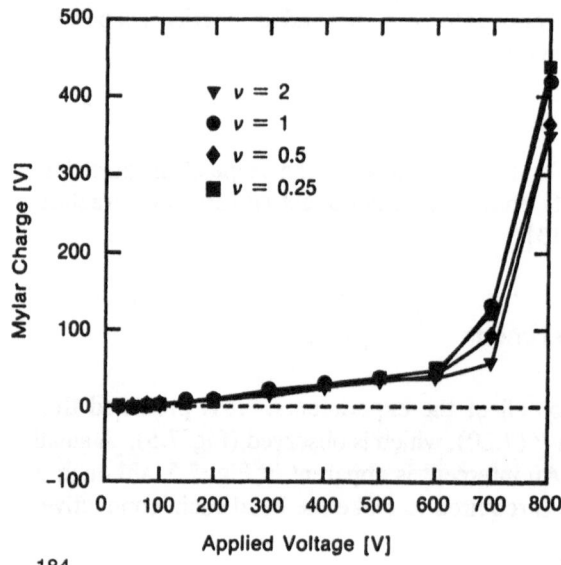

Fig. 7.15. Voltage on Mylar due to charges deposited during development as a function of the applied potential. The voltage increase is gradual until about 700 V. Beyond that point the mix appears to break down, passing charge to the Mylar surface [7.13]

does appear at higher voltages as a significantly increased Mylar charging. Such data have been published by *Lee* and *Beardsley* [7.9] and are shown in Fig. 7.15. Above 700 V the Mylar charge increases rapidly, and simultaneously spots, devoid of toner, appear on the photoreceptor approximately equal to the size of a carrier bead. Presumably during breakdown the charge on the photoreceptor is neutralized and no electric field remains to attract toner.

7.5 Line Development

Very little has been published on line development. Generally, line/solid ratios of about 1 are commonly observed (for equal optical reflection density input). The physical reason for this result is straightforward. If the counter electrode is closer to the photoreceptor surface (as Schein's model postulates) than the photoreceptor ground plane, then lines and solids have approximately the same electric field above the photoreceptor.

Scharfe [7.6] has published the only study of the relative density of 100 μm, 235 μm, 433 μm, and 981 μm lines with input densities of 0.4, shown in Fig. 7.16. As can be seen, the relative density appears to be flat to 1 line pair/mm. Above that, the relative density decreases. Whether the decrease is due to the falloff in electric fields (as assumed by Scharfe) or the decreased

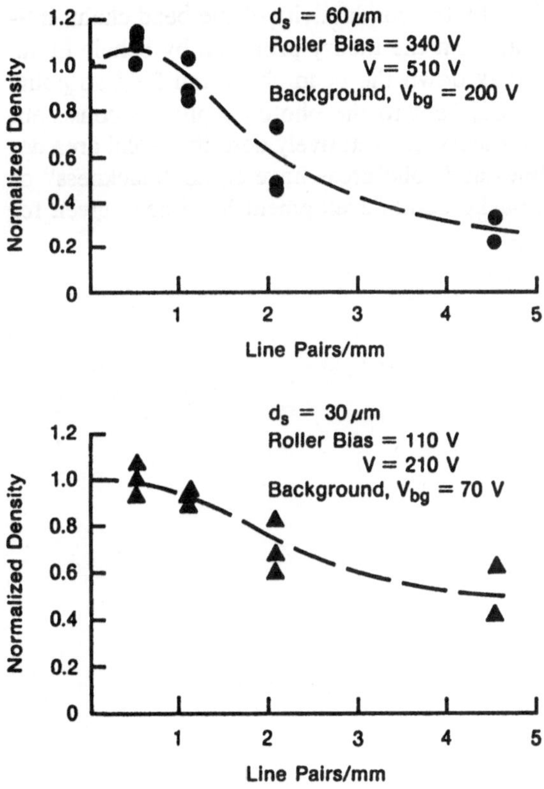

Fig. 7.16. Normalized developed density for a conductive developer as a function of line frequency and photoconductor thickness [7.6]

time for the brush to move the electrode to the carrier beads adjacent to the photoreceptor is not known. However, the falloff appears at a similar frequency as was observed for insulative magnetic brush (Fig. 6.27), suggesting Scharfe's assumption is correct.

7.6 Background Development

No information on background development specific for conductive magnetic brush development systems has been published. Background development is generally observed to be less than that observed in insulative systems. This could be due to the larger reverse electric fields in the background regions.

7.7 Summary

Conductive magnetic brush development has produced significantly improved copy quality, both by producing increased M/A 's for lines and solids, increasing the customer-perceived "blackness," and by making the optical reflection density of lines and solids equal.

Several theories have been proposed to account for this increased efficiency which can be characterized by the conductivity of the bead chain: infinite, partial, insulative. It appears that the theory proposed by Schein et al., which assumes infinite conductivity down the bead chain and field stripping of toner from the carrier beads adjacent to the photoreceptor, is consistent with all of the experiments. The theory quantitatively describes solid area development and explains why lines and solid areas have equal "blackness" or M/A. No data or theories of background development have been given for this development system.

8. Toner Charging for Monocomponent Development Systems

Just as in two component development systems, toner for monocomponent systems must be charged so that the electric field of the latent image can exert its Coulomb force. The problem which this chapter addresses is how to charge and transport toner when carrier is absent. The following chapter discusses the physics of the development process in which monocomponent toner is attracted to the photoreceptor.

Toner particles used in monocomponent development systems are similar to toner particles used in the two component systems (Sect. 4.4). They are polymer-based with carbon black added for colorant. Magnetic properties, if needed, are obtained by adding magnetite, γ-Fe_3O_4, or similar materials, with 50% loading not uncommon. Higher conductivity, when required, is obtained by adding additional carbon black to the bulk or surface. Triboelectric charging is enhanced by the addition of charge control agents. As for toner for two component development systems (Sect. 4.4.3), the patent literature is rapidly evolving.

Similar to the situation in two component development systems, the charge properties of the toner critically affect the performance of the development system. Wrong-sign toner can produce background, uncharged toner produces dust and the average charge-to-mass ratio determines character and solid area optical density. In addition, as in two component systems (Sects. 6.3.4, 7.3), the steepness of the M/A versus V curve, which partially determines the gray scale rendition, can also be affected by the charge properties (Chap. 9).

A surprisingly large variety of charging methods have been identified and incorporated into monocomponent development systems. Probably the simplest is induction charging of conductive toner, patented by 3M in the 1970s and currently being used in Océ copiers. Insulating toner can be charged by injection (3M) or contact electrification (Canon, Ricoh, Toshiba). Corona charging of insulating particles has been suggested (Xerox, IBM). Other methods of charging and transporting particles, not yet applied to the creation of new monocomponent development systems, also will be discussed.

8.1 Induction Charging

Perhaps the easiest method of charging particles is to make them conductive, contact them to a metal and impose an electric field E_{air}. Charge will flow from

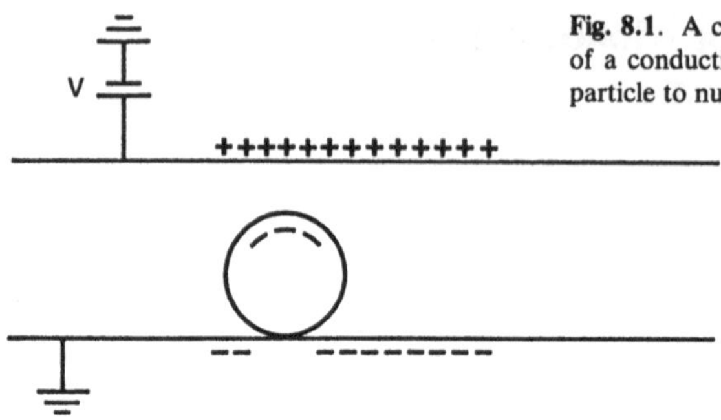

Fig. 8.1. A configuration for induction charging of a conductive particle. Charge flows into the particle to null internal electric fields

the metal to the particle to exclude the electric field from the interior of the particle.

A configuration for single-particle induction charging is shown in Fig. 8.1. A conducting particle sitting on the negative plate becomes negatively charged. Due to electrostatic repulsion, it is repelled from the negative plate. It then lands on the positive plate, loses its negative charge and becomes positively charged. It is then repelled from the positive plate. The particle will bounce between the two plates. *Choi* [8.1] has shown that the average electric field at the surface of a spherical particle on the plate is 1.65 times the electric field of the charging plates. Hence, the induced charge Q and the charge-to-mass ratio Q/M are

$$Q = (1.65\ E_{\mathrm{air}})(4\pi\varepsilon_0 r^2), \tag{8.1}$$

$$Q/M = 4.95\varepsilon_0 E_{\mathrm{air}}/r\rho_t, \tag{8.2}$$

where r is the particle radius, ρ_t is the particle's density, and ε_0 is the permittivity of free space. By allowing the particles to escape from a small hole in one plate, Choi characterized the charge and radius of inductively charged conductive particles. Quantitative agreement with (8.1) was obtained (Fig. 8.2).

Making the particles conductive provides a charging method, but a development system still requires a means of transporting the particles. *Kotz* [8.2] suggested using magnetic forces. By loading the toner with magnetic material such as magnetite, he could move the toner around a roller, see Fig. 8.3. Either the magnets or the outside roller can rotate. When the magnets rotate the toner moves in the opposite direction because the motion is determined by the erection and falling of toner chains, following magnetic field lines. In the development zone the toner is chained by the radial magnetic field between the roller and the photoreceptor surface, forming a conductive path. In the presence of the electric field due to the latent image, charge flows down the toner chains, charging the toner particles at the ends of the chains adjacent to the

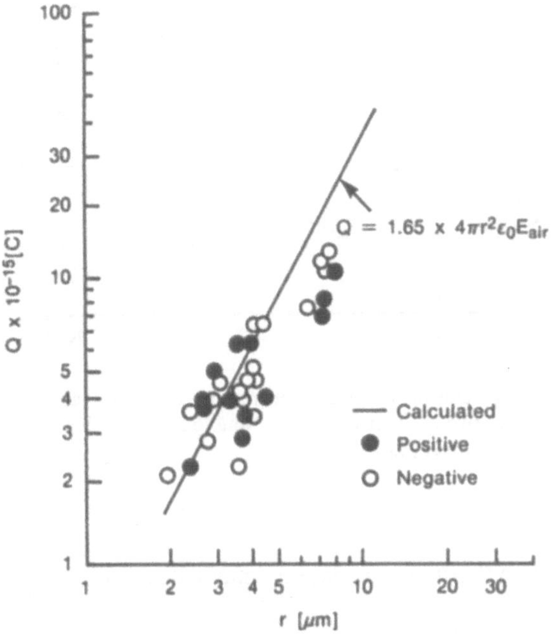

Fig. 8.2. Charge characteristics of carbonyl nickel particles charged inductively [8.1]

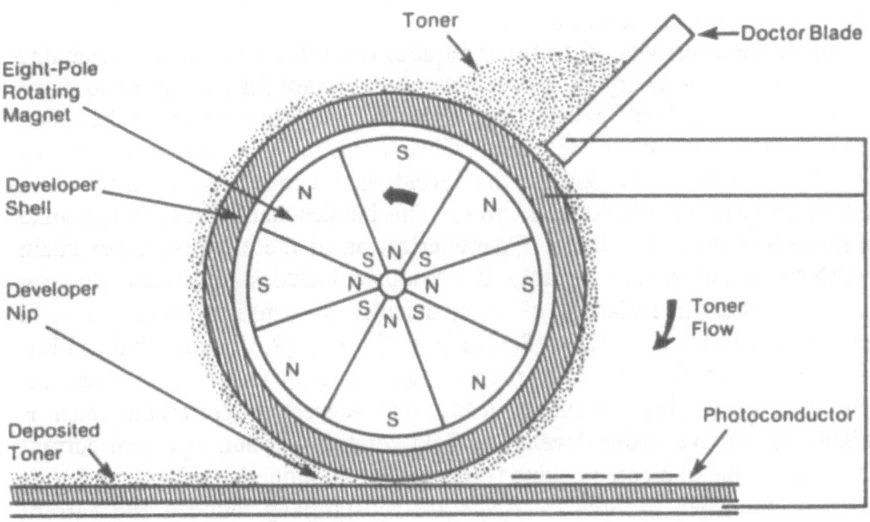

Fig. 8.3. Schematic of a development system that uses induction charging. The magnets rotate, causing the toner to move around the roller in the opposite direction. In the development zone, charge flows down the toner chains in response to the electric fields of the latent image [8.3] (© 1983 IEEE)

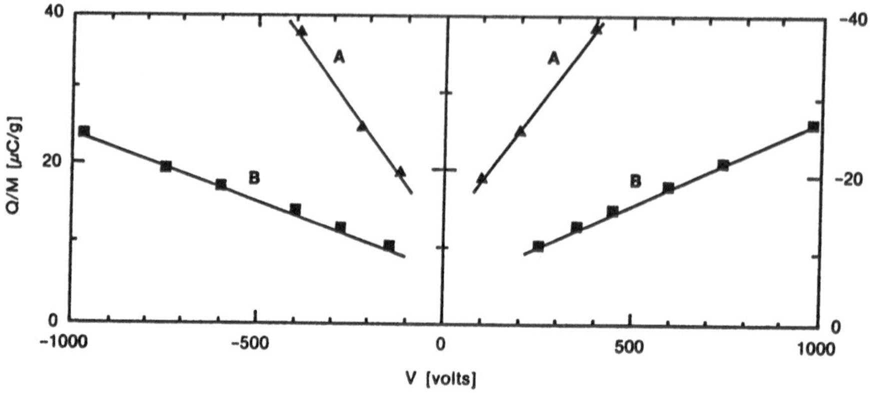

Fig. 8.4. Relationship between the magnitude of the developed toner's charge-to-mass ratio and surface potential of photoreceptor layer for two conductive toners. Note that Q/M is V shaped about zero volts, characteristic of conductive toner [8.4]

photoreceptor, which will then adhere to the latent image because of the Coulomb force.

Of course, either positive or negative charge can be induced on the toner. Hence a measurement of the magnitude of the developed toner's charge should be V shaped about zero volts. This has been shown by *Shimada* et al. [8.4] among others (Fig. 8.4). This result implies that reverse biasing to minimize background development (Sect. 2.1.3) is not possible with this system unless a diode characterization can be built into the toner. This may have been achieved by *Shimada* et al. [8.4].

Setting the resistivity of the toner requires careful consideration. Assuming a simple RC circuit charging model, the time constant for the charge to move down the chain is simply $\rho K_t \varepsilon_0$ where K_t is the dielectric constant of the toner, and ρ is the resistivity of the chain. Clearly this must be much smaller (say $10\times$) than the time the toner is in the development zone (20 ms for a nip width of 1 cm and a roller velocity of 5 cm/s). This implies that the resistivity should be much less than 10^{11} Ωcm. Characterization of the effective toner chain resistivity is not simple because it obviously includes interfaces between toners. Further, the resistivity of such particulate systems is often electric field dependent, as shown by *Faust* [8.5] (Fig. 8.5). *Kotz* [8.2] argues that too low a resistivity may cause the toner charge to leak onto the photoreceptor surface, effectively destroying the latent image, and suggests a preferable toner is "highly conductive under developing field conditions when electrical current flow is desirable to create imaging forces and less conductive prior to and after development when the electric fields are substantially reduced and current flow is not desired." Another consideration: a resistivity intermediate in value, such that $\rho K_t E_0$ is approximately equal to the development time, could lead to very erratic results because ρ can easily vary from chain to chain.

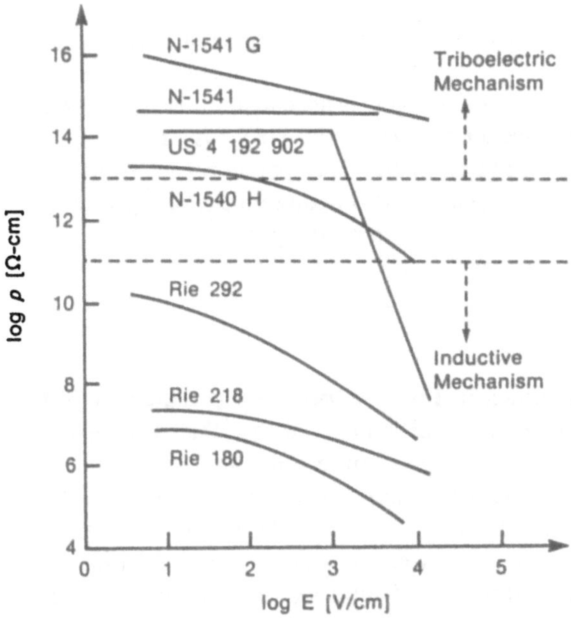

Fig. 8.5. Volume resistivity versus electric field for different mono-component magnetic toners [8.5]

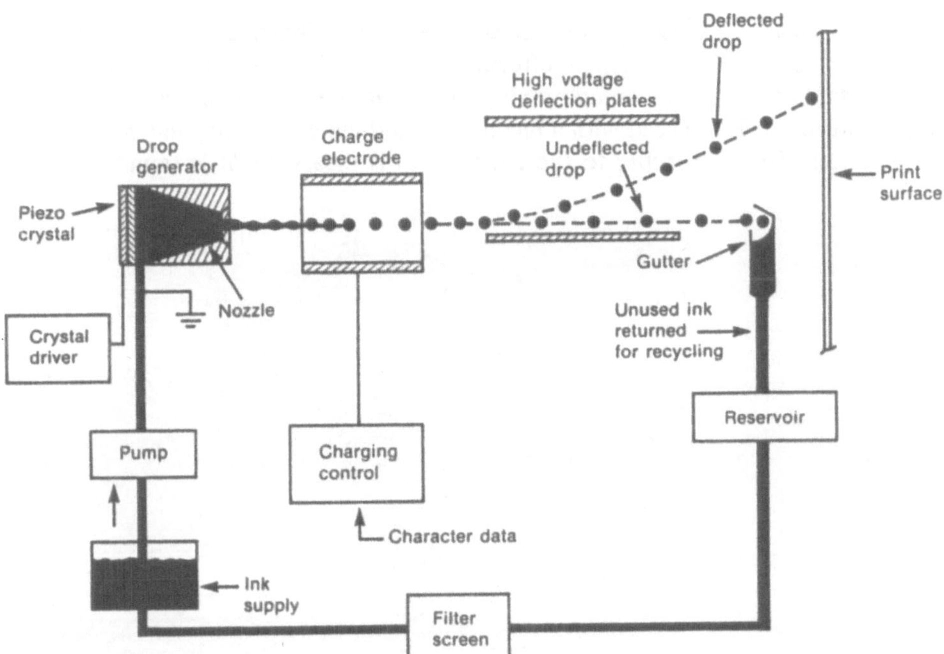

Fig. 8.6. One form of ink jet printing uses conductive ink which is inductively charged and then deflected with an electric field [8.6]. (Copyright 1977 by International Business Machines Corporation; reprinted with permission)

Development problems (Sect. 9.1) with this system caused by the high conductivity of the toner have led people to consider methods of charging insulating toner.

As an aside, this induction method of charging is used in another nonimpact printing technology, continuous ink jet printing (Fig. 8.6). As a stream of conductive ink breaks into drops, induced charge is trapped on each drop. A Coulomb force caused by an applied electric field acting on the charged drop is used to deflect the drop to the proper position on the paper.

8.2 Injection Charging

Injection charging methods as shown in Fig. 8.7 were also patented by 3M [8.7]. In this case insulating magnetic toner is metered to a thickness of approximately 250 μm (or about 25 toner layers) on a roller surface. Toner is transported around the roller by rotation of either the roller or the magnets. The insulating toner is charged by injection from the roller in the presence of an electric field. A critical aspect of this invention is the provision of a means to produce rapid, turbulent physical mixing of the toner particles so that uniform charging occurs.

The details of precisely how the turbulence contributes to toner charging remain unclear. That dynamic effects associated with the roller velocity are important is well documented by all groups working on this problem. If the photoreceptor is replaced with a metal plate and the current flowing in an applied field from the roller to the metal plate is measured, it is found to be

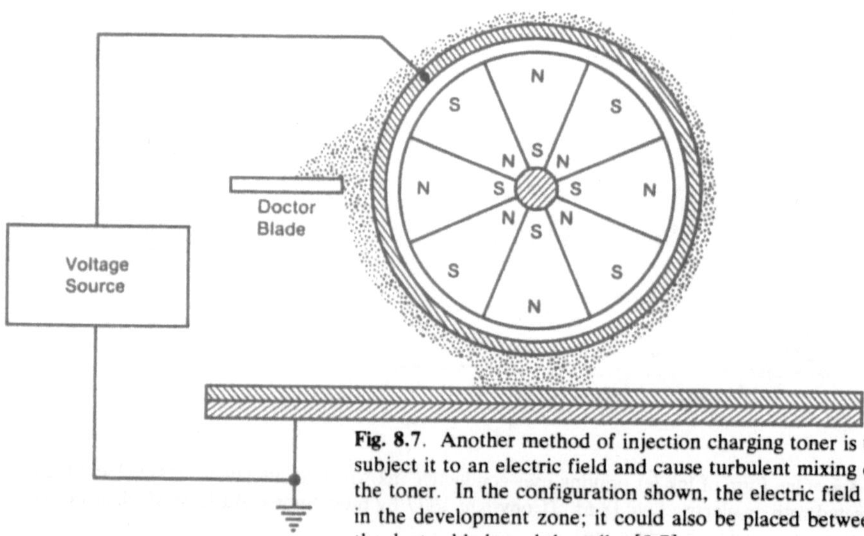

Fig. 8.7. Another method of injection charging toner is to subject it to an electric field and cause turbulent mixing of the toner. In the configuration shown, the electric field is in the development zone; it could also be placed between the doctor blade and the roller [8.7]

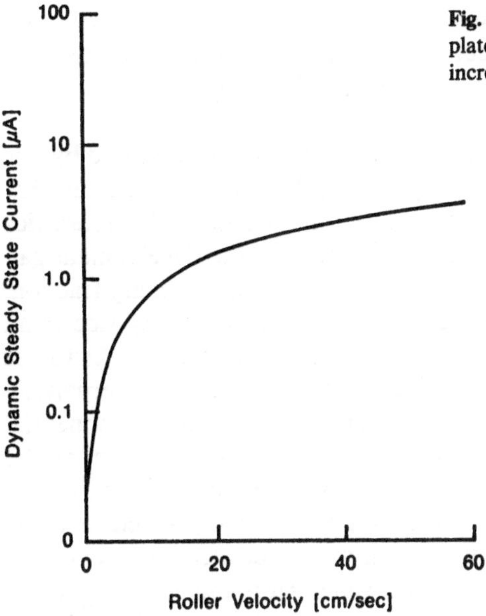

Fig. 8.8. Current between the roller and a metal plate placed in the position of the photoreceptor increases strongly with roller velocity [8.7]

strongly dependent on roller speed. Figure 8.8, which shows this effect, is taken from *Nelson*'s patent [8.7]. The same result has also been reported by *Field* [8.3], *Nakajima* et al. [8.8], and *Lee*, *Imaino* et al. [8.9–13]. Nelson suggests that the charge injected into a toner particle can be later transferred from one toner to another in the presence of an electric field. Nakajima et al. provide an interesting picture that depends upon chain rotation (Fig. 8.9).

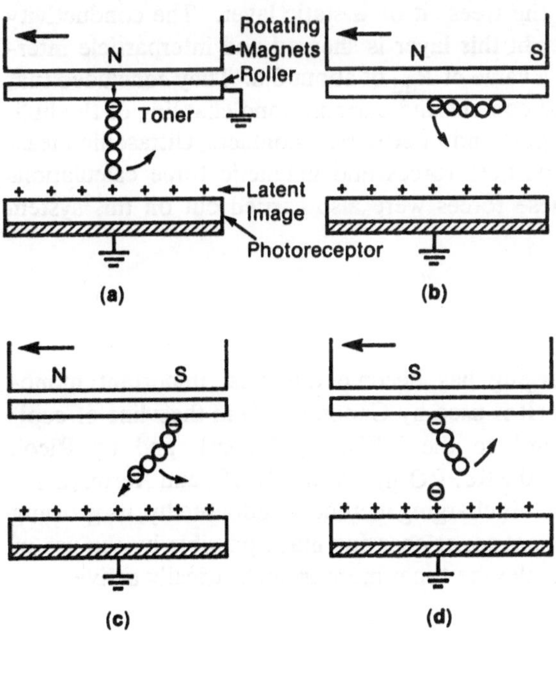

Fig. 8.9. A model for toner charging of high resistivity toner by injection from the roller [8.8]

Field argues that rapid toner movement is possible in the charging region. He postulates that charge injection occurs at the electrode-toner interface and then charged toner rapidly moves to the vicinity of the photoreceptor in response to the electric fields.

A study of the complications involved in this process was presented by *Lee* and *Imaino* and co-workers in a series of papers [8.9–13]. They showed that different current characteristics and presumably toner charging characteristics can be obtained by varying R_{dd}, defined as the ratio of the development gap to the doctor gap. At values of R_{dd} less than 0.9, i.e. high packing fraction, their toner behaved conductively, producing a dc current when exposed to an applied bias, with a metal plate substituted for the photoreceptor. At values of R_{dd} above 0.9, the dc current progressively decreased, i.e., the interparticle conductivity decreased, and they observed a transition to a simple exponential current-time relationship. They associated this with mass transport of the charged toner particles. These current-time curves also produced evidence for toner development. During the first revolution of the roller, the current was observed to decrease relatively rapidly, which was associated with toner development onto the electrode. This was observed at large values of R_{dd} which Nelson recommended for optimum development conditions. This work was coupled with direct observations of the toner flow through a transparent plastic window. Only at large R_{dd} is toner development observed. At smaller ratios, individual particles are not fixed with respect to their neighbors and the velocity near the electrode, i.e. the plastic window, is much smaller than that near the roller surface. When the roller rotates with the magnetic field fixed, toner filaments form larger lumps, or trees, that cartwheel along the surface. This is probably the source of the kinetically enhanced conductivity. An unexpected observation is that the trees sit on a static layer. The conductivity region occurs when the toner in this layer is sheared and interparticle interactions occur. This occurs at a value of R_{dd} of about 0.8. They concluded that charge is injected at the metal contact and currents (and charging of the bulk toner) result from mass transport and interparticle contact. Ultrasonic measurements which probe interparticle forces and magnetic force calculations which partially determine these forces were also carried out on this system [8.12, 13].

8.3 Contact Charging

Triboelectric or contact charging has become the most important monocomponent charging method. It is used by Canon [8.14] in their line of copiers, from the PC 10 (8 cpm) to the NP500 (50 cpm), and by Ricoh [8.15–17] in their new copier, the RePRO jr. Toshiba [8.18] and Xerox [8.19] have papers and patents using this charging method. Undoubtedly, its growing importance is due to the efficient toner charging made possible by the use of charge control agents which make the toner more triboelectrically active.

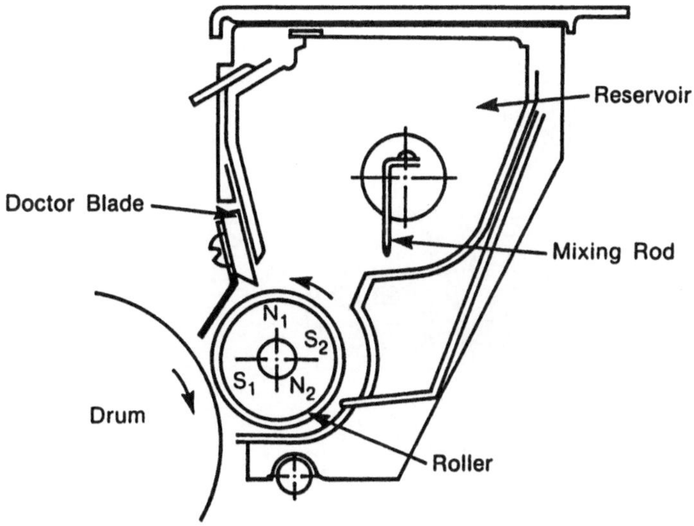

Fig. 8.10. Schematic of the Canon magnetic monocomponent development system. The roller rotates about stationary magnets, carrying magnetic toner past the doctor blade into the development zone [8.14]

The configuration used by Canon [8.14] is shown in Fig. 8.10. A roller (called a sleeve by Canon) rotates about stationary magnets in a reservoir (called a hopper by Canon) of magnetic, highly insulating, monocomponent toner. The toner is charged by contact with the roller and is carried out of the reservoir past a magnetic doctor blade. Evidence that the toner is charged against the roller was obtained by varying the coating material on the roller. Normally, for a metallic (aluminum) surface an electrostatic potential of the toner layer of −30 V was obtained. With different resins coated on the roller, values from +40 to −40 V were obtained. The doctoring process determines the amount of toner on the roller in the development zone. The magnetic doctor blade operates by splitting the toner chains where the spatial derivative of the magnetic field, which determines the toner-toner adhesion force, vanishes (Fig. 8.11).

Ricoh has discussed [8.15–17] a system, shown in Fig. 8.12, to charge nonmagnetic monocomponent toner triboelectrically. A supply roller made of foam pushes toner against a roller (called a development roller by Ricoh). Toner which adheres to the roller must pass under a spring-loaded metering blade. The source of toner charging has been determined by two experiments. First, *Demizu* et al. [8.15] determined that the counter charge is on the roller surface by measuring the electrostatic potential above the roller. Knowing the toner charge and the dielectric thickness of the roller, he predicted the electrostatic potential in the absence of the counter charge. If the counter charge is on the roller surface then the potential should be approximately zero. The data are consistent with the counter charge being on the roller surface, sug-

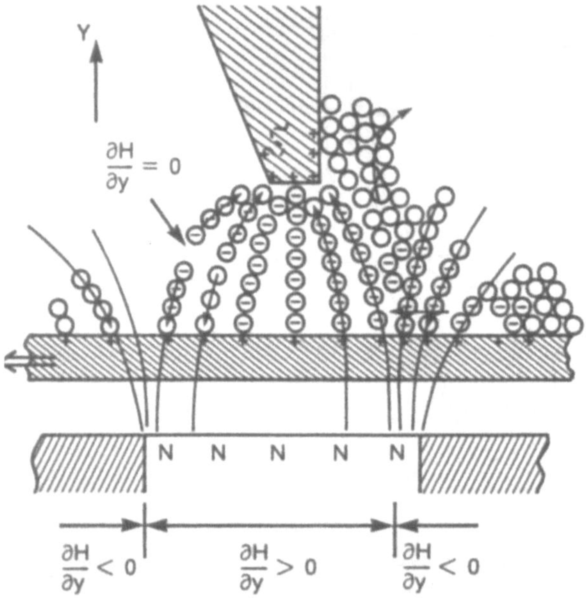

Fig. 8.11. The doctor blade in the Canon system is magnetic; the toner chains split where the force, proportional to $\partial H/\partial y$, vanishes [8.14]

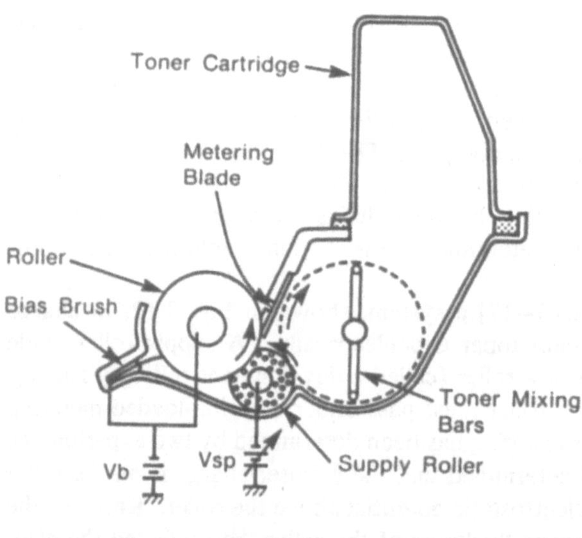

Fig. 8.12. Schematic of the Ricoh nonmagnetic monocomponent development system. Toner is pushed against the roller by the supply roller, where it obtains its charge. The toner is then metered before it contacts the photoreceptor (at approximately the 12 o'clock position) [8.15]

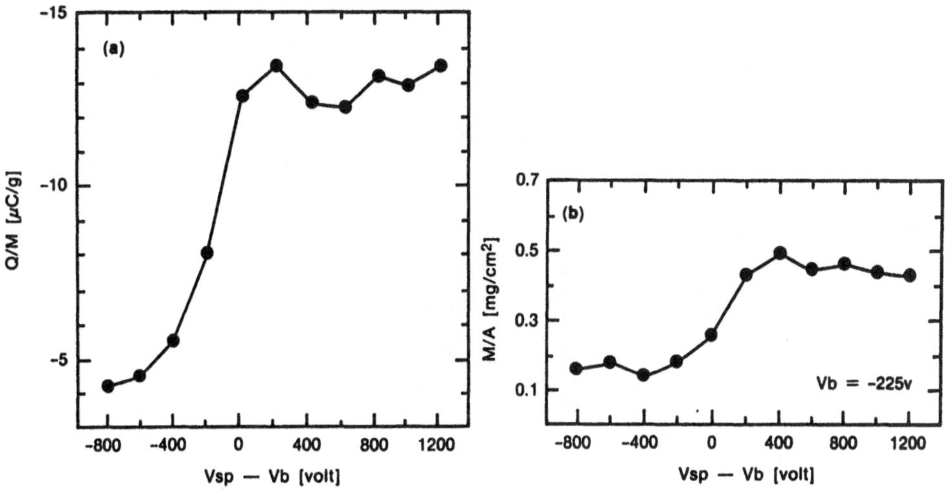

Fig. 8.13. The Q/M (a) and M/A (b) of the toner on the roller as a function of the potential difference between the roller and the supply roller [8.15]

gesting that the toner obtains its triboelectric charge by interacting with the roller surface. Second, the charge-to-mass ratio and M/A of the toner on the roller is measured as the potential between the roller and the supply roller is varied (Fig. 8.13). It is found that Q/M is independent of this potential difference and M/A increases between 0 and 400 V. Since M/A increases as the electric field increases it must be that the toner is charged and is probably being charged at the supply roller-roller interface. That Q/M is independent of bias suggests the charging process is independent of field, at least to the values used in the experiment. A method of measuring the individual toner particle's charge and radius appears to be available to the Ricoh group. Although the technique is not specified, data are shown (Fig. 8.14) which suggest that $Q \propto$

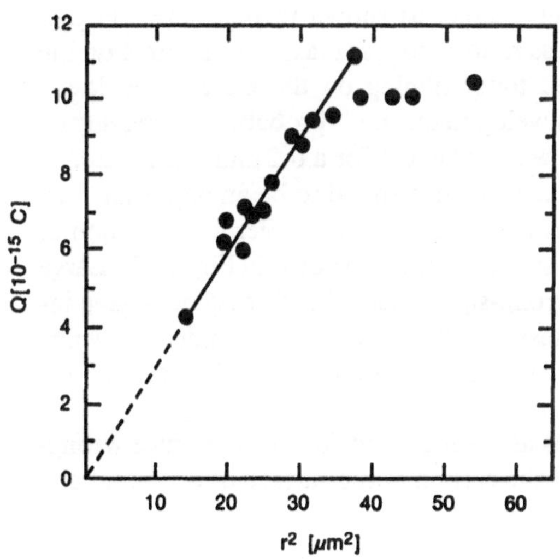

Fig. 8.14. Toner charge versus particle radius squared. Note $Q \propto r^2$ below $r = 6\ \mu m$ [8.15]

r^2 below $r = 6\ \mu m$. (Almost identical results were obtained by Terris et al., as discussed in Sect. 4.4.4). Above 6 μm radius, the toner charge tends to become independent of r. As the roller surface is insulating, means must be provided to discharge it. This is accomplished by a biased brush (Fig. 8.12) which contacts the roller after development occurs, at the 7 o'clock position.

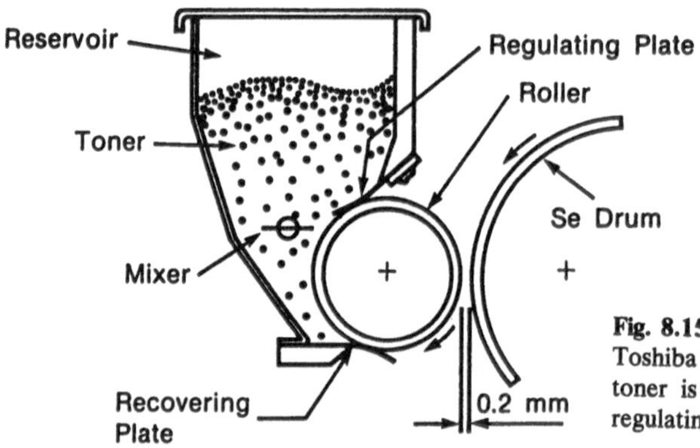

Fig. 8.15. A schematic diagram of the Toshiba development system. The toner is metered and charged by the regulating plate [8.18] (© 1985 IEEE)

At the same meeting at which Ricoh presented their results (see above) *Hosoya* et al. from Toshiba [8.18] presented their ideas for charging nonmagnetic insulating toner. Their apparatus is shown in Fig. 8.15. It is similar to the Ricoh design without the supply roller. A metering blade (called a regulating plate by Toshiba), by applying pressure to the roller, determines both the toner charge and the toner layer thickness. Detailed experimental analysis of the factors affecting toner charge were presented.

Both elastic metal and rubber plates were tried. Elastic metal such as stainless steel or phosphor bronze, 0.1–0.2 mm thick, was preferred because it did not deform with use and produced less wrong-signed toner. Plate pressure significantly affected the toner charge, as shown in Fig. 8.16. Higher pressures gave higher charge-to-mass ratios and less mass per unit area on the roller. Too high a pressure caused toner filming on the plate. Too low a pressure produced background development, i.e., probably wrong-signed toner. The recommended pressure was 100 g/cm for a 0.2 mm thick plate.

The roughness of the roller surface was also found to be an important variable. The toner charge-to-mass ratio and M/A on the roller as a function of mean roller roughness produced by sandblasting are shown in Fig. 8.17. Large roughness gave background, i.e., wrong-sign toner. Small roughness gave insufficient image density. A roughness of 0.7 μm was recommended. Toshiba maintained the surface roughness with use by catalytic nickel plating of the roller.

Carbon content of the toner was also varied and found to produce wrong-sign toner, due to "contact between the carbon at the surface of one toner and

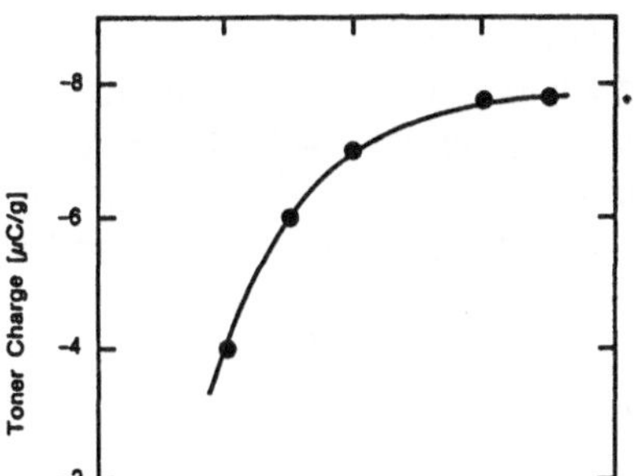

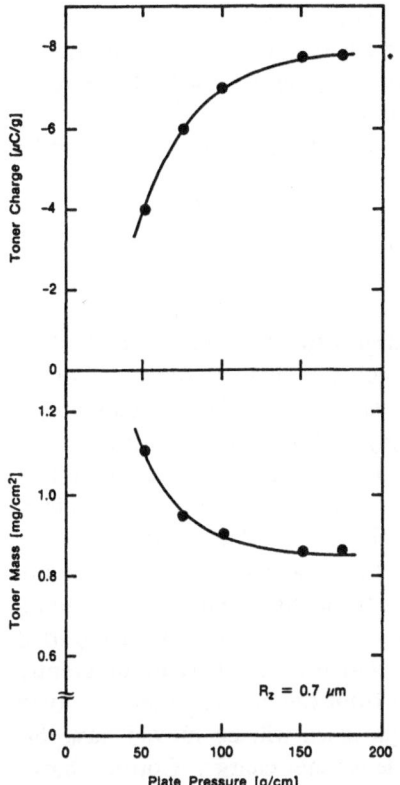

Fig. 8.16. Plate pressure affects the toner Q/M and M/A observed on the roller. Higher pressures increase Q/M and decrease M/A. The value R_z is the surface roughness of the roller (Fig. 8.17) [8.18] (© 1985 IEEE)

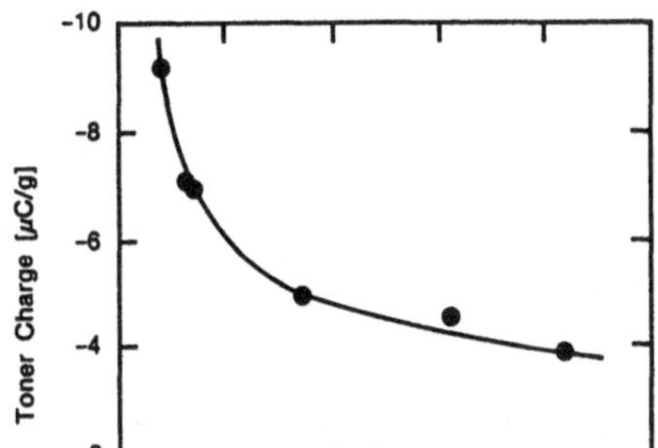

Fig. 8.17. The surface roughness of the roller also affects the toner Q/M and M/A on the roller. Larger roughness decreases Q/M but increases M/A [8.18] (© 1985 IEEE)

the polyester resin at the surface of another toner." Keeping the carbon content below 3% eliminated the background problem.

By measuring the current to the plate, it was determined that 30% of the toner charge is produced by friction with the plate and 70% by friction with the roller. Hosoya et al. observed a dc current which flows through the toner particles from the plate to the roller.

Patents have recently been issued for other nonmagnetic insulating monocomponent development systems, although none of these has been discussed in the scientific literature. For example, (1) *Gundlach* [8.19] of Xerox has recently suggested a triboelectric charging configuration in which the toner is forced through several wire screens of different meshes. (2) *Yoshikawa* of Canon [8.20] has suggested a modification to the Canon magnetic monocomponent system which converts it into a nonmagnetic system. Magnetic particles are confined between the roller and a magnetic doctor blade through which nonmagnetic toner flows, presumably charging against both the roller and the magnetic particles. (3) Fuji Xerox has claimed a contact metering blade which produces a uniform toner layer less susceptible to disruption by contaminants [8.21].

8.4 Corona Charging

Flint [8.22] patented a method of corona charging toner. He placed a corotron adjacent to a roller which attracted magnetically loaded toner. Corona charging was also used by *Chang* and *Wilbur* [8.23] in their studies of impression, i.e. contact, development. A supply of magnetic insulating toner is moved continuously around a roller past a corona charging device where ions generated by the corona are sprayed on the toner. Some obvious problems with this system are discussed by *Nelson* [8.7]. Corona devices are subject to toner contamination, especially by airborne toner, which will result in nonuniform ion emission along the length of the corona wire, and therefore nonuniform toner charging. Nonuniform ion emission is characteristic of negative corona (Sect. 2.1.1), which would make negative toner charging inherently nonuniform. Also continuous ion emission from the corona source coupled with the probability that individual toner particles will be moved past this source many times can result in time-dependent toner charge. Corona charging of particles and limitations on the amount of charge due to eventual repulsion of incoming ions has been discussed by *Hendricks* [8.24].

8.5 Charging Methods for Powder Coating

Electrostatic coating (painting) of surfaces with powders is a well-developed technology. The charging of insulating powders is accomplished during spraying by either corona charging or triboelectric charging (as the particles interact with the walls of the spray gun) [8.25]. A third potential source of charging, which may be operative in powder cloud development involves particle-particle charging. While the net charge may be zero, a distribution of charge about zero due to particle-particle contacts is likely.

8.6 Other Charging Methods

Hendricks [8.24] discusses other particle charging methods which have not yet been implemented in development systems. Particles can be charged by passage through an electron or ion beam. If the beam diameter is D_B and the particle velocity is v, then the particle will be in the beam a time D_B/v. For a beam current density of J, the particle charge will be

$$Q = \frac{r^2 J D_B}{v} \,. \tag{8.3}$$

This is the maximum charge. If the potential of the particle becomes comparable to the beam energy, the electrons or ions will be repelled and further charging ceases. Also, secondary emission of electrons from the particle surface during electron charging can limit Q.

While not usually observed at room temperature, thermionic emission can charge particles. When the component of electron velocity in a direction perpendicular to the material surface becomes great enough to overcome the image force, the electron will leave the surface. This effect is relatively well understood and is employed in hot cathode vacuum tubes.

If light falls on the surface of a particle, the light quanta can transfer sufficient energy on impact to eject electrons from the particle. The available energy on impact is the photon energy less the work function of the material. In the range of visible light, very few materials exhibit photoelectric charging. However, ultraviolet [8.26] and x-ray charging are much more efficient due to lower reflection coefficients, higher surface absorption, and greater interactions with bound electrons which are easier to eject. This method obviously yields positive particles since electrons are ejected.

It has long been known that the cleavage of mica and other crystalline materials leaves the fresh surfaces charged. If the crystals are cleaved under vacuum, electrons with energies of hundreds of kilovolts are emitted. It has been suggested that a charged double layer at the separation surface is the location of the charges. When the double layer is disturbed or separated, high electric field intensities are produced and field emission of electrons can occur. Similar effects can occur during a process in which small particles are broken or torn from a surface, e.g. the manufacture of toner particles.

8.7 Traveling Electric Fields

In 1987, a nonmechanical means of transporting toner was patented by *Schmidlin* [8.27] and discussed by *Melcher* et al. [8.28]. The toner is placed on a linear array of spaced electrodes which are electrically wired to produce a traveling electric wave. The use of traveling electric fields to transport charged macroscopic particles was pioneered by *Masuda* [8.29,30] and is also being studied for agricultural applications [8.31]. The traveling electric wave can carry the toner, either in hops across the surface, synchronously with the electric field, or airborne, asynchronously with the electric field [8.28]. Toner is charged either prior to being placed on the electrodes or by interacting with the surface which contains the electrodes. In the latter case, provision must be made to drain off the buildup of charge, by either surface or bulk conductivity whose value must be chosen so that the traveling electric field is not screened.

9. Monocomponent Development

While two component development (Chaps. 6 and 7) is today the predominant development system used in copiers and printers with speeds above 20 cpm (Table 1.1), it clearly is not the simplest system imaginable. For example, since toner is used but carrier is recirculated, means must be provided to sense depleted toner, and then to add and mix fresh toner. In addition, hardware must be built to recirculate a mixture of powders, of which only a small percentage of the weight (the toner) is directly used in creating marks on paper.

An obviously simpler development system is a monocomponent system in which the only powder component is toner. That concept was, of course, well known to the early inventors of electrophotography. The challenge is learning how to charge and transport the toner into the vicinity of the latent image. To date, at least four different monocomponent systems have been used in actual products (Sects. 9.1, 9.4–6). Because the relative simplicity compared to two component development implies fewer parts, smaller hardware, and lower manufacturing cost, monocomponent development systems have had their greatest application in low speed, low cost copiers. How far monocomponent development can penetrate the medium and high speed market is an important, unresolved question in electrophotography today.

Monocomponent development systems can be characterized by three independent choices: (1) The toner can be conductive or insulative. (2) The toner can be magnetic or nonmagnetic. (3) During development the toner, carried on a roller, can jump across a gap to the photoreceptor or be placed in simultaneous contact with the roller and the photoreceptor. Some people include a fourth choice in which the roller can run synchronously with or at a different speed from the photoreceptor. In this chapter we will consider the implications of these choices, as manifested in known development systems. Early work is discussed in Sects. 9.1 and 9.2. A theory of monocomponent development, which appears to be common to all published systems, is described in Sect. 9.3. In Sect. 9.4 the conductive toner system is described. Finally, the newest and most successful systems based on insulative toner are described in Sect. 9.5 (magnetic) and Sect. 9.6 (nonmagnetic). The current status of monocomponent development is summarized in Sect. 9.7.

9.1 Aerosol or Powder Cloud Development

One of the earliest monocomponent development systems that was used in a product was called aerosol or powder cloud development [9.1–7]. Work on this system was first mentioned in the mid-1950s [9.1–3]; a thorough study was reported by *Bickmore* et al. [9.4] in 1960 just at the dawn of the copier revolution and later by *Lewis* and *Stark* [9.5] in 1972. The objective of Bickmore et al. was to test the resolution capability of selenium photoreceptors. Resolutions in excess of 100 line pairs per millimeter were obtained using toner in the range of 0.1–0.8 μm diameter. Lewis and Stark's objective was to investigate the edge enhancement capabilities of this development system. The high resolution and the edge enhancement capabilities were exploited in the Xerox 125, an electrophotographic medical tool for obtaining x-ray pictures.

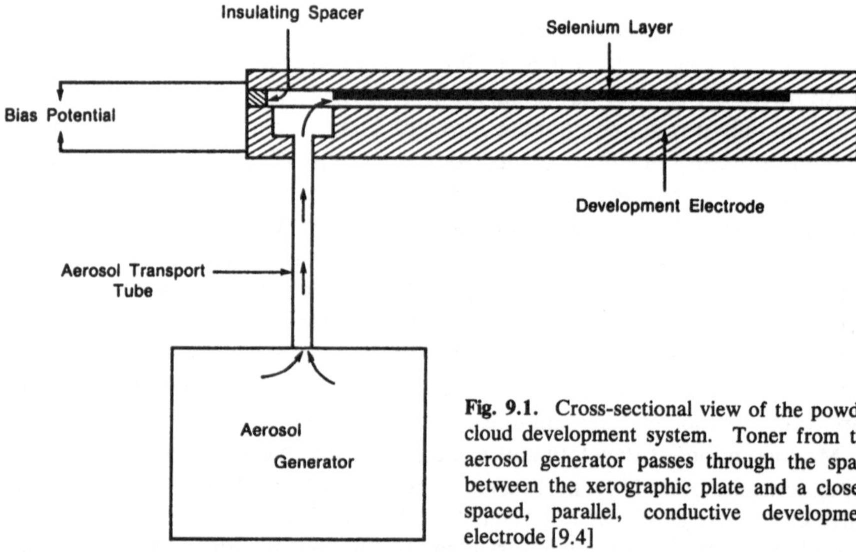

Fig. 9.1. Cross-sectional view of the powder cloud development system. Toner from the aerosol generator passes through the space between the xerographic plate and a closely spaced, parallel, conductive development electrode [9.4]

The system studied by *Bickmore* et al. [9.4] is shown in Fig. 9.1. In this configuration an electrode is present. Development to neutralization gave accurate reproduction of the gray scale, leading to copy quality similar to photographic film. Without an electrode or with a far-spaced electrode, toner responds to density gradients and is therefore especially sensitive to edges of solids and lines. It is this aspect of powder cloud development that was studied in [9.5].

In the system shown in Fig. 9.1, an aerosol generator lifts toner into a transport tube. The toner, suspended in air, is introduced into one end of a channel formed by a photoreceptor and an electrode. If the air flow is laminar then in the absence of electric fields the toner will not strike the walls. While the toner is probably close to electrically neutral on average, toner particles possess charge equally distributed between positive and negative polarities. In the presence of an electric field due to the latent image, one polarity is attracted to the photoreceptor. This is one of the powder coating charging mechanisms discussed in Sect. 8.5.

The work reported by *Bickmore* et al. [9.4] was a detailed study of various effects of parameters on development. They found that a small electric field applied across the development zone was crucial. With no field the toner in the proximity of the photoreceptor was insufficient to cause complete development of edges. They found that 10–20 V across the few centimeters between photoreceptor and electrode usually led to complete development with "tolerable" background. Above 20V, background increased with little or no improvement in image completeness. By varying the development time (between 20 and 100 s) they showed that images were essentially completely developed in this system only when the fractional neutralization was 80% or higher (80 s). They also showed that the development rate was proportional to the potential difference in the development zone. Surprisingly, it was found that increasing the aerosol concentration by a factor of 100 had little effect on development time. Furthermore, higher aerosol concentrations produced images of poorer quality, containing missing sections, background, and streaks. The charge-to-mass ratio of the developed toner was found to significantly decrease, from 250 μC/g to 60 μC/g when the aerosol concentration was increased from 1 mg/l to 130 mg/l. It was suggested that toner space charge may be a factor in these effects.

The remarkable feature that was investigated by *Lewis* and *Stark* in 1972 [9.5] was the edge enhancement capability of this development system, of obvious practical importance in assisting doctors detect pathologies in x-ray pictures. What was observed was a white gap at the edge of lines and solids. The system investigated appears similar to the one studied by *Bickmore* et al., except the toner radius used was larger, 2 μm, leading to lower charge-to-mass ratios, 3–5 μC/g, and a far-spaced electrode (3.8 cm) was used. Cloud mass densities remained the same, approximately 6 mg/l.

The white gap (minimum in optical density) is shown in microdensitometer traces in Fig. 9.2 for a 30 V step. Data for a 3 V step were also given and it was claimed that even a 1 V step could be detected by eye. It was suggested that the source of this step is the geometry of the electric field, shown in Fig. 9.3, a phenomenon discussed by *Sullivan* and *Thourson* [9.8] in the cascade development literature (Sect. 5.3.1). Near a step the electric field lines arch back to the photoreceptor, penetrating only a small volume of space above the photoreceptor, unable to capture much toner. The width of the zone in which the field arches back on itself, the "forbidden zone", is compared with experiment in Fig. 9.4. Both figures show good agreement between theory and experiment.

As indicated above, this development system is in fact used commercially in the Xerox 125, an electrophotographic x-ray copier (Fig. 9.5). It was

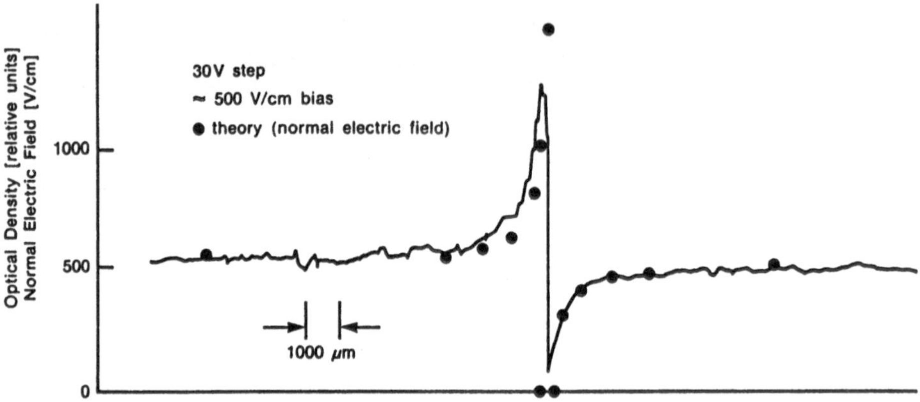

Fig. 9.2. A density trace across a 30 V step. The filled circles are a theoretical calculation of the normal electric field [9.5]

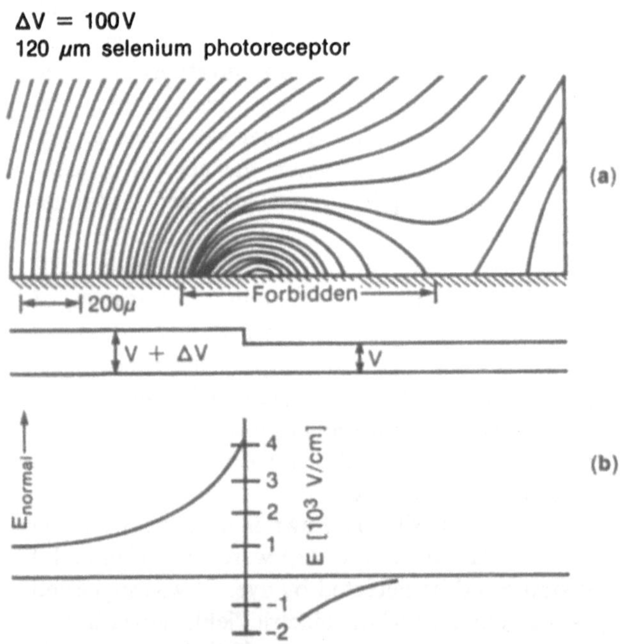

Fig. 9.3. The electric field pattern over a step in plate charge density **(a)** and **(b)** a plot of the normal component of electric field [9.5]

pointed out [9.6] that xeroradiography produces images with contrast sensitivities and resolutions at least equal to film, has short exposure-to-viewing time (about 30 s), and does not need a darkroom. However, its slow speed and high background characteristics make it unsuitable for high speed, high quality copiers.

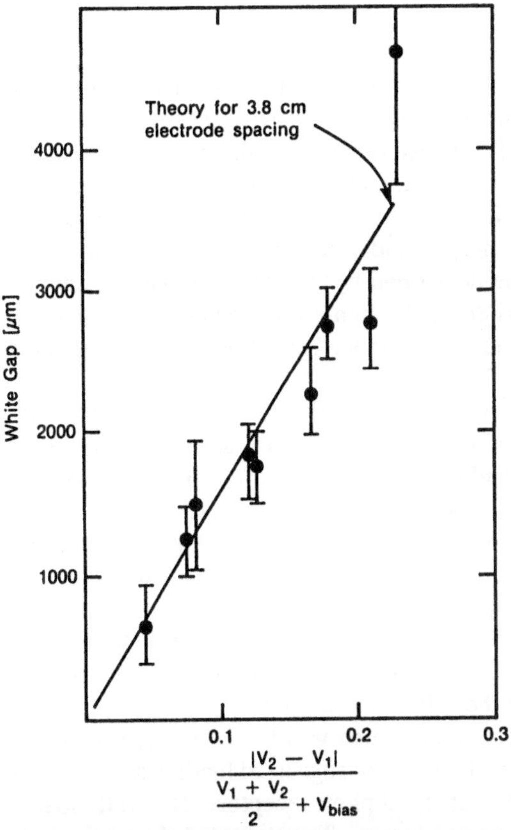

Fig. 9.4. Measured values of the white gaps at the edges of charge steps together with the results expected from a solution of the electrostatics problem. V_1 and V_2 are the voltages on either side of the step and V_{bias} is the potential on the counter electrode [9.5]

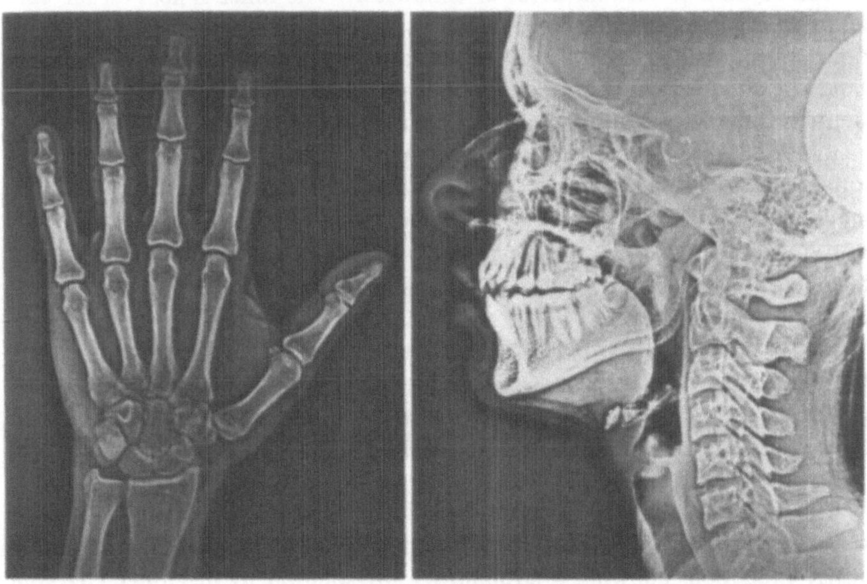

Fig. 9.5. Examples of Se plates exposed to x-rays developed with a powder cloud development system without an electrode (Courtesy of H. Bogdonoff of Xerox Corporation)

9.2 Early Work

Work on monocomponent development was begun in the early 1950s at Battelle as part of the effort to find a viable development system for an automatic electrophotographic copier [9.9]. An initial embodiment was a sheet of specially coated paper or rubberized cloth, called a donor, which was coated with a thin layer of toner by rubbing the toner over the surface with a cotton or fur pad. This triboelectrically charged and spread the toner in a thin layer. A charged photoreceptor could be developed by "touching down" the donor sheet on the plate momentarily (hence, the name touchdown development). Several such touchdowns were needed to produce high density images because of the sparce toner coating on the donor.

Thicker coatings of toner on the donor were obtained by various techniques. However, they usually produced background development, most likely because the toner was not sufficiently charged. Two new concepts were proposed to solve this density/background problem [9.9]. First, the donor was spaced 25–50 μm from the photoreceptor surface. Toner will jump across the narrow air gap to charged areas of the photoreceptor, but not to uncharged areas. This was given the name spaced touchdown. Second, a metal donor cylinder (which we call a roller in this book) was loaded with toner by powder cloud deposition techniques (Sect. 9.1) and then corona charged. This remained a slow process because of the roller loading step.

Work on monocomponent development also was being pursued at RCA. In 1957 a patent was issued to *Greig* [9.10]. He suggested loading a roller with toner which is then contacted to the charged photoreceptor. The roller material can be hard or soft, insulator or conductor. The toner is held on the surface either electrostatically or by tackiness. If the roller is conductive, then the toner can be corona charged; if an insulator, by triboelectricity. The toner should be a material with "good dielectric properties, such as sublimed sulfur." Mention is made of doctoring the toner on the roller.

In 1958, Wilson, president of Haloid Xerox (the name of the Xerox Corporation at the time), suggested using insulative magnetic toner in a monocomponent system [9.11]. The magnetic toner is picked up from a toner reservoir by many circular magnets which are adjacent to each other, forming magnetic brushes. The magnets rotate, continuously bringing fresh toner to the latent image.

In 1959, *Mayo* [9.12] of Battelle suggested using a belt which passes through a reservoir of toner as a development "roller". He assumed triboelectric charging of the toner will occur and suggested coating the belt with known carrier-coating materials. He recommended contact development and speed ratios greater than one.

The first conductive toner monocomponent development patent was issued to *Gundlach* [9.13] of Xerox Corporation in 1965. The charging method is clearly induction, allowing a straightforward solution to the toner charging problem. The toner is loaded onto the conductive donor surface by directing a powder cloud at the surface. The toner will adhere either due to van der Waals forces or electrostatically if charged by, for example, "feeding at turbulent rates through fine tubes." The conductive toner on the conductive donor surface is then pressed into contact with the latent image. The toner is charged inductively by the electric field of the latent image and then adheres to the latent images. It was pointed out that this system should have no threshold voltage for development.

In 1966, *Willmott* [9.14] of IBM was issued a patent which claims development from a belt which is spaced a distance from the photoreceptor. This work follows from an earlier patent by *Lowrie* of IBM [9.15] in which a spaced touchdown system is described. Toner is attracted across the space only when an electrostatic image exists on the photoreceptor. Where no electrostatic image is present, no toner touches the photoreceptor, leading to low background. Toner is loaded onto a belt which is dipped into a reservoir of toner. Speed ratios less than one are suggested.

Battelle continued work on what they called spaced touchdown development. Apparently Mayo's concept of running a belt through a toner reservoir did not adequately load and charge toner since *Andrus* et al. [9.9] of Battelle reported at the Second International Conference on Electrophotography in 1974 on a microfield donor (which we will call a roller). This roller consisted of an aluminum cylinder coated with 25 μm thick insulating enamel over which a copper screen was created. The 150 mesh copper screen pattern was created by standard photoresist and etching techniques. By impressing 200 V between the aluminum and copper, small microelectric fields were created which attracted toner of both signs from a reservoir of fluidized toner. This toner layer was then corona charged to give the toner the same polarity and charge level. The roller was spaced 25–50 μm from the photoreceptor with shims at the ends of the roller. Reverse biasing to decrease background was used. It was observed that the corona charging step adds sufficient charge to the insulating squares between the copper screen to bring the whole roller substrate to an equipotential, preventing further reloading. Hence, a neutralizing corona discharge of the roller before reloading is necessary. The Battelle group report excellent solid areas (remember, they were comparing their results to cascade development!), good resolving power (8–10 line pairs per millimeter), but a high gamma, i.e., high contrast, making the process not suitable for continuous tone images. Studies of the sensitivity of copy quality to spacing revealed that below 25 μm unsatisfactory images were produced because the toner loading was generally not sufficiently uniform. Above 75 μm spacing, image density dropped and the ability to reproduce fine lines and dots was reduced. This clearly created a challenging tolerance problem.

At the same meeting in 1974, *Chang* and *Wilbur* [9.16] of IBM reported on another version of monocomponent development, which they called impression development. In this system they attempted to coat a roller with a material (a copolymer of vinyl chloride and vinyl acetate) which triboelectrically charged the toner. To allow the coating to discharge, it was loaded with 30% carbon to give it appropriate conductivity. The carbon also gave the surface a roughness of $\approx 4\ \mu m$ which the authors suggested enhanced triboelectric charging on the toner. The toner was held in a reservoir against the roller and was doctored with Teflon blades to 2 or 3 monolayer thickness. A corona recharge was required to "improve the uniformity of the charge distribution." This roller was then contacted to the photoreceptor at synchronous speeds. "Good" resolution (5 line pairs per millimeter), solid areas and line copy were reported. As reported by Battelle, high gammas were observed.

The interested reader can find a more complete list of monocomponent patents up to 1972 in *Schaffert*'s book, *Electrophotography* [9.17].

9.3 Theory of Monocomponent Development

In Sects. 9.4–6, monocomponent development systems which have been successfully implemented in automatic copiers are described. In reviewing this literature it became apparent that concepts used by the various authors could be combined in a universal theory of monocomponent development, which is described below.

As with theories of two component development systems, our goal is to express the developed mass per unit area as a function of the measurable hardware and material parameters of the system. We will assume for this discussion that the toner has been uniformly doctored and charged on a roller (Chap. 8) and brought into the vicinity of the latent image. After proposing this theory, it will be applied in the following sections to available monocomponent development experimental data. The theory might be called "field stripping from a finite source".

In a monocomponent development system, the toner adheres to the roller by electrostatic F_{es} and magnetic forces (for magnetic toner) F_M. This adhesion must be overcome by the Coulomb force caused by the electric field E_{air} of the latent image (Fig. 9.6a). Therefore a threshold electric field E_{th} is predicted (Fig. 9.6b) below which development is zero:

$$Q_t E_{th} = F_{es} + F_M \,. \tag{9.1}$$

Expressions for all three terms have been derived before or are well known. The electric field acting on toner in air is related to the electrostatic potential of the latent image V, any bias potentials V_{bias} on the roller, the dielectric thickness of the photoreceptor d_s/K_s, the air gap L, and the developed toner dielectric thickness d_t/K_t,

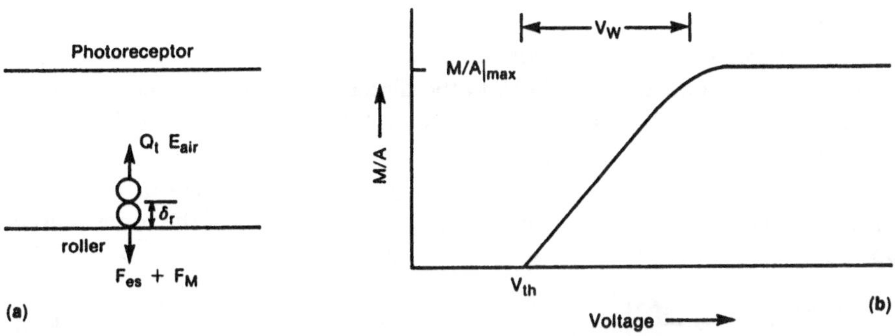

Fig. 9.6. **a** A universal theory of monocomponent development is proposed based on the forces experienced by toner. **b** This theory predicts a voltage threshold and a maximum mass per unit area. The voltage width V_W of the rising portion of the curve is determined by the toner adhesion distribution

$$E_{air} = \frac{V - V_{bias}}{d_s/K_s + L + d_t/K_t}. \tag{9.2}$$

The electrostatic adhesion of toner to the roller is the image force

$$F_{es} = \frac{1}{4\pi\varepsilon_0} \frac{Q_t^2}{(2r + \delta_r)^2} \tag{9.3}$$

plus any van der Waals adhesive forces. In (9.3) we have allowed for the fact that the bottom of the toner particle may be spaced a distance δ_r above the roller surface because it is above the first monolayer (Fig. 9.6a). The magnetic adhesion is given by [9.18]

$$F_M = \frac{\mu - 1}{\mu + 1} r^3 \overline{\boldsymbol{H}} \cdot \nabla \overline{\boldsymbol{H}}, \tag{9.4}$$

where μ is the permeability of the toner, r is the toner radius, $\overline{\boldsymbol{H}}$ is the magnetic field and $\nabla\overline{\boldsymbol{H}}$ the spatial derivative of the magnetic field. Solving for $V - V_{bias}$ and labeling it as a threshold voltage V_{th} gives

$$V_{th} = \left(\frac{d_s}{K_s} + L + \frac{d_t}{K_t} \right) \times \left(\frac{1}{4\pi\varepsilon_0} \frac{Q_t}{(2r + \delta_r)^2} + \frac{\mu - 1}{\mu + 1} \frac{r^3 \overline{\boldsymbol{H}} \cdot \nabla \overline{\boldsymbol{H}}}{Q_t} \right). \tag{9.5}$$

As V is increased above V_{th}, toner is field stripped from the roller. If all of the parameters were a single value, a step function development curve would result. Clearly the toner charge and radius are distributed parameters.

Furthermore δ_r, the distance of the bottom of the toner from the roller surface, can vary if more than one monolayer is present. The value of $\bar{\boldsymbol{H}} \cdot \nabla\bar{\boldsymbol{H}}$ can vary in the development zone both in the direction toward the photoreceptor and along the development width. These effects produce a width V_w of the development curve (Fig. 9.6b).

As the roller is usually loaded with only a few monolayers of toner, the developed mass per unit area reaches a maximum when all of the toner is used up:

$$\frac{M}{A}\bigg|_{max} = \frac{M}{A}\bigg|_{roller} \nu, \qquad (9.6)$$

where ν is the ratio of the roller to the photoreceptor velocities and represents the length of roller that contacts a unit length of photoreceptor (6.63).

The predicted development curve, i.e., mass per unit area versus voltage, is shown in Fig. 9.6b. This theory of monocomponent development might be called field stripping (since the electric field strips toner from the roller) from a finite source (the source of toner is limited to the few monolayers on the roller).

Inspection of Fig 9.6b immediately indicates that monocomponent development systems have nonlinear development characteristics that will affect the gray scale reproduction characteristics. For example, under the condition that the voltage width is small compared to the threshold, the development curve approaches a step and only whites and blacks are reproduced. This characteristic is usually described by gamma, the slope of a D_{out} (output density) versus D_{in} curve. Gamma differs from 1, the ideal for perfect gray scale rendition, due to nonlinear transfer functions in a system, such as given by the development curve shown in Fig. 9.6b.

In these systems an ac voltage V_{ac} is sometimes superimposed on the dc voltages of the latent images. This adds an additional time-dependent force to (9.1). One proposal for the effect of V_{ac} [9.19] predicts a change in slope of M/A versus V. As the effect of V_{ac} appears to affect the threshold voltage more strongly (see below), the following alternative picture is proposed. The force balance on toner attracted to the roller and photoreceptor is shown in Fig. 9.7. In Fig. 9.7a is shown the half cycle in which V_{ac} causes a force towards the photoreceptor. If the force pulling the toner towards the photoreceptor overcomes the adhesion force, toner will develop (the Canon [9.19] literature uses the word "project") towards the photoreceptor. During the next half cycle, the force due to the ac voltage is in the opposite direction (Fig. 9.7b). The same logic applies. Hence, for high enough frequency the toner will "project" back and forth.

As the roller rotates and L increases past the center of the development zone the forces due to the applied voltages will decrease. Where the toner ends up will depend upon which force condition is largest when projection stops. If

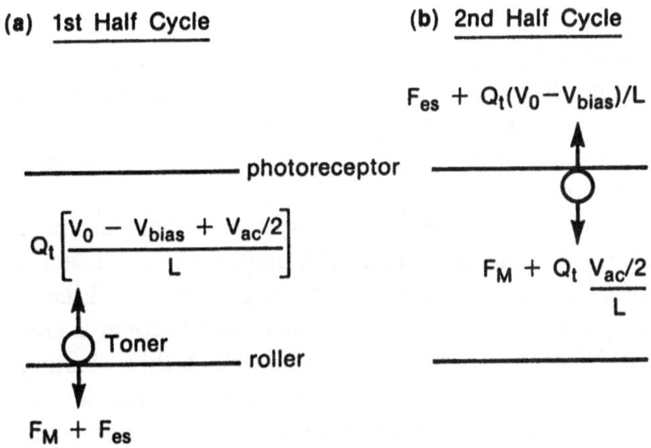

Fig. 9.7. The effect of ac bias on development characteristics is to introduce a time-dependent force which "projects" the toner from both the photoreceptor and the electrode. Shown are the forces on toner particles during each half cycle of the ac voltage

$$Q_t\left(\frac{V_0 - V_{bias} + V_{ac}/2}{L}\right) - (F_M + F_{es}) > \left(F_M + \frac{Q_t V_{ac}/2}{L}\right) - \left(F_{es} + \frac{Q_t(V_0 - V_{bias})}{L}\right) \tag{9.7}$$

then toner ends up on the photoreceptor. This is just

$$\frac{Q_t(V_0 - V_{bias})}{L} > F_M, \tag{9.8}$$

a condition independent of V_{ac}. It is of course possible that the imposition of V_{ac} increases the width of the development zone, moving the last projection to regions where the magnetic force is lower. This makes F_M a function of V_{ac} and can lead to increased development,

$$\frac{Q_t(V_0 - V_{bias})}{L} > F_M(V_{ac}). \tag{9.9}$$

Such an effect suggests that the precise shape of the magnetic field and its spatial distribution at the edges of the development zone could be important in determining the development characteristics. Note also that F_{es} is missing from (9.8). Because the toner is now airborne most of the time, it is far from the roller and photoreceptor, making the electrostatic adhesion force zero. This predicts that the threshold voltage, in the presence of a V_{ac} large enough to cause projection, will be reduced, i.e., $F_{es} = 0$ in (9.1).

It is mentioned by both Canon [9.19] and Toshiba [9.20] that use of V_{ac} in development systems with a gap improves the edges of lines. A reason for this may be that F_{es}, the toner image force to the roller, is eliminated from the physics by the "projection" condition. Development of toner onto edges from a powder cloud (generated during projection) should be more uniform than from a roller in which adhesive bonds need to be broken.

Line to solid density ratios for these development systems should approach 1 as the roller-photoreceptor spacing approaches the photoreceptor dielectric thickness and the line to solid area electric field ratio approaches 1. Line to solid area ratios will also approach 1 if the available toner on the roller is used up. *Sakamoto* et al. [9.21] discuss a novel geometry in which thin, floating, metal electrodes are placed on the roller surface. They argue theoretically that this enhances the electric fields for line development, although no supporting experimental data are given. It is shown in [9.21] that solid area density decreases as the distance to the roller increases, as expected.

Discussions of mechanisms of background development for monocomponent systems are not available. It is obvious that the magnetic force (for magnetic toner) and reverse biasing (for correct sign toner) aid in removing background toner.

9.4 Conductive Toner

The first monocomponent development system for an automatic copier was introduced in a product by 3M in 1971 [9.22]. It used magnetic, conductive toner that was charged inductively in the development zone. A schematic of this system, taken from *Kotz*'s patent [9.22], is shown in Fig. 9.8. The magnetic toner is metered onto a roller and held by magnetic forces. The toner is moved around the roller, spaced approximately 750 μm from the photoreceptor, by rotating either the magnets or the roller. In the development zone the toner contacts the photoreceptor. The charge of the latent image initially creates an electric field in the conductive toner. Charges flow from the roller down the toner chains (formed in response to the magnet fields) to neutralize this field. With sufficiently conductive toner, the charge moves to the monolayer of toner immediately adjacent to the photoreceptor.

The above qualitative discussion is sufficient to identify the threshold voltage and the maximum development [from (9.1) and (9.6)]. The threshold voltage is found by finding the threshold electric field from the force condition

$$Q_t E_{th} = F_M . \tag{9.10}$$

The electrostatic adhesion of toner to the adjacent toner layers is ignored because it is cancelled to first order by the same force to the photoreceptor. Since the toner charge per unit area σ_t equals the photoreceptor charge per

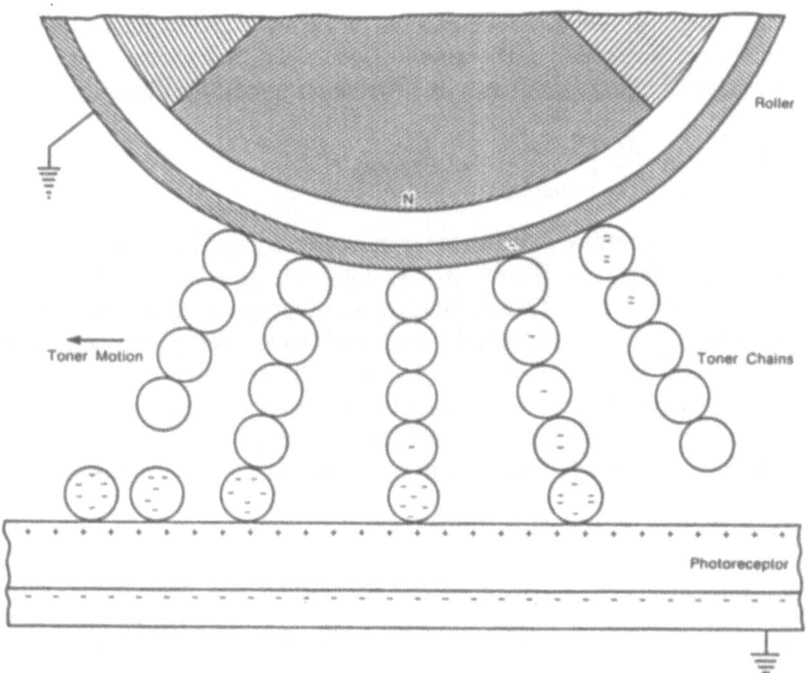

Fig. 9.8. Conductive toner is charged by induction. Charge flows down the toner chains in response to the electric field of the latent image [9.22]

unit area σ_p in an inductively charged system,

$$\sigma_t = \sigma_p, \tag{9.11}$$

$$\frac{Q_t}{\pi r^2} = \frac{V\varepsilon_0}{d_s/K_s} \quad \text{and} \tag{9.12}$$

$$E_{th} = \frac{V_{th}}{d_s/K_s}, \tag{9.13}$$

we obtain

$$V_{th}^2 = \frac{F_M(d_s/K_s)^2}{\varepsilon_0 \pi r^2} \propto r, \tag{9.14}$$

since $F_M \propto r^3$, (9.4). Kotz claims $F_M \approx 10^{-4}$ dynes, giving a threshold voltage of 36 V for $d_s/K_s = 10 \mu m$ and $r = 10\ \mu m$. Larger diameter toner particles are used in conductive systems because the threshold voltage depends on their radius [see (9.14) and below]. (For the reader who chooses to check the calculations in this chapter it is useful to recall that there are 10^5 dynes per newton and 1 N = 1 CV/m.)

The maximum development is a monolayer of toner and is independent of speed ratio since any toner in the second layer which has charge will immediately transfer its charge to toner in the first layer to null the electric field:

$$\left.\frac{M}{A}\right|_{\max} = \left.\frac{M}{A}\right|_{\text{monolayer}} = \frac{4}{3} r\rho_t p_t , \tag{9.15}$$

where r is the toner radius, p_t is the toner mass density and ρ_t is the toner surface packing. For $r = 10\,\mu\text{m}$, $\rho_t = 1\,\text{g/cm}^3$, $p_t = 0.6$, (9.15) gives $M/A|_{\max} = 0.8\,\text{mg/cm}^2$. That $M/A|_{\max}$ is independent of the speed ratio, and that $M/A|_{\max}$ is approximately $1\,\text{mg/cm}^2$ have been demonstrated by *Nelson* [9.23] and *Shimada* et al. [9.24,25] (Figs. 9.9 and 10).

When the potential difference across the development zone is below V_{th}, i.e. in the background regions, toner will not develop as long as F_M for all the toners is kept large enough. This begins to define the control required on the

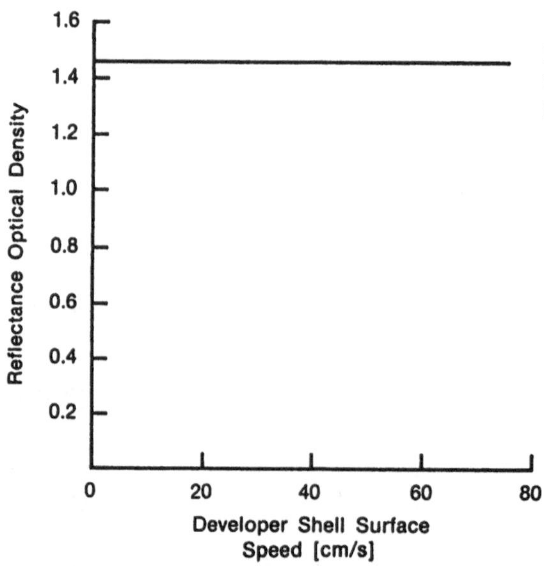

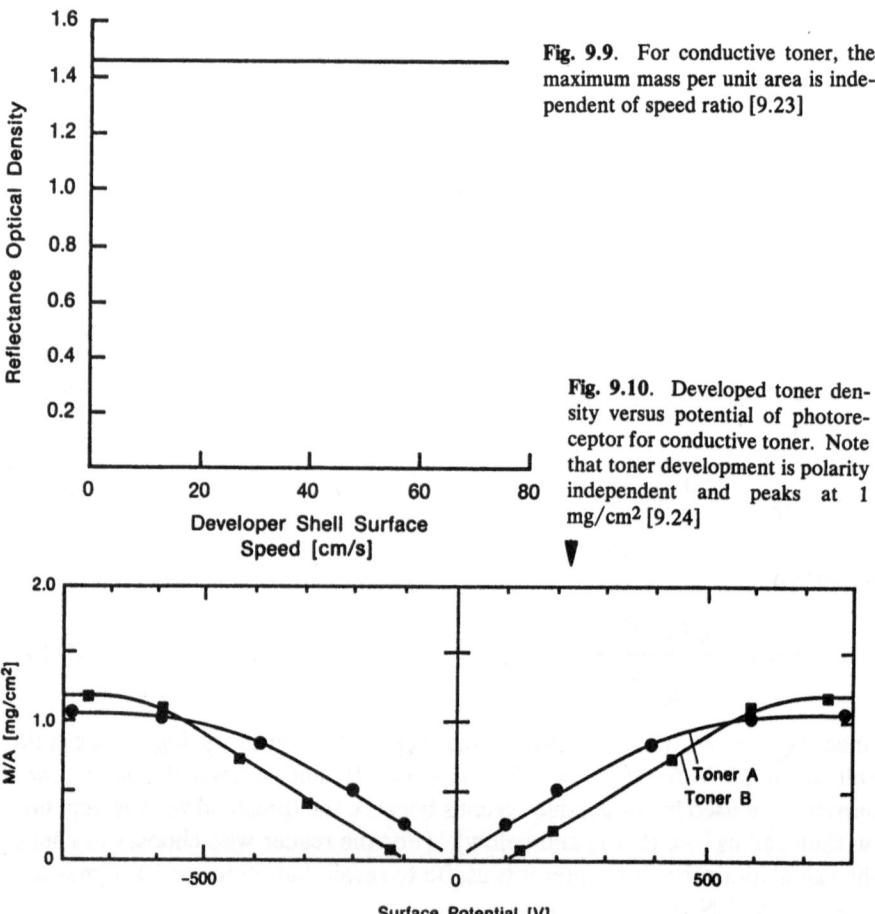

Fig. 9.9. For conductive toner, the maximum mass per unit area is independent of speed ratio [9.23]

Fig. 9.10. Developed toner density versus potential of photoreceptor for conductive toner. Note that toner development is polarity independent and peaks at 1 mg/cm² [9.24]

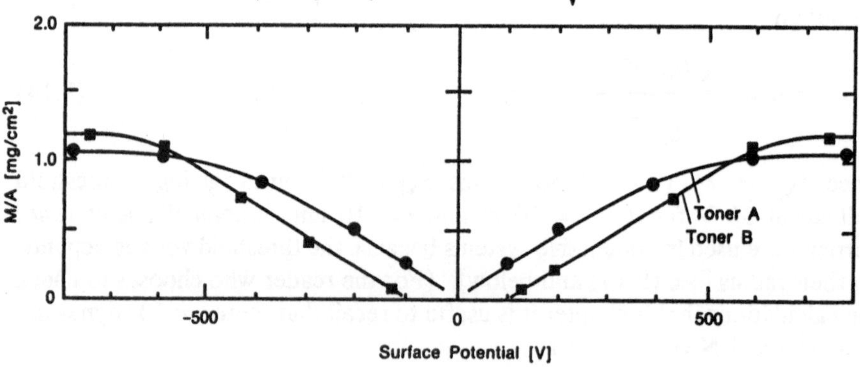

amount of magnetite in each particle and the size distribution allowed. Clearly, small particles will be a potential source of background toner (9.14).

The question of conductivity control is an interesting one. *Field* [9.26] states that the toner should charge in a few microseconds, requiring conductivities of $10^7\ \Omega$ cm. Such short charging times compared to development times (milliseconds) limit development to one monolayer. One could increase development by making the charging time approximately equal to the time the toner is in the development nip, so that some charge remains on toner above the monolayer adjacent to the photoreceptor. However, due to variation in conductivity down each chain, uniformity and reproducibility would probably suffer.

Another prediction for inductively charged toner is that development should be independent of polarity. The full development curve for both polarities was published by *Shimada* et al. [9.24,25]. Their data are shown in Fig. 9.10. Note M/A saturates at about 1 mg/cm^2 in agreement with (9.15), the developed characteristics are independent of the polarity of the surface potential, and a small threshold voltage is observed for toner B.

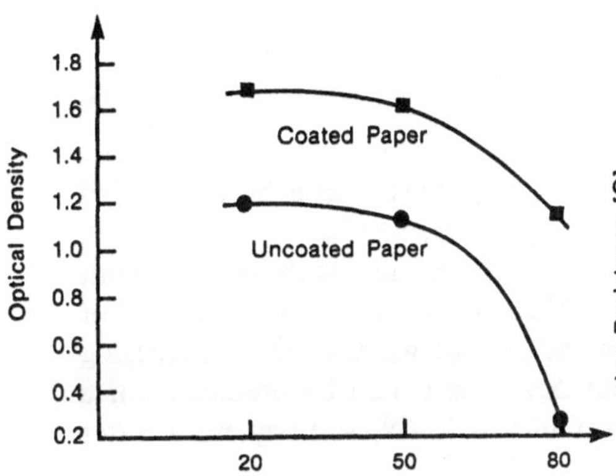

Fig. 9.11. Line density with conductive magnetic toners (volume resistivity: $10^8\ \Omega$ cm) versus relative humidity [9.27]

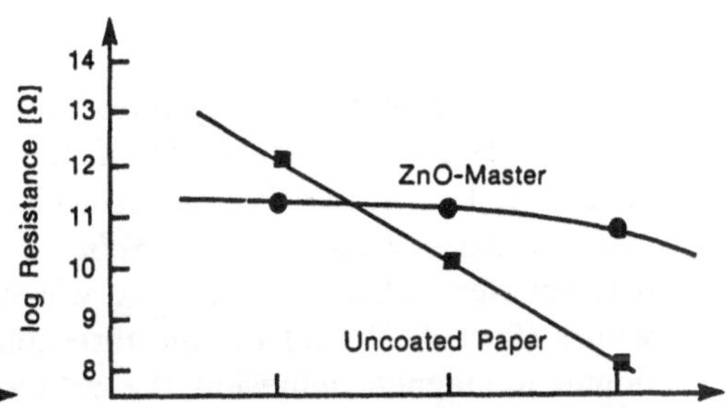

Fig. 9.12. Surface resistance of various papers (at 1 kV) versus relative humidity [9.27]

The two problems with conductive toner are now well known [9.27]. First, as discussed above, only a monolayer of toner can be developed on the photoreceptor for sufficiently conductive toner. Second, and perhaps more seriously, under high relative humidity conditions, it is found that the optical density decreases (Fig. 9.11). This occurs because under high relative humidity conditions paper becomes conductive (Fig. 9.12) and the toner charge, but not the toner, is transferred to paper at the transfer station.

Conductive monocomponent development is currently being used by two companies, Delphax and Océ. Delphax makes printers based on ionography, as discussed in Sect. 1.4.2. They transfer and "fix" the toner in one step, a high pressure "transfix" station. The toner is mechanically transferred and pressed into the paper, eliminating the need for electrostatic transfer. Océ uses a double transfer system in their copiers in which the toner is transferred to a warm intermediate belt before being transferred to paper. A thermal transfer is used in place of an electrostatic transfer.

9.5 Magnetic, Insulative Toner

For this system, we assume uniform charging and doctoring of a few monolayers of the toner on a roller has been accomplished prior to the toner entering the development nip region (Sect. 8.3). Since Q_t is fixed, we can directly use (9.5) and (9.6):

$$V_{th} = \left(\frac{d_s}{K_s} + L + \frac{d_t}{K_t}\right)\left(\frac{1}{4\pi\varepsilon_0}\frac{Q_t}{4(r+\delta_r)^2} + \frac{F_M}{Q_t}\right), \quad (9.16)$$

$$\left.\frac{M}{A}\right|_{max} = \left.\frac{M}{A}\right|_{roller} \nu. \quad (9.6)$$

The first commercial implementation of this process was achieved by 3M in 1977 [9.23]. Discussions of the charging mechanism were given in Sect. 8.2. Unfortunately, few results have been published on the development characteristics of this system. *Nelson* [9.23] shows M/A increases with ν, but it is not clear whether $M/A|_{max}$ was measured, or whether Q_t is increasing with ν (Sect. 8.2), shifting the threshold down, as would be predicted for a dominant magnetic adhesion, the last term in (9.16). (Nelson argues the latter is occurring.)

Published data from the impression development work [9.16] verify the general features of (9.16) and (9.6). Copy reflection density versus contrast voltage has a threshold (of about 25 V) and M/A saturates above 150 V.

By far the most successful implementation of the magnetic, insulating toner, monocomponent development system is practiced by Canon [9.19]. This system is used in their whole line of copiers from the 6 cpm personal copier all the way up to a 70 cpm copier (Table 1.1). A schematic is shown in Fig. 8.10. As discussed in Sect. 8.3, charging and doctoring are done by frictional contact with the roller and with a magnetic doctor blade.

In the development zone the photoreceptor is spaced 300 μm from the roller surface so the toner has to "jump" across an air gap to develop onto the photoreceptor. The speed ratio is set at 1. A development curve is shown in Fig. 9.13. It has the now familiar voltage threshold and saturated optical density. The copy density saturation, about 1.2 OD, corresponds to (from

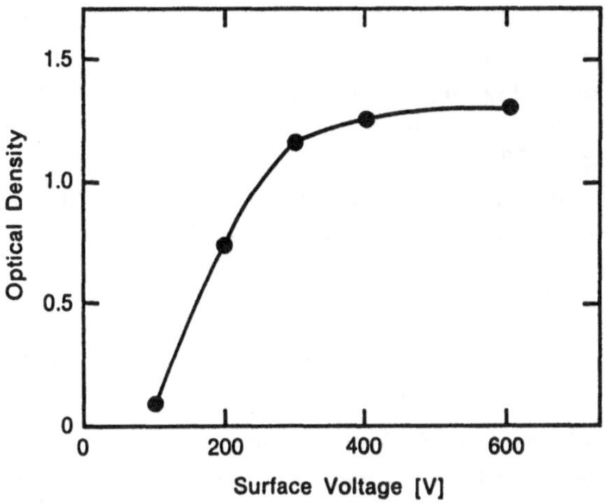

Fig. 9.13. Developed solid area density as a function of surface voltage without ac bias voltage for the Canon magnetic insulating monocomponent development system [9.19]

[Ref. 9.19; Fig. 7]) 0.5 mg/cm^2, very close to what would be expected for one monolayer of 5 μm radius toner, see (9.6), a speed ratio of 1, and $p_t = 0.6$:

$$\left.\frac{M}{A}\right|_{max} = \left.\frac{M}{A}\right|_{roller} \nu \rightarrow \frac{4}{3} r \rho_t p_t = 0.4\ \mathrm{mg/cm^2} \quad \text{(monolayer)} \tag{9.17}$$

It is indicated in [9.19] that in the development zone the toner layer thickness is 100 μm, which must represent the height of the magnetic brushes. The toner particles at the top of the bristles are very far from the roller surface and the primary attractive force between toner and roller is the magnetic adhesion. The threshold voltage is then given by, see (9.16),

$$V_{th} = \left(\frac{d_s}{K_s} + L + \frac{d_t}{K_t}\right)\left(\frac{F_M}{Q_t}\right). \tag{9.18}$$

For $F_M = 0.34 \times 10^{-4}$ dynes, $Q_t = 1.5 \times 10^{-15}$ C, and $L + d_s/K_s + d_t/K_t = 300\ \mu$m as given in [9.19], (9.18) predicts $V_{th} = 70$ V, very close to the observation.

In the Canon paper the voltage width (V_w in Fig. 9.5) is estimated by assuming that all of the toner particles have one charge and one radius. Further, it is assumed that the last toner particle to be developed, a toner particle immediately adjacent to the roller, experiences an image force due to the image charges of all the toner originally on the roller, i.e., the whole layer is developed at once. In this case the additional electrostatic adhesion of a toner in the first monolayer adjacent to the roller is proportional to Q_t times the charge per unit area σ_t of the rest of the toner, or

$$F_{es} = Q_t \frac{\sigma_t}{K_t \varepsilon_0} = Q_t \rho_{tv} d_t / K_t \varepsilon_0, \tag{9.19}$$

which is the second term in [Ref. 9.19; Eq. (11)]. For $K_t = 1$ (because toner packing is so low when $d_t = 100\,\mu$m), $Q_t = 1.5 \times 10^{-15}$ C and $\rho_{tv} = 4 \times 10^{-2}$ C/m^3 this is about 2 times F_M, suggesting an additional voltage of 2 times V_{th}, or 140 V, is required to develop all the toner. This corresponds closely to the observation (Fig. 9.13).

Another approach to this calculation is to assume the last toner particle experiences only the electrostatic adhesion due to its own image charge. This adds

$$\frac{1}{4\pi\varepsilon_0}\frac{Q_t^2}{4r^2} \tag{9.20}$$

to the adhesion force, which equals 0.2×10^{-4} dynes (for $r = 5\,\mu$m), requiring an additional 40 V to completely develop the toner. The experiments indicate approximately 200 V extra is needed, possibly indicative of the existence of higher charge or smaller radius toner or larger magnetic forces near the roller surface, all reasonable possibilities.

In addition to the dc bias, Canon uses ac biases to assist in development. Data are shown in Fig. 9.14 for $V_{ac} = 800$ and 1400 V and two frequencies, 500 and 1000 Hz. The lower scale is the total dc potential across the gap, $V_0 - V_{bias}$. It can be seen that the addition of V_{ac} has substantially lowered the threshold (from 100 V, Fig. 9.13, to zero for $V_{ac} = 800$ and -100 V for $V_{ac} = 1400$ V at 500 Hz), as predicted qualitatively in Sect. 9.3.

Frequencies too high for the toner to follow obviously will be less effective. The higher threshold at 1000 Hz compared to 500 Hz probably reflects this since, as *Takahashi* et al. [9.19] note, the toner transient time T across the gap L is

$$T = \sqrt{\frac{2M_t L^2}{Q_t V}}\,, \tag{9.21}$$

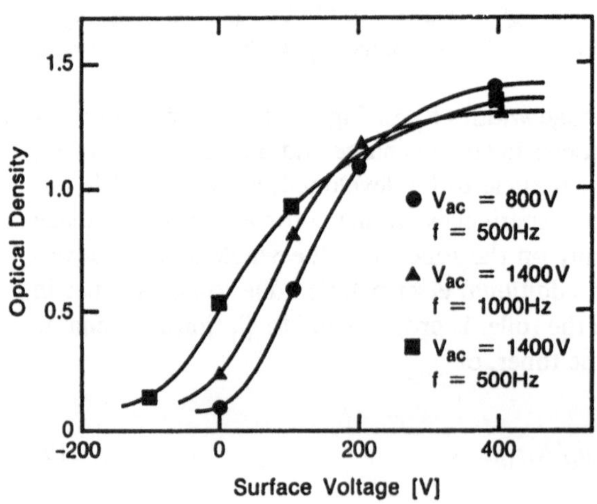

Fig. 9.14. Developed solid area density as a function of surface voltage for different ac bias conditions (V_{ac} is the peak-to-peak voltage) for the Canon magnetic insulating monocomponent development system The voltage axis has been recalculated including the bias potential of 200 V [9.19]

which equals 0.21 ms for $L = 300\,\mu m$, $Q_t = 1.5 \times 10^{-15}$ C, $V = 1400$ V and $M_t = 5.4 \times 10^{-10}$ g.

No data characterizing line development, line/solid ratios or background development have been published on this system.

9.6 Nonmagnetic, Insulative Toner

In 1985 at the IEEE-IAS Conference in Toronto both Ricoh [9.21,28,29] and Toshiba [9.20,30] announced a *nonmagnetic* insulating toner monocomponent development system. In the Ricoh system the roller with toner contacts the photoreceptor. In the Toshiba system the toner must jump a gap. Charging in the Ricoh system was described in Sect. 8.3. No development data for this system have been published by Ricoh. This system is commercially available in the colored toner cartridges for the Ricoh RePRO jr. copier (8 cpm) and in their PC Laser 6000 printer (6 ppm).

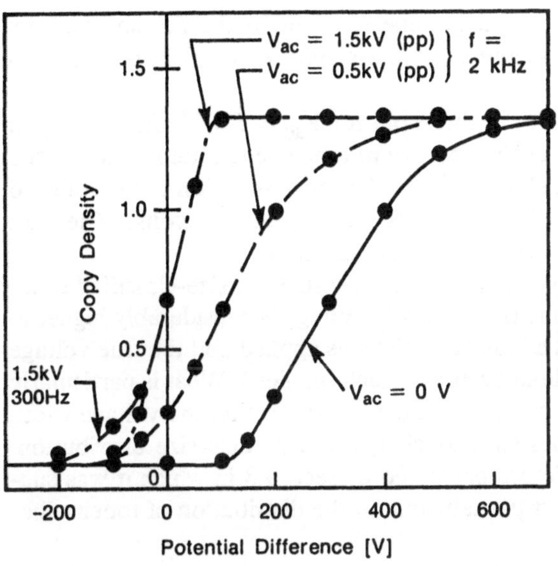

Fig. 9.15. Developed solid area density as a function of potential difference between the roller and the photoconductor for various ac bias conditions for the Toshiba nonmagnetic insulating toner monocomponent development system [9.20] (© 1985 IEEE)

The Toshiba system is shown in Fig. 8.15. The hardware variables which control toner charging were discussed in Sect. 8.3. The development characteristics are shown in Fig. 9.15. With $V_{ac} = 0$, the now familiar voltage threshold, rise, and saturation are observed. With no magnetic force, the threshold is now entirely due to the Coulomb force required to overcome the electrostatic adhesion, i.e., from (9.5),

$$V_{th} = \left(\frac{d_s}{K_s} + L + \frac{d_t}{K_t} \right) \frac{1}{4\pi\varepsilon_0} \frac{Q_t}{4r^2} \approx \frac{L}{4\pi\varepsilon_0} \frac{Q_t}{4r^2}, \tag{9.22}$$

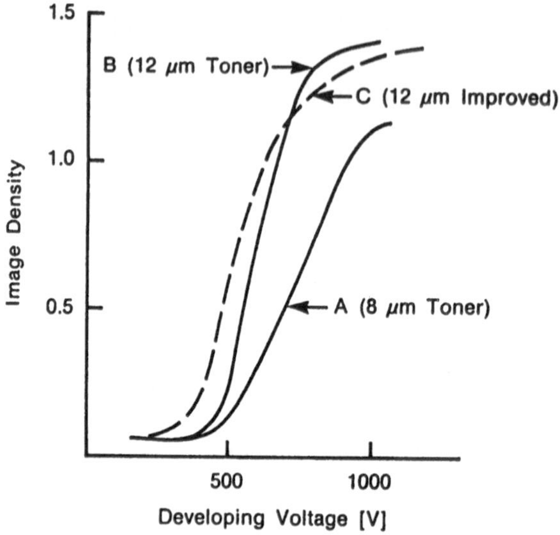

Fig. 9.16. Development curve for the Toshiba nonmagnetic insulating toner monocomponent development system for different size toner particles [9.30]

which is predicted to be 66 V for the parameters given by the Toshiba workers ($L = 0.2$ mm, $Q/M = 7\ \mu$C/g, and we will assume $r = 5\ \mu$m) in good agreement with the observations.

The effect of V_{ac} is to shift the threshold for $V_{ac} = 500$ V and to steepen the curve for $V_{ac} = 1500$ V. In the absence of a magnetic retaining force, the application of a large ac field should reduce the toner adhesion to the roller to zero, see (9.22), suggesting all toner should develop at zero volts. The data taken with $V_{ac} = 1500$ V appear to approach this limit.

Figure 9.16 shows the development characteristics for size-classified toner for the Toshiba system. (While the threshold voltage is considerably higher in this figure, it is probable that a bias potential was applied and the true voltage difference at which the threshold occurs is much smaller.) What is particularly interesting about this set of data is that it demonstrates that the voltage width can vary with toner properties such as charge and particle-size distribution. This is consistent with the picture presented in Sect. 9.3 in which it was suggested that the voltage width is partially due to the distribution of toner adhesion.

Finally, we show data [9.31] taken on an experimental setup partially based upon components from a commercially available magnetic monocomponent development system which demonstrate and compare the effects of magnetic and nonmagnetic toner and the application of ac voltages. The development system was mounted so that M/A of the developed toner could be measured. This was done (1) in the standard configuration (magnetic monocomponent development), (2) with the magnet opposite the development zone removed, simulating nonmagnetic monocomponent development, and (3) with and without V_{ac}. The data, shown in Fig. 9.17, demonstrate several interesting effects. The threshold clearly shifts to lower values with the magnets removed.

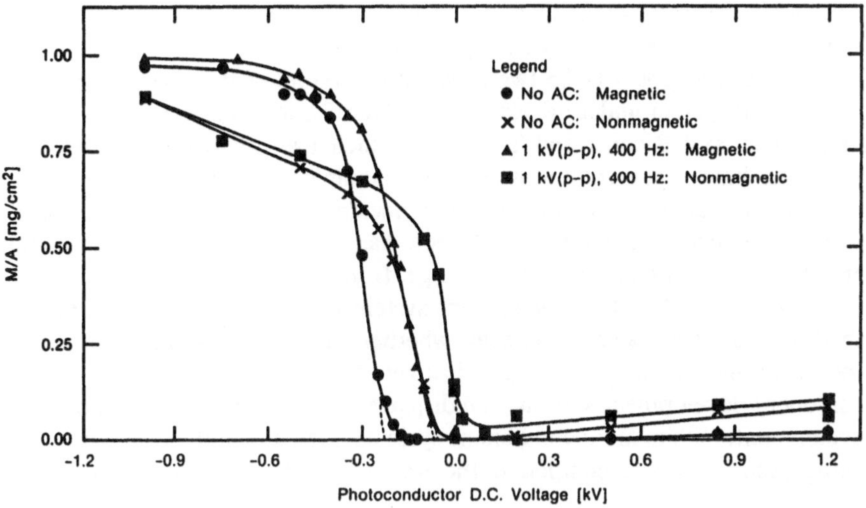

Fig. 9.17. Data for M/A taken on an experimental monocomponent development system. The effect of nonmagnetic toner is simulated by removing the magnets from the development zone. Also shown are data with and without the ac field [9.31]

Also the application of an ac voltage lowers the threshold. Both effects are predicted by the model given in Sect. 9.3. Unexpected is the effect of removing the magnets on the maximum toner development. It decreases the maximum development, at least over the range of voltages measured. One could imagine that the magnets may play a role in breaking the toner-roller adhesive bonds. As the toner moves past the poles, reforming chains, its contact with the roller surface is momentarily broken, decreasing adhesive forces. The magnets also play an important role in preventing the development of wrong sign toner in reverse bias. Note M/A increases with positive bias when the magnets are removed. Measurements of toner charge (not shown) indicate this is due to wrong sign toner.

9.7 Summary

It has always been clear to electrophotographers that the monocomponent development system is simpler than two component, yet successful implementation was not achieved until 1980 with Canon's magnetic insulative development system. The solution to the problem of charging and transporting toner in a monocomponent development system slowly evolved, with many ideas preceding the Canon work. Today this area of electrophotography represents one of intense activity, as can be judged by recent papers and patents. It is an area like many in electrophotography in which success is achieved only with a truly interdisciplinary approach. The increased contact charging sensitivity of the toner, brought about by the introduction of charge control agents, was clearly a crucial element in producing a viable system.

Today, there appear to be two successful variants of the monocomponent development system under development. Canon's, which uses magnetic insulating toner, and Ricoh's and Toshiba's, which use nonmagnetic insulating toner. All have made compromises. For example, the Canon toner is more complicated than two component toner because it requires the addition of magnetic material. Nonetheless, Canon has successfully used this system commercially in an enormous speed range, from 6 to 70 cpm (Table 1.1). So far Ricoh has produced a development system that works only at a very low speed, 8 cpm. It remains to be seen whether the system is viable at higher speeds. Perhaps most interesting, it has yet to be established whether monocomponent development systems can displace two component magnetic brush development. While monocomponent systems have a manufacturing cost advantage, they must also compete in the areas of reliability and copy quality.

10. Liquid Development

The most prevalent method of liquid development uses the phenomenon of electrophoresis. In electrophoretic development systems, charged particles, suspended in a nonconductive dielectric liquid, move in response to the electric fields of the latent image. Other liquid development techniques besides electrophoresis are described in Sect. 2.3.4, but none have been commercially implemented.

Electrophoresis has been used as an electrodeposition technique since the early part of this century. Its application to electrophotography was suggested independently by *Metcalfe* [10.1,2] and *Mayer* and co-workers [10.3]. In 1955, Metcalfe described liquid development of selenium plates exposed with x-rays. Xeroradiography, a medical x-ray diagnostic tool, was identified early as a potential application of electrophotography (Fig. 9.5). Metcalfe argued that liquid development combined the best features of the other two known development systems: the high contrast, sharp line images of cascade development (Chap. 5) and the excellent halftone images of electroded powder cloud development (Sect. 9.1). In 1957, *Straughan* and *Mayer* [10.3] used a liquid development system in the first electrophotographic printer based on selenium plates addressed by a cathode ray tube. This work was followed up [10.4] at Battelle in 1958 (for preparing transparencies) and RCA in 1960 (for making five color maps on Electrofax paper). One of the first copiers to use liquid development was introduced by SCM corporation. It made 10 copies per minute on zinc-oxide-binder paper. The most commercially successful copiers using a liquid development system were those manufactured by Ricoh (and sold in the U.S. as Savin copiers) that resulted from an international collaboration between Metcalfe's group in Australia, Ricoh of Japan, Kalle of Germany, and Nashua and Hunt Chemical of the U.S. (An interesting account of this collaboration can be found in [10.5]). Ricoh was successful (Sect. 1.1) because they produced a slow but small, inexpensive and reliable copier, a market ignored by Xerox during the 1970s. Liquid development has also been used to develop images created by ionography (Sect. 1.4.2). An example is the Versatec configuration in which the latent image is formed on dielectric paper by applying a voltage sufficient to cause air breakdown to the styli of a multi-stylus array. This printing process has been given a special name (electrography or direct electrophotography) because light and the photoreceptor are eliminated from the electrophotographic process. As mentioned in

Sect. 2.3.4 and shown in Fig. 3.11, liquid development has the capacity to produce high resolution images. This and its color capability has led Eastman Kodak to introduce recently a pre-offset electrophotographic color proofing system based on liquid development [10.6,7].

The phenomenon of electrophoresis is actually one of four electrokinetic effects associated with colloidal suspensions that are well known to physical chemists [10.4,8]. They are all based on the electrical double layer discovered by Helmholtz that exists at a solid-liquid boundary: one sign of charge resides on the surface of the solid and the opposite charge is distributed within the liquid. Since movement of the solid particles relative to the liquid can separate the charge on the particles' surface from the counter charge in the liquid, electrokinetic effects can occur. If mechanical force is applied, one of two effects occurs, depending on which phase is stationary. If the liquid is forced through a porous plug or capillary tube, a potential gradient known as the streaming potential is set up at the ends of the tube. If the solid particles are allowed to fall down a tall column of liquid, a potential gradient called a sedimentation potential develops down the tube. If an electrical force is applied, two other effects occur. If the liquid moves and the solid phase is stationary, the effect is called electro-osmosis. If the solid phase moves, the effect is called electrophoresis, the one of interest to us.

10.1 Material Requirements

For electrophoretic liquid development to occur (1) the toner particles suspended in the liquid must be charged and (2) the liquid must be nonconducting and must not chemically interact with the photoreceptor.

10.1.1 Toner Charging

The problem of charging toner particles in liquid developers is just as complicated and empirical as in dry toner systems (Chaps. 4 and 8). That particles in liquid quite often acquire charge was first noted by Helmholtz, as mentioned above. He characterized the charging behavior by the zeta potential, which is approximately the potential difference between the charge on the particle and the countercharge in the liquid.

Predicting and controlling this charge is the difficult problem [10.4]. Some material properties obviously drive the charge exchange. "Contact potential differences" between the liquid and solid phase have been invoked. Coehn and Metcalfe (see discussion in [10.4]) have suggested that the higher dielectric constant material becomes positive. Early it was found that surface modification could affect the charge. For example, the polarity of carbon black particles is sensitive to small variations in the mode of preparation of the solid particle, its heating history, gases to which it was exposed and mode of dispersion in the liquid. In fact the charge of most materials is found to depend upon the environment of the particle at the time of formation and the method of preparation. Surface sensitivity led to the use of "surface active agents" or "control agents" such as films of alkyd resin and linseed oil (suggested by Metcalfe). *Stotz* [10.9] suggested the use of surfactants, e.g., metallic salts of organic acids with sufficiently long aliphatic chains. By predominant absorption of one ionic species, the particles receive a net charge. The amount of charge can be regulated simply by changing the additive concentration until the absorption becomes saturated. Polarity can be determined by the choice of surfactant. For example, a suspension of carbon black in liquid isoparaffins becomes negatively charged by calcium petroleum sulfonate and positively charged by calcium disopropylsalicylate. *Halfdanarson* and *Hauffe* [10.10] published a study of the effect of varying charge control concentration on developer properties and copy quality. *Novotny* [10.11] characterized a commercial toner with radius 0.1 μm and charge of 35 positive unit charges. He states "The origin of this high charge is of interest. The charge control agent in this colloid is a metal stearate which adsorbs and dissociates at the solid-liquid interface. The metallic ion stays preferentially on the particle surface and the organic moiety" goes into the liquid. "...other mechanisms of charging such as charge transfer complexes and acid-base interactions involving charge control agents, polymer and solvent could also be operative in non-aqueous media". Such discussions are similar to the ideas considered for dry toner systems (Sect. 4.4.3).

Charge stability with time is also an important consideration. It has been shown [10.4] that impurities which enhance conductivity, such as water, can cause charge exchange, particle chain formation, irregular particle movements and complete reversal of the toner charge leading to regular transits between the two electrodes!

10.1.2 Liquid Properties

The most important property of the liquid medium is high volume resistivity ($>10^{10}\ \Omega$ cm) so that it will not destroy the latent image. There are many liquids that satisfy this resistivity requirement, such as gasoline, kerosene, turpentine, benzene, carbon tetrachloride, and cyclohexane [10.4]. Nonflammable liquid developers have been based on fluorinated hydrocarbons such as Freon and silicone. If restrictions on flammability, toxicity, volatility, and resistivity are applied, one is limited to fluorinated hydrocarbons, silicones and to those aliphatic hydrocarbons with high flash points (one commercially available is called Isopar). Isopar is used in the Ricoh (Savin) copiers.

Another restriction is that the liquid does not destroy the photoreceptor. The aliphatic hydrocarbons appear to interact with a-Se [10.4] and organic photoreceptors with polycarbonate-based transport layers. These photoreceptors could be covered with an overcoat. In the Ricoh copiers, zinc-oxide-binder paper is used and chemical interactions are too slow to affect the copying process.

Liquid recovery is another requirement, especially in today's world. There are threshold limit value (TLV) concentrations allowed in the office environment and Environmental Protection Agency (EPA) permitted total emissions per year per site for organic solvents. The difficulty in meeting these requirements can be illustrated with two calculations. For example, assume a total allowed emission of 1000 kg per year per site. For a high speed, high volume application where 6 million copies are made each month, the total solvent carryout per page can only be 10–15 mg, to be compared with typical values of 100 mg/page. For a low speed application, the TLV limit is difficult to achieve. For Isopar, the TLV is 300 ppm. If the amount of Isopar vapor emitted from each page is 100 mg and the copier operates continuously at 10 page/min in a 40 m^3 room which has its air completely refreshed each hour, an equilibrium concentration of approximately 300 ppm is produced.

10.2 Development Theories

Development theories which account for first-order effects were proposed in the early 1970s (Sect. 10.2.1). Descrepancies between theory and experiment led people to an appreciation of the complexities of this process. Theories to account for some of these complexities have been proposed (Sects. 10.2.2, 10.2.3).

10.2.1 First-Order Effects

We again address the question of predicting the developed mass per unit area M/A as a function of known or measurable hardware and material parameters. A theory of development along with supporting experimental data was first

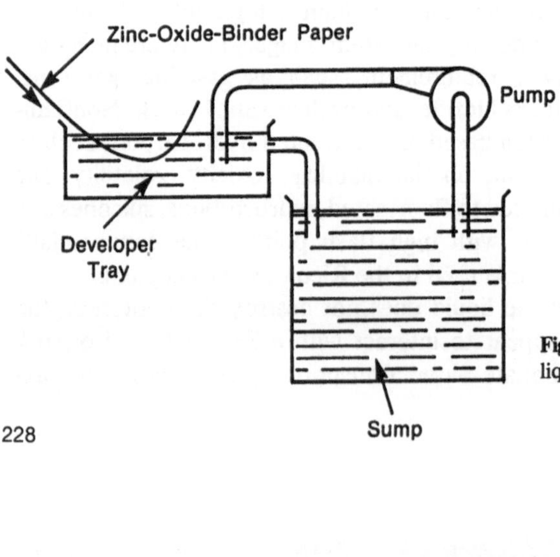

Fig. 10.1. Schematic of an immersion liquid development device [10.4]

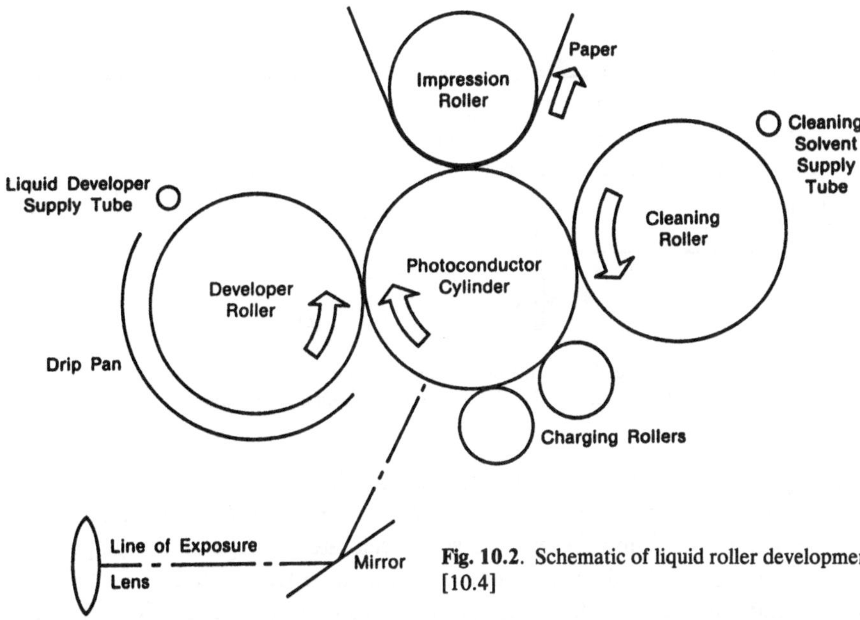

Fig. 10.2. Schematic of liquid roller development [10.4]

described in the early 1970s by *Stark* and *Menchel* [10.12]. Two development configurations are shown in Figs. 10.1 and 10.2. In Fig 10.1, immersion development is shown: the zinc-oxide-binder paper is immersed (or flooded) with liquid developer. As in open cascade development (Sect. 5.2) only fringe field development occurs and consequently solid areas are not developed. Figure 10.2 shows roller development. In this system, the liquid is applied to the surface of a roller in the form of a thin film before contacting the photoreceptor. Spacing a conductive roller 125–500 μm from the photoreceptor capacitively couples electric field lines into the liquid, producing solid area development.

Much of the early experimental work was done with zinc-oxide-binder paper. While that represented experiments on "real" systems, it introduced several severe problems associated with the electric field. It was found that the corona charge does not remain on the surface but migrates (and is trapped) in the bulk [10.13-17]. The solvent can migrate into the bulk of the paper, changing its dielectric constant [10.13,14,18], and current is observed to flow through pinholes [10.13,18]. To minimize such effects, which complicate specification of the electric field, Stark and Menchel did their liquid development experiments onto Mylar with a voltage V applied between the back of the Mylar and the roller (see below). The theory, proposed by *Stark* and *Menchel* [10.12], which was similar to one proposed by *Kurita* [10.19], was constructed to illustrate the first-order physics occurring during solid area development while carefully pointing out the simplifications. Experimental data, taken on Mylar, were given, which provided a test of the theory.

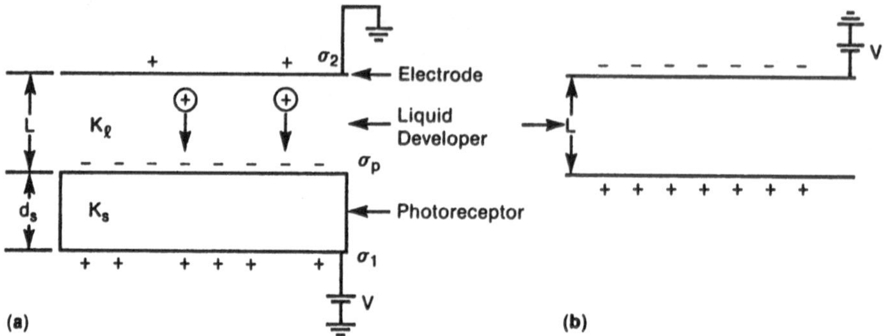

Fig. 10.3. Schematic sketches illustrating the (**a**) development configuration and (**b**) electrophoretic cell configuration for the plate-out charge-to-mass measurement

Stark and Menchel assumed that the latent image is neutralized by charged toner which moves in response to the electric field of the latent image. The electric field in the liquid E_ℓ due to a solid area latent image of charge per unit area σ_p can be obtained by techniques identical to those used in Sect. 6.2.1. To maintain charge neutrality, charge densities σ_1 and σ_2 must flow onto the photoreceptor ground plane and the electrode (Fig. 10.3a) where

$$\sigma_1 + \sigma_2 = \sigma_p \ . \tag{10.1}$$

Since the total voltage drop between the two ground planes must be zero,

$$\frac{\sigma_1 d_s}{\varepsilon_0 K_s} - \frac{\sigma_2 L}{\varepsilon_0 K_\ell} = 0 \ , \tag{10.2}$$

where d_s is the photoreceptor thickness, K_s is the photoreceptor dielectric constant, L is the gap between the photoreceptor and the electrode and K_ℓ is the liquid dielectric constant. Solving for the electric field in the liquid E_ℓ, using (10.1 and 2), gives

$$E_\ell = \frac{\sigma_2}{K_\ell \varepsilon_0} \approx \frac{d_s}{L K_s \varepsilon_0} \sigma_p \ , \tag{10.3}$$

where it has been assumed that $L/K_\ell \gg d_s/K_s$. Taking into account (Sect. 6.2.1) the (first-order) decrease in the surface charge density on the photoreceptor caused by the development of toner with charge density σ_t gives

$$E_\ell = \frac{d_s}{L K_s \varepsilon_0} (\sigma_p - \sigma_t) \ . \tag{10.4}$$

This electric field causes the toner to move. The current density j due to the motion of the toner charge density σ_t is proportional to the conductivity of the liquid σ and the electric field in the liquid:

$$j = \frac{d\sigma_t}{dt} = \sigma E_\ell \ , \tag{10.5}$$

giving a differential equation for σ_t

$$\frac{d\sigma_t}{dt} = \frac{\sigma d_s}{LK_s\varepsilon_0}(\sigma_p - \sigma_t) \tag{10.6}$$

with the well-known solution

$$\sigma_t = \sigma_p [1 - \exp\ (-t/\tau_\ell)] \ , \tag{10.7}$$

where

$$\tau_\ell = \frac{\varepsilon_0}{\sigma} \frac{LK_s}{d_s} \ . \tag{10.8}$$

Equation (10.7) can be written in more familiar terms by recalling (6.5) that σ_p can be related to the electrostatic voltage V above the photoreceptor:

$$\sigma_p = \frac{V\varepsilon_0}{d_s/K_s} \ . \tag{10.9}$$

This voltage V can also represent (6.8) the voltage placed across the ground planes with zero charge on the photoreceptor (or the photoreceptor replaced with an uncharged insulator such as Mylar). In addition

$$\sigma_t = \frac{M}{A}\frac{Q}{M} \ , \tag{10.10}$$

where M/A is the developed toner mass per unit area and Q/M is the average charge-to-mass ratio of the developed toner (identical to Q_t/M_t since for this simple theory a single value for the toner mass M_t and charge Q_t are assumed). Using (10.9) and (10.10) in (10.7) gives

$$\frac{M}{A} = \frac{V\varepsilon_0}{(Q/M)(d_s/K_s)}[1 - \exp(-t/\tau_\ell)] \ , \tag{10.11}$$

That is, M/A approaches the neutralization condition (Sect. 6.3.1) exponentially in time with a time constant related to the geometry of the development system (LK_s/d_s) and the conductivity of the developer (σ). Stark and Menchel assume the conductivity of the liquid is due entirely to the toner particles (ignoring the presence of counter- and excess ions), giving

$$\sigma_s = N_\ell Q_t \mu_t \ , \tag{10.12}$$

where N_ℓ is the number of toners per unit volume, Q_t is the charge of the particles, and μ_t is their mobility.

Several experiments were carried out to check (10.11). It was shown that at constant time t, $M/A \propto V$, as predicted. By integrating the current measured in an external circuit, Stark and Menchel showed that this charge equaled (within 10%) the toner charge deposited on the Mylar (although it is unclear why the counterions did not contribute to the current). This allowed measurement of external currents to characterize the development process. Differentiating (10.7) with respect to time gives

$$j = \frac{d\sigma_t}{dt} = \frac{\sigma_p}{\tau_\ell} \ \exp\ (-t/\tau_\ell) \tag{10.13}$$

$$= \frac{V\sigma}{L} \ \exp\ (-t/\tau_\ell) \ . \tag{10.14}$$

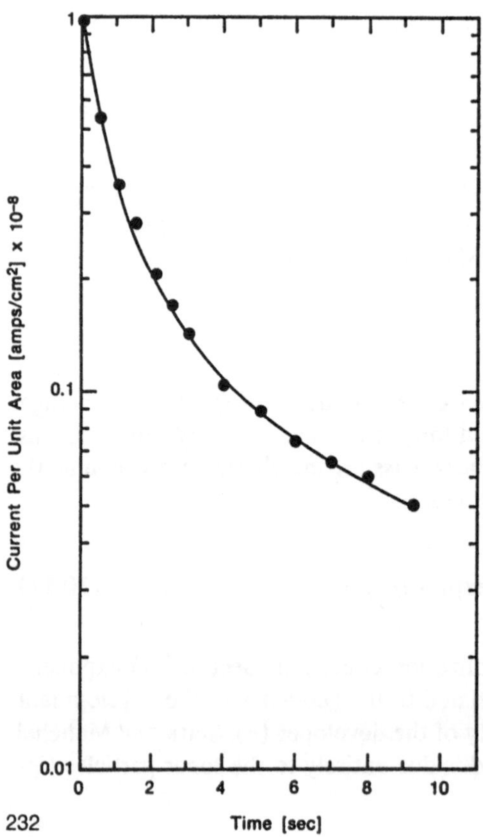

Fig. 10.4. Decay curve showing current density versus time during development onto a Mylar layer [10.12]

Experimental data, shown in Fig. 10.4, do not indicate a simple exponential dependence of current on time, but a time constant that gets smaller as time increases. It was pointed out that this could be due to several effects. The conductivity of the liquid developer was observed to have some field dependence. The effect of the countercharge in the liquid has been ignored; *Kurita* [10.19] showed the correction to τ_ℓ taking into account ionic conductivity σ_i is

$$\tau_\ell = \frac{\varepsilon_0}{(\sigma + \sigma_\mathrm{i})} \frac{LK_\mathrm{s}}{d_\mathrm{s}} \,. \tag{10.15}$$

Finally toner depletion just above the latent image will slow down development. This produces a time-dependent conductivity and is suggested (by Stark and Menchel) to be the primary cause of the nonexponential dependence seen in Fig. 10.4.

The initial slopes of curves such as Fig. 10.4 are plotted as a function of L/d_s in Fig. 10.5. The slope gives a conductivity of 3.3×10^{-12} $(\Omega\,\mathrm{cm})^{-1}$ close to the observed values $(2.7-4) \times 10^{-12}$ $(\Omega\,\mathrm{cm})^{-1}$. However, the intercept is not predicted by the theory. Also no attempt was made to relate the observed conductivity to the toner properties.

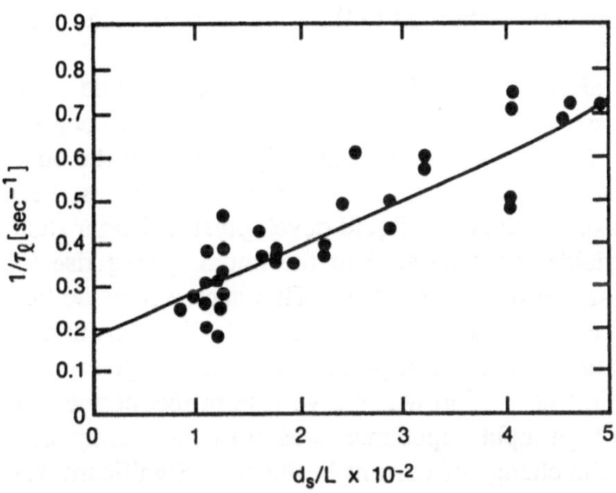

Fig. 10.5. Initial slope of decay curves such as shown in Fig. 10.4 versus d_s/L [10.12]

Mohn [10.20], in the following year (1971), generalized these equations to take into account toner depletion and conductivity of the liquid due to the counterion. In our notation, he allowed for a time dependence of N_ℓ, the number of toner particles per unit volume of liquid, which makes σ in (10.6) time dependent and produces another differential equation. With just this simple modification, a solution is not possible in closed form, but implicit and approximate solutions were given. Experimental current-time curves were semiquantitatively described by their equations; the predicted time constant was smaller (about 20%) than observed. The difficulty was attributed to inhomogeneous initial particle distribution and turbulence. Nonetheless, Mohn showed that the changing time constant, first reported by Stark and Menchel, most likely has its origins in the changing number of toner particles per unit volume.

Stark and Menchel mention turbulence in the fluid during the first 1–2 s of development. This caused difficulty in estimating the initial currents. It is indicative of other phenomena which were ignored in their simple theory and which were more thoroughly studied later by others. Also ignored in the theory is the space charge of the remaining countercharges and excess ionic charge in the liquid. This would tend to slow down development by reducing the electric field. Later work was directed at more fully exploring these and other aspects of liquid development.

10.2.2 Complexities

A more complete theory of liquid development must contend with a wide range of physical and chemical complications. These include considerations of space charge effects, macroscopic fluid flow including turbulence, toner depletion effects, the complicated electric field associated with lines, particle interactions, and the effect of other ions in the solvent. Some of these complexities are reduced when lateral liquid flow occurs, as in the roller development configurations shown in Fig. 10.2.

Clear demonstrations of space charge effects and macroscopic fluid flow have been given by *Stechemesser* [10.21] and others [10.22,23]. Space charge effects can occur when the charge per unit area of moving charges in the bulk of the liquid approaches the charge per unit area on the electrodes, i.e., $K_\ell \varepsilon_0 E_\ell$. This is in fact the usual situation in liquid development. Under such circumstances, nonuniform fields are generated in the liquid, giving rise to nonuniform toner velocities and development rates. That such effects do occur can be demonstrated by measuring the potential across the liquid from one electrode to the other. Deviations from linearity indicate space charge effects. Such deviations can be seen in Fig. 10.6 in which a voltage probe, connected to an electrometer with very high input impedance, was drawn across the cell in a time short compared to the change of current in the cell. Significant departures from linearity are observed at both electrodes.

That significant macroscopic fluid flow occurs during liquid development was also graphically demonstrated by *Stechemesser* [10.21] by using transparent electrodes. In liquid development systems, the black toner particles arrange themselves to give a directly visible pattern. The patterns observed were dependent on time, voltage and the conductivity of the mix (Fig. 10.7). The honeycomb structures arise from electrohydrodynamic effects similar to macroscopic flow patterns generated by temperature gradients. At high enough voltages even turbulence was observed as an irregular fog. At the edge of a voltage step, a whirl was observed, its speed of growth dependent on ap-

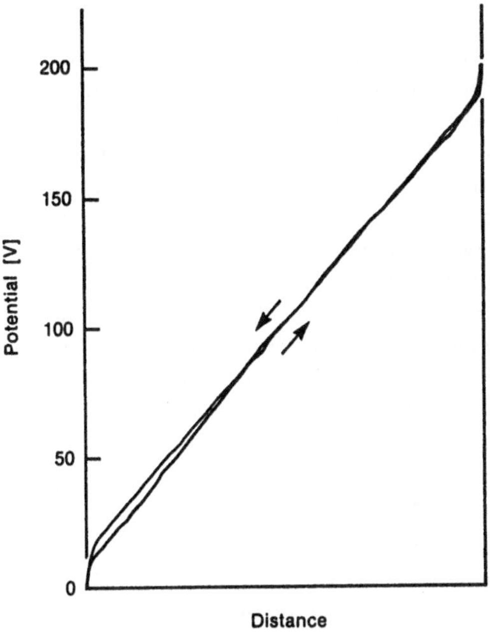

Fig. 10.6. Potential versus distance in a liquid developer. Recording was started 5 s after switching on the voltage. The probe was moved from the grounded electrode to the negative electrode (separated by 3 cm) during 16 s, and then retraced. At voltages above 300 V, vigorous streaming on the surface close to the negative electrode could be observed [10.21]

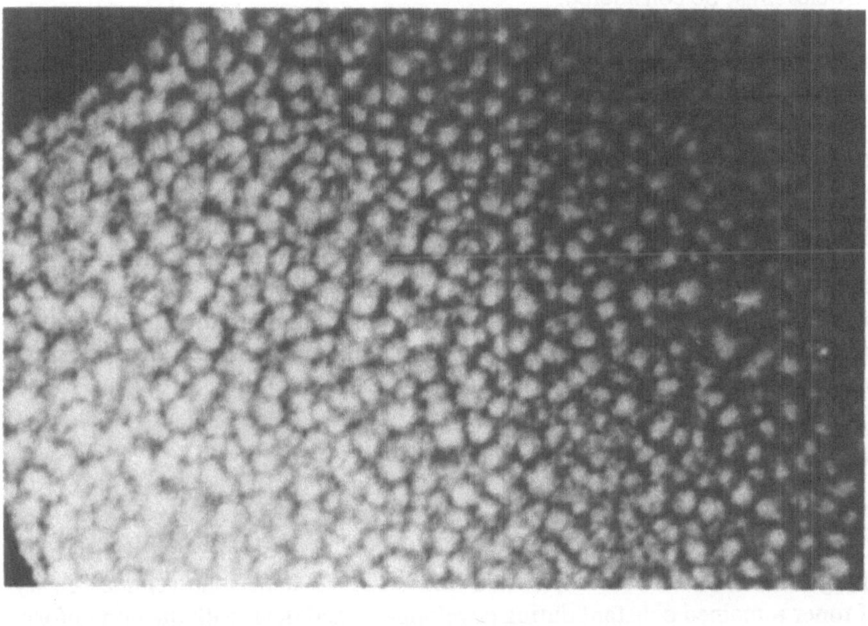

Fig. 10.7. A photograph of toner being developed between transparent electrodes 1 mm apart. The diagonal of the picture is 2.5 cm in the original. The toner is deposited on the negative electrode, giving rise to a directly visible pattern. The size of the honeycomb patterns is approximately equal to the electrode separation. Above 500 V, turbulence occurs giving rise to an irregular fog. Similar patterns are observed when one electrode is covered with ZnO coated paper [10.21]

plied voltage. It was argued that these macroscopic flow patterns are important for understanding heavy development near the boundaries of charged areas: the circulating liquid flow, caused by the strong electric field near the edge, brings toner from a large region into the immediate vicinity of the edge.

Stark and *Menchel* [10.12] suggested and *Mohn* [10.20] demonstrated that time-dependent time constants had their origin in toner depletion effects. This occurred despite the fact that a screen electrode was used. It was expected that once a particle entered the development region through the screen, it would move away from the screen with a velocity given by its mobility times the electric field (ignoring space charge effects). However, Stark and Menchel showed, by solving the diffusion equation, that the diffusional flow of fresh toner into the development zone is significantly slower than the flow of toner due to development. It may be that the use of roller development (Fig. 10.2), which continually brings in fresh toner, decreases these effects.

The electric field associated with line latent images has been shown to be quite complicated, decreasing exponentially into the fluid, dependent on the line width and photoreceptor thickness (Sect. 3.2).

Interactions between particles have been considered by *Reed* and *Morrison* [10.24] They showed that for separation distances of less than three particle diameters, double-layer interactions and applied field distortions between particles must be considered.

Other ionizable species besides toner can significantly complicate characterization techniques [10.25]. If the electrodes are not covered by blocking layers, electrochemical processes such as charge injection and even charge reversal can occur at the electrodes and contribute to background conductivity. In addition, dissociation of neutrals into charged species and recombination of ionized species can take place in regions where the charge concentration has been altered from the equilibrium value. Dissociation and recombination can be field dependent. Uncharged particles in nonuniform electric fields with dielectric constant different from that of the liquid will move in field gradients by dielectrophoretic motion. In addition, small amounts of dissolved water significantly affect the conductivity of the liquid and hence current measurements [10.26].

10.2.3 Better Development Theories

The first group to attempt a more complete theory of liquid development was that of *Junginer* and *Strunk* [10.27–29]. They ignored toner depletion, space charge effects and macroscopic fluid flow. They assumed the volume density of toner remained constant during development and dealt with the nonuniform electric fields associated with the latent images of lines and steps. Their procedure was to solve for the electric field pattern, given a charge latent image, at zero time. Some toner and counterions are attracted to the latent image changing, the charge distribution on the photoreceptor. The electric field is then recalculated as additional toner development occurs. An important result is that higher frequency components develop faster. This causes a change with time of the charge profile. A numerical example of the development of a charge step is given along with high speed motion pictures of the deposition of toner. Toner depletion effects are observed to complicate the experiments.

Numerical solutions which incorporate the effect of the space charge effects, toner depletion, and nonuniform surface charge were presented by *Crofoot* and *Cheng* [10.30]. Ignored were macroscopic fluid flow and spatial dependences in the space charge (which could be caused by spatial variations in the surface charge). This model allowed clarification of the range of validity of the previous treatment and a better estimate of development rates. It is shown that space charge effects are large and, even in the limit of small space charge, the equations do not reduce to those of Junginger and Strunk. Nonetheless, Junginer and Strunk's analysis may be valid if there is sufficient turbulence in the liquid so that local charge neutrality is maintained. This model also predicts a slower rate of development (Fig. 10.8 shows the decrease of the charge on the photoreceptor with time for two frequencies for this theory and for Junginger and Strunk's) for the obvious reason that toner depletion reduces the supply of toner to the photoreceptor surface.

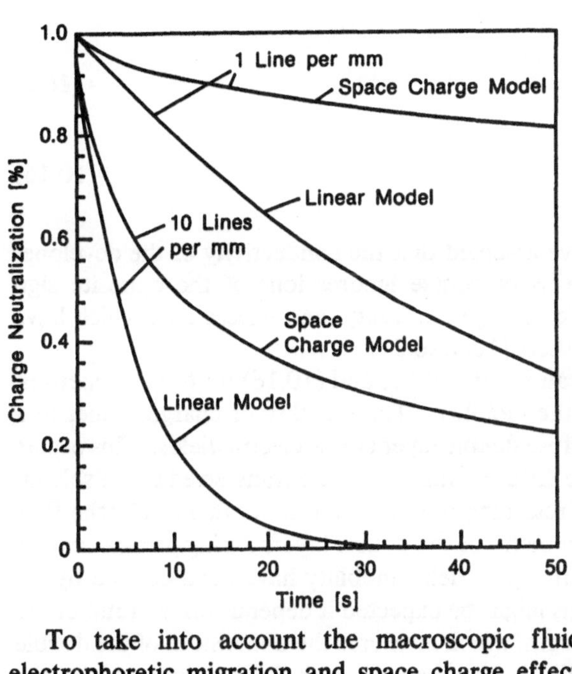

Fig. 10.8. Development of different spatial frequencies as predicted by the linear model of Junginger et al. and the space charge model of Crofoot et al. [10.30]

To take into account the macroscopic fluid motion as well as the electrophoretic migration and space charge effects requires solution of the laws of hydrodynamics (Navier-Stokes and continuity equations for an incompressible fluid), electrostatics (Poisson equation), and charge transport (continuity equation for the charges). For unipolar conduction, i.e., charge injected at one electrode, such a comprehensive model has been developed and solved by the Grenoble group [10.31,32]. Obviously, the resulting equations are complicated and nonlinear, and such results are only the first step towards understanding the liquid development problem in which charge carriers of both signs exist in the bulk of the fluid.

10.3 Toner Characteristics

10.3.1 Optimized Properties

It is clear from the above discussion that toner properties partially determine the characteristics of the development system. Identifying optimum properties is clearly information that toner formulation people would find useful. We follow the first-order theory of *Stark* and *Menchel* [10.12] and discussions by *Stotz* [10.33] and *Kohler* et al. [10.34].

In Sect. 10.2.1 it was shown that the developed mass per unit area is

$$\frac{M}{A} = \frac{V\varepsilon_0}{(Q/M)(d_s/K_s)}[1 - \exp(-t/\tau_\ell)] \ , \qquad \text{where} \tag{10.11}$$

$$\tau_\ell = \frac{\varepsilon_0 L K_s}{\sigma d_s} \ , \tag{10.8}$$

$$= \frac{\varepsilon_0 L}{N_\ell Q_t \mu_t} \frac{K_s}{d_s} \ , \tag{10.16}$$

using (10.12), where we have assumed that the conductivity of the developer is due entirely to toner. This of course ignores ions of the opposite sign (10.15), which must exist for charge neutrality, and excess ions which have been reported by several authors [10.9,26,34].

To understand the implications of (10.11) and (10.16) for toner properties, we need an expression for the mobility. The mobility of charged toner in a solvent is affected by the diffuse double layer at low electric fields. This occurs because forces on ions in the diffuse double layer are transferred to the solvent molecules, which exert two retarding forces on the toner due to (1) the flow of the liquid and (2) the asymmetry of the double layers during particle motion [10.9,35]. Expressions for the (low field) mobility have been derived by several authors [10.9,35,36]; as might be expected it depends on the ratio of the toner radius to the Debye length, which measures the dimensions of the double layer. This low field mobility is less than the high field mobility where the double layer is stripped off and the mobility is simply obtained by equating the Stokes and Coulomb forces,

$$Q_t E_\ell = 6\pi r \eta v \ . \tag{10.17}$$

Since by definition

$$\mu_t = \frac{v}{E_\ell}, \tag{10.18}$$

$$\mu_t = \frac{Q_t}{6\pi r \eta}. \tag{10.19}$$

Stotz [10.9] has shown that the low field mobility can be 30 times smaller than the high field mobility given by (10.19). The field at which the stripping of the double layer occurs (observed to be between 10^5 and 10^6 V/cm) depends primarily on the developer conductivity; it occurs if, during the electrical relaxation time of the liquid (dielectric constant times resistivity), the particle travels a distance which exceeds the double-layer thickness.

Using the high field expression for the mobility (10.19) gives for τ_ℓ (10.16)

$$\tau_\ell = 6\pi\varepsilon_0\eta \frac{r}{N_\ell Q_t^2} \frac{LK_s}{d_s}, \tag{10.20}$$

where we have separately displayed the dependence of τ_ℓ on liquid properties (η), toner properties ($r/N_\ell Q_t^2$) and geometry (LK_s/d_s). Equations (10.11) and (10.20) indicate tradeoffs exist in this development system: Lower toner charge gives higher M/A but longer development times. As pointed out by *Kohler* et al. [10.34], if (10.11) is solved for time t and dt/dQ is set equal to zero, one can show that for a desired optical density or M/A on paper, there is an optimal value of Q_t at which the development time is a minimum. In principle, the development time can be reduced by increasing the number of toners per unit volume, N_ℓ. But according to *Dahlquist* and *Brodie* [10.37] this leads to higher background development. This is the same problem encountered in two component development, and may have a similar cause: higher toner concentration leads to more wrong-sign toner. (This reference, by the way, is the only discussion of background development in liquid development literature known to the author.)

The stability of colloids has been considered by *Stotz* [10.33]. He applied more general theories to low dielectric constant, nonpolar media. The concept is that the total potential energy V_p of two particles should exceed $10kT$ where k is Boltzmann's constant and T is absolute temperature. Stotz shows that for solvents used in liquid developers, the potential energy V_p is given to a good approximation by the electrostatic repulsion of two spheres minus the van der Waals force of attraction:

$$V_p = \frac{4\pi K_\ell \varepsilon_0 r^2 \psi_0^2}{x} - \frac{A}{6}\left[\frac{2}{(x/r)^2 - 4} + \frac{2}{(x/r)^2} + \ln\left(\frac{(x/r)^2 - 4}{(x/r)^2}\right)\right], \tag{10.21}$$

where x is the center-to-center distance, $A = 10^{-19}$ J and ψ_0 is the potential at the surface of a particle (which is the zeta potential to a first approximation):

$$\psi_0 = \frac{Q_t}{4\pi\varepsilon_0 K_\ell r} . \tag{10.22}$$

The potential energy V_p has a maximum as a function of x. (x is determined by the particle concentration.) Setting the maximum equal to $10kT$ (assuming $10kT$ is a sufficient barrier in the presence of thermal energy to keep the particles apart) one obtains the minimum charge (or charge-to-mass ratio) required to keep the colloid stable, which depends on ψ_0 and the particle radius. For example from [Ref. 10.33; Fig. 3], at $\psi_0 = 70$ mV, 0.1 μm particles require Q/M's of at least 200 μC/g to be stable; 1 μm particles require Q/M's of 2 μC/g. Smaller particles, which give higher resolution, require higher Q/M's for stability and therefore develop less M/A, see (10.11).

Application of the more sophisticated development theories to the question of toner property optimization will more finely tune the above conclusions. Nonetheless, the above discussion clearly indicates that knowledge and control of toner properties is important for making a successful liquid development system.

10.3.2 Determination of Toner Properties

Dahlquist and *Brodie* [10.37] were among the first to suggest techniques for characterizing toner properties. They determined the radius and charge on the toner particles by several complementary methods for commercially available toners manufactured by Hunt Chemical Corp. The toner radius was determined by electron microscopy to be between 130 and 250 Å with the vast majority close to the average of 215 Å. The average charge-to-mass ratio was determined by placing the colloidal suspension between two large area closely spaced electrodes and applying an electric field for a sufficient time to plate out all the toner (Fig. 10.3b). This was later called the plate-out technique. The charge and mass accumulated on the plate was measured. Assuming $r =$ 190 Å gives the number of charges per particle to be 1. The charge was obtained by monitoring the current as a function of time (Fig. 10.9). Two components are observed, a slow and a fast component. The fast component was attributed to higher mobility counterions present to assure charge neutrality. The slow component was attributed to the toner. The time extrapolated from the second region for the current to reach zero is the transit time τ_t of the toner as it traverses the cell of width L, giving the particle drift velocity v_t and mobility μ_t

$$v_t = \mu_t E_\ell = \frac{L}{\tau_t} . \tag{10.23}$$

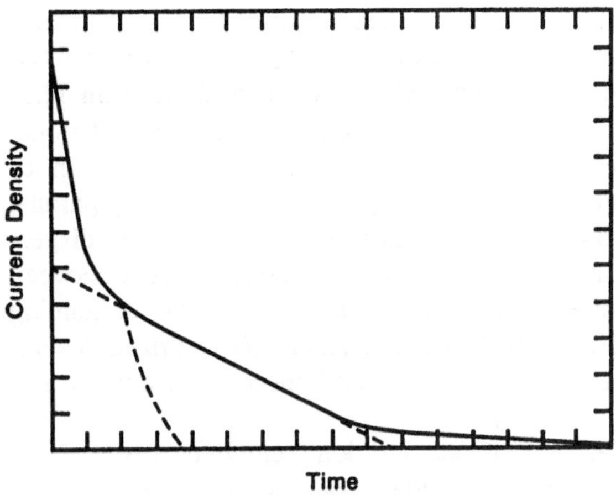

Fig. 10.9. Current transient versus time showing the "fast" and "slow" decay of the current. The slow component due to toner is extended to $t = 0$ and evaluated at that point. The waveform shown is for the case of 152 V applied across 190 μm. The abscissa is 2 s full scale, the ordinate, 20 μA [10.37]

(This electronic means of measuring μ_t replaced earlier photographic observations of particle displacement under pulsed electric field conditions [10.38]). The initial current density

$$j_0 = N_\ell Q_t \mu_t E_\ell = \sigma E_\ell \ , \tag{10.24}$$

where N_ℓ is the number of toner particles per unit volume, Q_t is the charge per particle, and σ is the conductivity due to the toner. It was shown that j_0 is linear in E_ℓ and N_ℓ, as predicted. Using the high field expression for the mobility (10.19) gives

$$\sigma = \frac{N_\ell Q_t^2}{6\pi r \eta} \ . \tag{10.25}$$

Measurements of the conductivity also gave 1 electron per particle and $r = 204$ Å. An increase in conductivity was noted at electric fields above 5×10^4 V/m. This was later shown by *Stotz* [10.9] to be due to the stripping of the ionic double layer (as mentioned above).

Mohn [10.20] suggested similar characterization techniques. He pointed out that, in the measurement of the toner charge-to-mass ratio, it was necessary to have an electric field E sufficient to plate out all the particles, i.e.,

$$E_\ell > \frac{N_\ell}{\varepsilon_0 (K_\ell / L + K_s / d_s)} \ . \tag{10.26}$$

Similar characterization techniques and results are reported by *Stotz* [10.33], *Kondo* and *Yamada* [10.39], *Van Engeland* et al. [10.40] and *Kohler* et al. [10.34]. *Stotz* [10.9], *Novotny* [10.41], and *Kohler* et al. [10.34] suggested techniques for separating the toner current from current due to other like-charged ions in the liquid. *Stotz* [10.9] suggested measuring Q/M as a function of the number of toner particles per unit volume. At high concentrations the correct value is obtained; at low concentrations, contributions from like-charged ions make the observed Q/M larger than Q/M of the toner. *Novotny* [10.41] suggested a modification of this idea: allowing sedimentation to produce a variable toner density and probing this with a cell with many parallel electrodes. *Kohler* et al. [10.34] suggested centrifuging the developer dispersion to remove the toner and then measuring the difference in current between twin electrophoresis cells, one containing developer and one containing toner-free developer. Using such techniques, Kohler characterized several liquid developers. One showed aging properties and one was sensitive to relative humidity.

A review of optical and electrical characterization techniques was given by *Novotny* [10.25] in 1981. Adding to the electrical transient and electrical plate-out measurements, he discusses quasi-elastic light scattering and optical transient measurements. It is suggested that electron microscopy is unreliable because agglomerates in microscopy samples may be quite different from those in the dispersion. He suggests in situ light scattering is more reliable. Nonsphericity and polydispersity complicate the interpretation of these measurements, but an average spherical diameter can be obtained. Values of Q/M are obtained by electrical plate-out measurements. The optical transient, associated with the sweepout of scatterers from the dispersion, obtained by collecting light from a laser which is shown perpendicular to the electrodes, can be analyzed to give the particle mobility when below space-charge-limited conditions. Furthermore, just after the application of the electric field, evidence for particle orientation in the optical transient (of nonspherical particles) can be observed in the optical transients. The particle mobility can also be obtained from the current transient (such as Fig. 10.9); it also decreases to zero as the charged particles are swept out of the cell and plated on the electrode. Measurements of the current density at initial time gives the initial conductivity. Knowing the ionic concentration and the particle charge (Q/M from plate out, r from light scattering) gives another electrical determination of the particle mobility. A liquid developer characterized by Novotny had $2r = 0.19\,\mu$m, $Q = 56 \times 10^{-19}$ C (or 35 charges), $\mu = 0.4 \times 10^{-8}$ $m^2V^{-1}s^{-1}$ at concentration of 0.48 g/l or $42 \times 10^{18}\,cm^{-3}$.

Even with complete characterization of toner properties, it remains essential to test the liquid developer in actual hardware. This "final" test can in fact be implemented into hardware as a screening tool [10.42]. Such hardware should require relatively small volumes of toner, an important consideration when toner design is largely empirical in nature and requires many iterative steps on experimental samples.

10.4 Recent Developments

The reduction of solvent carryout on the paper was mentioned as an important goal in Sect. 10.1.2. One way to reduce the amount of solvent carried out on the paper is to prevent direct contact of the developer liquid and the paper surface. Landa has suggested several techniques to minimize this contact. One [10.43] is to spray 50 μm diameter balls on the photoreceptor after liquid development and before electrostatic transfer. The spacer balls should be separated by about 3–4 mm to be effective. Another method [10.44] is to add the spacer balls to the liquid developer. If two types of spacer balls are added which have both signs of charge, some will go to image areas and some will go to nonimage areas. However, the spacer balls with charge opposite to that of toner will attract toner, putting black dots in nonimage areas. When they become covered with toner they will settle out of the liquid. Also, negative and positive spacer balls may coagulate. An improvement on this idea [10.45] is to use spacer balls with the same charge sign as the toner. These balls (20–70 μm in diameter) must have a dielectric constant greater than the liquid so that dielectrophoretic forces can be used to keep them on the photoreceptor in the background region despite the imposition of an electrostatic force drawing them away from the photoreceptor.

Another approach for dealing with the solvent carryout is to try to eliminate the solvent above the paper in the form of vapors. Vapors of Isopar can in principle be combusted to water and CO_2. One suggestion is to use a heated chamber at 200°C and catalyze the fumes using types of alumina [10.46]. Such treatment is claimed to reduce the concentration of Isopar in a confined space from 200 to 25 ppm. In another patent [10.47], the heat released during the catalytic oxidation is used to dry and fix the image on paper. Such a "fixer" allows the design of a high-speed liquid-developer copier (30 ppm) with fast start-up time, low power requirements, and with no need for a hooded exhaust system to handle vapor emissions.

Recently, *Landa* [10.48] described a new liquid developer which he calls ElectroInk. He points out that a limitation of liquid inks is that the toner is severely squashed under the pressures of electrostatic transfer, leading to edge roughness. This is solved by devising ink particles which have morphological structure resembling that of "jacks", i.e., particles which have tentacles or extended spines which enable them to interdigitate or interlock when in close proximity to one another. These particles resist deformation or squashing under pressure (during the transfer step) and yet redisperse upon the application of shear (necessary so that untransferred ink can be redispersed and returned to the development station). *Levy* et al. [10.49] and *Niv* et al. [10.50] measured the mobility and development characteristics of these ElectroInk particles.

10.5 Summary

Liquid development is a technically viable alternative to powder development and has several distinct advantages, including sharper images (Fig. 3.11) and the potential of high quality color. In addition, its hardware simplicity has certainly been demonstrated in the many commercially available low cost, low speed special paper copiers.

For solid area line development first-order theories exist that have been validated experimentally. A qualitative or semiquantitative understanding of other physical and chemical complexities also exists, including consideration of space charge effects, macroscopic fluid flow, toner depletion effects, and particle interactions. From these theories, one can make reasonable guesses of optimized toner properties, and toner characterization techniques of varying sophistication have evolved.

However, the technical difficulties in implementing this development system are clearly materials related. Additives are required to charge the toner. Maintaining the charge over time and keeping conductive contaminants out of the liquid developer are challenging problems that have obviously been solved in existing machines, but are not discussed in the literature. Furthermore, methods of keeping the background development low while producing good black images are required, just as in powder development. And just as in powder development, almost no information is available on the mechanisms of background development.

Perhaps the most challenging problem for liquid development involves the carryout of the liquid on the paper. As these liquids are usually organic solvents, TLV (threshold limit values) and EPA permitted total emission limits must be met. Such limits can be met with very slow speed machines (as in the Eastman Kodak pre-offset proofing system) or with occasional use of higher speed machines. Advances in this area would significantly increase the use of liquid development systems.

11. Color Electrophotography

Color electrophotographic copiers have finally arrived. Only since the end of the 1980s have these copiers become generally available within reprographics departments in large corporations and in copy stores in cities. What delayed the introduction of this technology?

Many of us remember discussing why the first color copier, the Xerox 6500 introduced in 1973, was not a more successful product. Was it because the copy quality was not photographic? Or perhaps because it was too slow, (4 pages per minute)? Was the cost per copy too high? Or maybe the problem was the lack of, or difficulty of creating, originals? I did find a use for the copier: making birth announcements for our first son. Apparently other people found similar trivial uses because the color copier was finally taken out of the Xerox Webster Research Center by the administration in one of their cost cutting moves.

The delayed introduction of color copiers and printers was not only the subject of lunch-time conversations. The various consulting firms which annually predict the growth of the copier and printer markets were clearly frustrated during the late 1970s and into the 1980s by the consistent failure of the color copier market to evolve, despite their continual optimistic predictions. With hindsight, it is now clearer how the need for solutions of many technical problems delayed the introduction of color copiers. To achieve acceptable color quality, many companies have introduced digital copying, in which the image information is first scanned in, digitized, and then printed using a laser printer. Having the information in electronic form allows manipulation of the information to produce acceptable gray scales and color corrections. Nonuniformities in virtually all of the electrophotographic subsystems had to be minimized to eliminate copy quality defects. The three- or four-colored toned images can be accumulated on the paper, on the photoreceptor, or on an intermediate surface; all three architectures have been introduced in commercially available products. The choice affects the physical size of the copier (accumulation on the photoreceptor can be the smallest), the number of times the image must be transferred, technical difficulties associated with interactions among the subsystems, and the amount of misregistration in the final image (accumulation on paper has the most). For all architectures, the physical size of the development system had to be reduced, since three (or four) systems are now needed in approximately the same space where one existed before. These needs (improved quality, image accumulation, and minimization of physical size) required the invention of many new development systems.

The source of colored originals was, and remains, a pervasive problem. Aside from a few hand-colored originals and blow-ups of 35 mm slides, many of the originals come from color printers based on a variety of technologies, as discussed below. This would suggest that there is a market for color electrophotographic printers. Since it is relatively straightforward, at least conceptually, to convert a digital color copier to a color printer, it would seem that the color electrophotographic printer market should quickly evolve. But a "chicken-and-egg" problem has existed: the market for color printers needs a means of reproducing the prints, i.e., a color copier, and the market for color copiers requires originals that color printers could provide. One would therefore expect that the color copier and color printer markets will evolve together.

However, the color printer market has some additional complications. Color printing requires significantly more data than monochrome printing, increasing the electronics costs. Further, a color electrophotographic printer, which uses subtractive colors, must convert RGB (red, green, blue) data into CMYK (cyan, magenta, yellow, and black) before printing. Maintaining color fidelity is a challenge. Standards for the data stream representation are yet to be agreed upon, making the exchange of color image files difficult [11.1]. This creates the need for a device-independent software description of color images, a concept receiving attention today [11.2,3]. The infrastructure to support color printing is still evolving, accounting for the additional delay in the emergence of color electrophotographic printing [11.1]. The proliferation of color work stations and color scanners in recent years has generated a need for color printers independent of the color copier market and has helped to build this market.

Color electrophotographic copiers and printers will be used in engineering/scientific applications, graphic arts, and home/education environments [11.1], but the largest market is expected to be in the office, in word processing, presentations, and electronic publishing. Consider the requirements for these applications. People want high quality, equivalent to magazine quality. People want the copier to produce both monochrome, i.e., black, and color images on "plain" paper. They want the copy to look like the original, a requirement not demanded of either photography or offset lithography. They want multiple copies that look alike, requiring significant consideration be given to stabilizing the electrophotographic process well beyond anything achieved for black and white. They want overhead transparencies, which is not a trivial extension because light scattering from partially fused $\approx 10\ \mu m$ diameter toner particles turns colored images into "muddy" or dark color. And, of course, people would like all this in a highly reliable, low maintenance engine at a modest price above black and white. The success in providing such low prices will obviously impact the market growth of this technology.

Another use for color copiers, which probably occurs to everyone upon first seeing this technology, is counterfeiting money. Despite the signs by every copier that copying certain documents, including money, is illegal, some people apparently cannot resist the temptation. In response to this problem the United States Bureau of Engraving and Printing convened a task force in 1987 to recommend the addition of a covert feature to paper money which cannot be copied. Their recommendation, which should appear shortly, is the addition of a security thread. This thin flat thread, which will be positioned vertically and off center on paper money, can be seen in transmission only (by looking through the bill at a light source). On it will be printed the domination. Since it will be invisible to the reflected light used in a copier, it cannot be copied.

Other technologies [11.4] provide color copying and printing capabilities today, but none match the overall potential of electrophotography. Photography and offset lithography produce color prints which are the standards by which other technologies are judged. But the "first copy out" time is slow (several hours to several days) and the cost per copy is expensive for a small number of copies. Color pen plotters and wire matrix printers can make lines, but are too slow for solid fill. Ink jet technologies can provide excellent quality, but can be expensive (continuous ink jet), and their quality can be dependent on paper characteristics. Thermal transfer technologies also provide excellent quality, but are generally slow (1 page per minute), require special smooth paper, and have high supply costs for the color ribbons.

There are two other analog color processes available today [11.1]. Mead Cycolor uses paper coated with microcapsules filled with leuco dyes and photoinitiators which polymerize upon selective exposure to light wavelengths associated with the three subtractive colors. When pressure is applied, the dyes in the microcapsules not photopolymerized are released to the paper. Printer implementation is a problem, the light sensitivity is several orders of magnitude less than that of electrophotographic photoreceptors, and the materials are expensive. The 3M dry silver process uses colored light to create an image directly on a special substrate, which is then heated to develop and fix the image. The supplies cost is inherently high and the color quality needs improvement [11.1].

This chapter begins with a technical history of color electrophotography (Sect. 11.1). The unique technical challenges that arise in electrophotography due to the color quality requirement are discussed in Sect. 11.2, emphasizing the new gray scale techniques that had to be invented for quality color electrophotography. Color science, the methods of quantifying colors, is dealt with in many other books, to which the reader is referred [11.5–7]. The technical issues and trade-offs associated with image accumulation of the colored toner images are discussed in Sect. 11.3. The new development systems which were invented in response to the color requirements are described in Sect. 11.4. Finally, the color market and currently available products are discussed in Sects. 11.5–7.

11.1 History

Color copying was an opportunity for electrophotography that was considered from its inception. *Carlson* himself experimented with colored toners [11.8]. Full color electrophotographic images using the "new" cascade development system were made from Kodachrome transparencies at Battelle in 1947 [11.9]. Support for the early continuous-tone work at Battelle during the 1950s was partially funded by the U.S. Army with a long term interest in copying color military maps [11.10].

During the 1950s the U.S. Army also funded RCA, where the Electrofax (Sect. 1.1) process was invented. Two approaches were adopted, one using the "new" magnetic brush development system [11.11] and one using liquid development [11.12]. The latter printed five-color maps at a rate of 41 cm/s. Several groups attempted to deal with the problem of the first toner layer altering the electrostatics of subsequent images by making the toner in the second and third layers photoconductive [11.11,13]. A second-generation multicolor liquid Electrofax copier was made for the U.S. Army by Harris Intertype and was mounted in a truck. It developed five-color maps in sequence from 70 mm microfilm at the rate of two thousand 57 cm × 76 cm sheets an hour [11.14]! Metcalfe's group in Australia was also experimenting with liquid development. Apparently some of these prints were "virtually indistinguishable from commercial color printing" [11.14]; this work evolved into the REMAK pre-proofing system described below. Another early use of color electrophotography, recently discussed by *Hays* [11.8], is in the area of film animation. Walt Disney Studios recognized around 1956 that electrophotography could be used to copy artists line drawings to make the color overlays, which were then colored by hand. By 1960, colored powder toners were supplied to Walt Disney by Xerox for inking the colored objects, a process still used today: the recent film "The Little Mermaid" lists credits for electrophotographic staff members.

The availability of colored toners at Xerox provided the impetus for Xerox to study color copying. This work led to the introduction in 1973 of the first automatic full color analog copier, the Xerox 6500, which sold for $39K. The strategy that was adopted in designing the Xerox 6500 was to change an existing engine (the famous Xerox 914 copier) as little as possible to minimize development and manufacturing costs, in view of the uncertain color market. The amusing result, according to those lunch-time conversations, was that the only part common to the two machines was the casters! (Anyone who has tried to modify an electrophotographic engine will sympathize.) A schematic diagram, shown in Fig. 11.1, begins to indicate why. The cascade development had to be replaced with three development systems, one for each subtractive color. This required using a more compact magnetic brush development system. After each image was developed on the photoconductor, it was transferred to paper, which required replacing corona transfer with roll transfer.

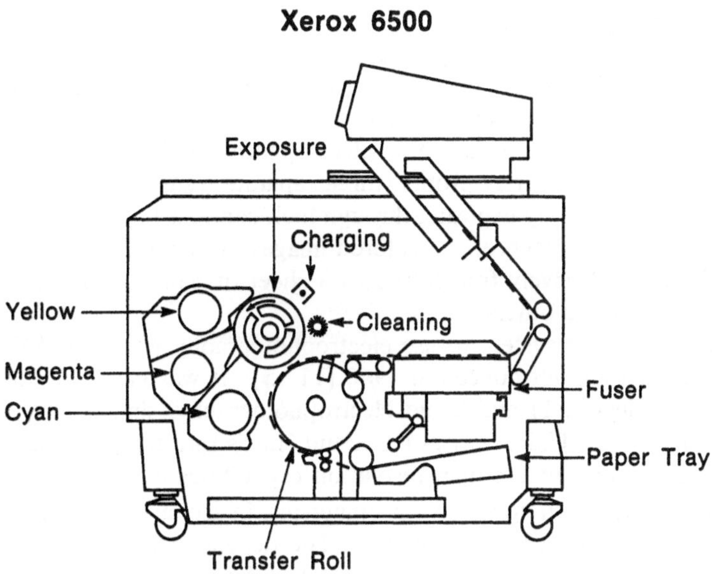

Fig. 11.1. Schematic diagram of the Xerox 6500 color copier, introduced in 1973 [11.8]

There is another amusing story about the Xerox 6500. It is well-known that an independent marketing assessment requested by Xerox in 1959 predicted that perhaps a dozen (monochrome) copiers could be sold in the United States. Clearly, this was a major marketing underestimate, given the success of the first copier, the Xerox 914. The market prediction for the Xerox 6500? You can guess: everyone will want one. Again, this turned out to be a major miscalculation, as discussed above.

Canon began work on color electrophotography in the 1970s and in 1978 introduced the Canon NP Color copier. It had the same architecture as the Xerox 6500 and was an analog process using three subtractive colors with magnetic brush development and accumulation of the toner images on paper.

Both of these copiers produced color copies which were obviously inferior to photography and offset printing and, partially for that reason, did not do well in the marketplace. The next generation of color copiers, introduced in 1987 and 1990 by Canon and Matsushita (Panasonic) respectively, were digital copiers. In this design, the light imaging lens is replaced with CCD (charge coupled device) scanners sensitive to the different colors. With the color information now in electronic form, color correction and gray scales can be greatly improved. Since then, the marketplace has greatly expanded. Color copiers can now be found in many copy stores in major cities and in reprographic centers in large corporations. A discussion of today's commercially available products is given in Sect. 11.6.

The potential for high quality using liquid development has been pursued by several companies for use in prepress proofing systems for offset lithographic presses. Considerable cost savings are possible if proofs are made to check the colors and the overall appearance before the expensive lithographic presses are involved. Non-electrophotographic "off-press" proofing systems based on overlay proofs (three separate transparent sheets, each made photographically by exposing through high contrast film, are combined) and integral proofs (each of the colored images, which are developed photographically, are adhesively transferred to a sheet of paper) have been offered by 3M, DuPont and others.

The first color proofing system to use electrophotography was REMAK, developed in Australia and introduced in 1960 [11.15]. It was based on the Electrofax process (Sect. 1.1) and liquid electrophotography, similar to the map project funded by the U.S. Army at RCA and Harris Intertype. Contact exposure was used to exploit the high resolution capabilities of the system. The development of the four toners, with immersion in Isopar-based liquid developer (Chap. 10), took about 8 minutes. The dependence of the ZnO binder photoreceptor's charge acceptance on relative humidity and variable residual potential resulted in inconsistent toner development.

In 1978 Coulter-Stork introduced the KC-color proofing system, which is also based on liquid electrophotography [11.15]. The photoreceptor is 0.35 μm thick polycrystalline CdS, which is less sensitive to temperature and relative humidity variations than a ZnO binder photoreceptor and has essentially zero residual potential. It produces a 50.8 cm $\times$ 61 cm, four-color proof in 10 min. Resolution of 1000 line pairs/mm is readily achieved. The photoreceptor is charged to only 35V and requires about 40 erg/cm^2 for discharge.

In 1985 Eastman Kodak introduced its Signature Color Proofing System [11.16,17]. Also based on liquid toner, it uses an organic photoreceptor. The four-color images are developed on the photoreceptor at the rate of three 76 cm $\times$ 100 cm proofs an hour, which are transferred to paper using heat and pressure. Exposure of each image with the halftone separation is through the ground plane of, and in contact with, the photoreceptor, eliminating problems associated with exposure through already developed toner layers.

In 1989 3M introduced the first digital prepress proofing system, Digital Matchprint, which is also based on liquid development electrophotography. The electronic information, used to make the four halftone separations, is used to write directly on the photoreceptor, simplifying the process and the equipment. It makes one 46 cm $\times$ 68 cm sheet in 16 min. The gray scale is produced with 1000 dots per cm and up to 69 picture elements per cm, giving 256 gray levels.

11.2 Image Quality

In black and white printing, image quality requirements are clear and few: the blacks should be black, and the whites should be white. Such qualities are easily quantified and measured, usually by a reflection optical density for the black ($\geq$1.4 optical density is typical nowadays) and a particle count or percent reflectance for the white. Once these goals are achieved, second order image quality requirements such as graininess of the solids and edge raggedness and sharpness of lines can be considered.

Color requirements introduce another dimension. While black can be saturated in black and white printing, producing a full range of colors requires repeated and uniform production of a continuous gray scale for each color. The difficulty of producing analog gray scales (by varying the exposure of the photoreceptor) can easily be demonstrated by copying, on any black and white copier, a piece of gray paper (Fig. 11.2). If the grayness is chosen so that the final copy is gray, nonuniformities will be obtained on the final copy on all size scales. Virtually every subsystem contributes to the blotches and streaks. Hence, for analog gray scales every subsystem must be significantly improved. Most people have decided instead to implement halftoning methods for achieving quasi-continuous gray scales. The logical evolution of halftoning and the new inventions required for electrophotography are described in Sect. 11.2.1.

Other challenges must be considered in order to achieve and maintain image quality. For example, uniformity and repeatability are significant challenges. Making copies of copies is one of the factors that enhanced the market for the original Xerox 914 copier and can be expected to be of increasing importance in the color copier market. These and other challenges are discussed in Sect. 11.2.2.

11.2.1 Gray Scale

Producing good gray scales is crucial for obtaining good color quality. But producing good gray scales in electrophotography has some unique difficulties. It was only when these difficulties were overcome by introducing digital halftoning that the market for color copiers finally developed.

Consider the most straightforward approach, analog gray scales, used in the Xerox 6500 and Canon NP Color Copiers. In this case, light reflected from a document is used to expose the photoreceptor. Among the various steps of expose, develop, and transfer, nonlinearities occur, stretching or shrinking the original gray scale. The most significant nonlinearity is in the discharge characteristics of the photoreceptor. This nonlinearity steepens the input versus output gray level curve (sometimes called the gamma of the sys-

Fig. 11.2. A copy of a uniform gray color on a typical black and white copier. The nonuniformities have the signature of almost all the electrophotographic subsystems

tem), reducing the possible number of gray levels. Most people would agree that the maximum number of gray levels possible in analog electrophotography is about 10.

A much more serious problem with analog gray scales is image defects that are accentuated in light gray print with normal electrophotography. Figure 11.2 is a copy of light gray color made on a typical black and white copier. Many types of nonuniformities are apparent, including horizontal and vertical lines and "blotchiness" on several size ranges from millimeters to several centimeters. This unacceptable analog gray "level" has the signature of most of the electrophotographic subsystems. The lines in the process direction (the direction the paper comes out) are usually due to dirt on the corona wire causing reduced charging in those areas. The lines perpendicular to the process direction may be due to nonuniform velocities or to varying gaps in the development system (as the not-perfectly-round rollers rotate). Large blotches may be due to transfer or development nonuniformities. Millimeter or smaller size blotches may be due to brush marks produced in a magnetic brush development system by bead chains. Lines approximately 5 cm apart may be due to one rotation of the development roller in a monocomponent development system. Such defects are usually covered up in black and white printing by over-developing the blacks.

Let us assume that it is desired to produce color copies equivalent to magazine quality. We will assume this requires approximately 64 gray levels per color and a picture element size of approximately 250 μm (1/40 cm or 1/100 inch). Clearly, analog gray scale approaches cannot begin to produce the requisite number of gray levels, so alternative techniques must be sought.

Analog halftoning techniques are well-known from the offset printing industry. In this technology, the gray scale is produced by printing circular dots of varying sizes. This creates the illusion of continuous tone images from binary (on-off) picture elements. The distance between the dots is called the line screen. By varying the area of the dot, grayness is achieved. It is a successful technique, as judged by the excellent image quality in almost any magazine or advertisement.

An implementation of this analog halftoning technique in electrophotography has been discussed by *Goren* [11.18] and implemented in the Eastman Kodak ColorEdge color copier [11.19]. In the ColorEdge implementation, a high-contrast screen is built into the photoreceptor structure underneath the ground plane. When the "photo" mode is chosen, the photoreceptor is exposed to a uniform diffuse light through the screen, in addition to the normal exposure from the image. This breaks up the latent image into small dots, separated by approximately 200 μm (1/133 inch), which produces fringe electric fields. The developed mass per unit area is now determined by the development of fringe electric fields of varying magnitudes. For a threshold development system (as used in offset lithography), this produces dots of varying size. Development in electrophotography is more complicated, but screening the image increases the number of gray scales that can be produced, reduces the sensitivity to development artifacts and improves the uniformity [11.19].

Another way to implement halftoning techniques is to scan in the information to be copied (using a CCD scanner) and then to print this information with a laser printer. Unfortunately, direct use of the variable dot size analog halftoning technique cannot be used in electrophotography because it is not known how to vary the spot size. Dots, of course, can be produced. In any laser printer, a laser beam scans the photoreceptors, creating dots the size of the laser beam. As the laser beam is fixed in diameter and the image data rate is very high, there are difficulties in making dots of variable size, as is done in a typical halftone picture.

With a fixed dot size, digital halftoning, also called spatial dithering [11.20], can be used to create the gray scale. In this case, the picture element is subdivided into pixels. The pixels which combine to make the picture element belong to a cell (sometimes called the threshold array or superpixel). An example of the 10 gray levels possible with a 3 × 3 cell is shown in

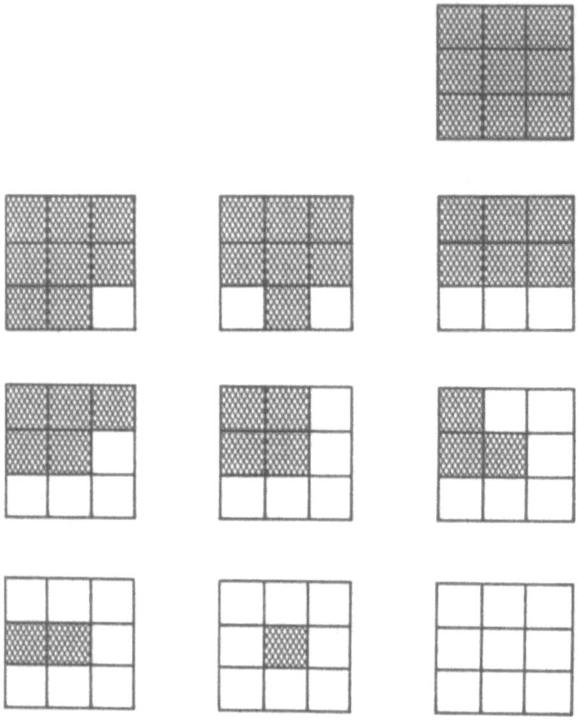

Fig. 11.3. The 10 gray levels that can be created with a 3 × 3 threshold array at 0°

Fig. 11.3. If the pixel size is 75 μm, the cell size is 225 μm, and there are 44 cells per cm, an acceptable number. But the number of gray scales is only 10, unacceptably below the desired 64. (Usually the laser beam diameter is chosen so that solid fill occurs if a diagonal line is written, suggesting setting the 50% power point at the pixel size times $\sqrt{2}$, or 106 μm in this example.)

It has been found that the human visual system is less sensitive to spatial frequencies oriented at 45°, as compared to 0 or 90°. Hence it is common practice to repeat the cell or threshold array, shown in Fig. 11.3, at a 45° angle, as shown in Fig. 11.4. This provides 9 more gray levels, which are numbered sequentially in Fig. 11.4. The threshold values of each gray level must be calibrated from the final print. If M is the number of pixels across the repeat pattern in the cell (M=3 in Fig. 11.4), and a is the size of the pixels in cm, then the number of gray levels is $2M^2+1$ and the number of lines per cm (line screen) is $(Ma\sqrt{2})^{-1}$, measured along the 45° angle. For the cell shown in Fig. 11.4, assuming $a = 8.5 \times 10^{-3}$ cm (1/300 inch), there are 19 gray levels with a line screen of 28 cm^{-1}, which is below magazine quality. Figure 11.5 shows a continuous gray scale ramp halftoned using this cell. Because the cell creates a shape in which the dots are clustered together, this is called a clus-

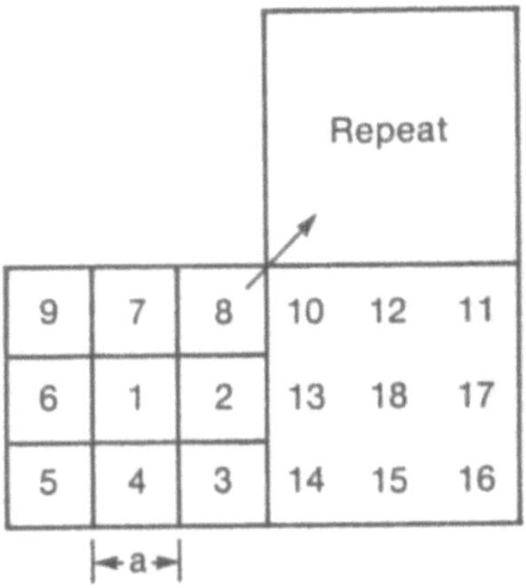

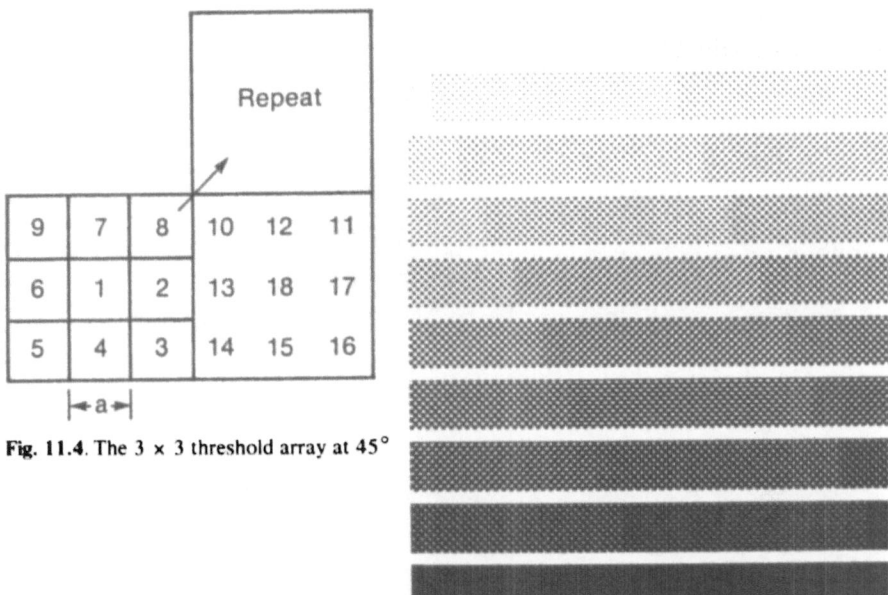

Fig. 11.4. The 3 × 3 threshold array at 45°

Fig. 11.5. A continuous gray ramp halftoned using the threshold array given in Fig. 11.4 [11.20] (Copyright © 1987 Massachusetts Institute of Technology; reprinted by permission)

tered dither. In the case in which the order of the numbers in the cell or threshold array creates a continuously growing circle, this digital halftone is sometimes called a supercircle. The threshold array does not have to be square and the threshold array pattern can be dispersed, although this is not optimum for electrophotography because of the difficulty of developing small dots (Sect. 11.2.2). Clearly, given the size of the pixel, there is a trade-off between picture element resolution (line screen) and the number of gray levels. For example, Panasonic, in their FP-C1 color copier, made the following choices in the "photo" mode. They chose the size of their pixel to be 62 μm (400 per inch), a line screen of 28 cm^{-1}, and a 4 × 4 array at 45° giving 33 gray levels. In this case, the number of gray levels and the line screen are somewhat lower than magazine quality, as defined above. A black and white blow-up of a copy is shown in Fig. 11.6b.

So why not make the laser beam smaller? That is possible, but major optical and electronic memory improvements are necessary. On a given page of 600 cm^2 area, there are approximately 16 Mbits (2 MBytes) of information at a resolution of 62 μm. Increasing the resolution by a factor of 2 increases the information required by a factor of 4, thus increasing the memory requirements and decreasing the amount of time allowed for information processing by this factor.

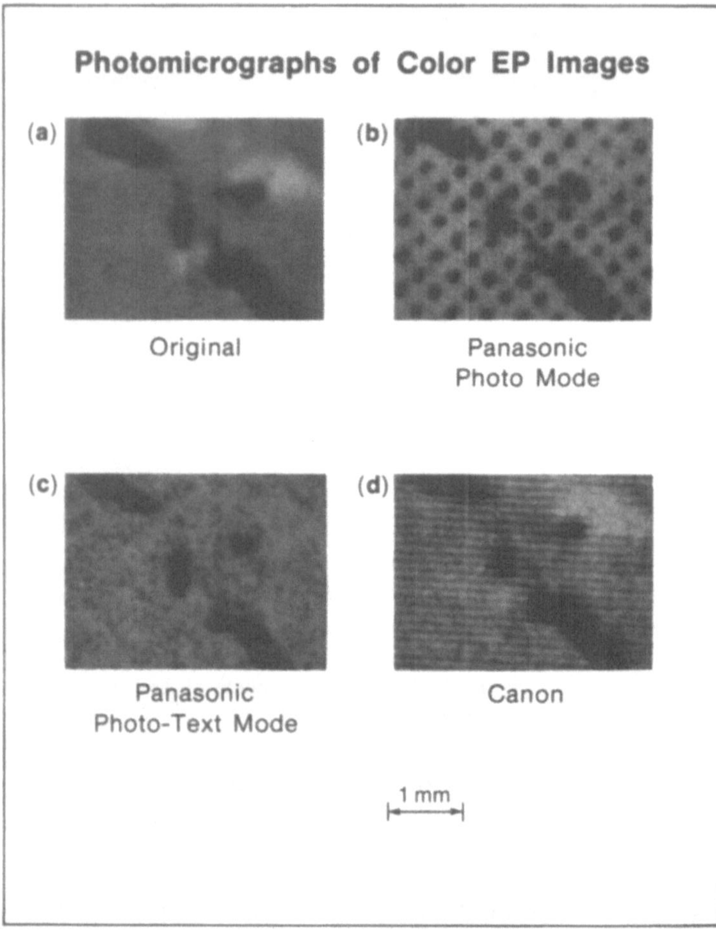

Fig. 11.6. The detail of the top of a tree against a light blue sky. **(a)** Original photograph and copies made on **(b)** the Panasonic FP-C1 color copier in the "photo" mode, **(c)** in the "photo-text" mode and **(d)** on a Canon CLC copier

Another technique known in digital halftoning, called error diffusion, is used by Panasonic in their "photo-text" mode [11.20]. It is not necessary to strictly limit the gray scale information to a picture element. One could "share" the gray scale information over several picture elements if the picture elements do not have enough latitude to express the gray scale information. This is done in practice by keeping track of the error between the gray scale information desired and printed (determined by the threshold gray scale values in the cell), and diffusing the error to nearby picture elements. The "resolution" of the gray scale information becomes lower than the resolution of the picture element. With an infinitely large area, error diffusion can

$$\left(\frac{1}{42}\times\right) \quad \begin{matrix} & & \bullet & 8 & 4 \\ 2 & 4 & 8 & 4 & 2 \\ 1 & 2 & 4 & 2 & 1 \end{matrix}$$

Fig. 11.7. An example of the error diffusion algorithm using the Stucki filter. The dot is the origin and the error is diffused to future pixels by the ratio given in the figure. Application of this algorithm to a continuous gray slide ramp is shown. Note the artifacts in some of the gray levels [11.20]

produce an infinite number of gray scales. Naturally, over a small distance, the number of gray scales is more limited. One can imagine that there are many conceivable algorithms for diffusing the error, each of which has its own special characteristics. So far, all algorithms proposed have artifacts, especially in the light gray areas, that give the image a structured or "wormy" look. An example of applying the Stucki filter to a gray scale ramp is given in Fig 11.7. Examples of other filters can be found in [11.20]. A photomicrograph of the Panasonic "photo-text" mode in which error diffusion is used can be see in Fig. 11.6c.

How about mixing digital and analog techniques? For example, if one or two gray levels could be added to each pixel in the 3 × 3 superpixel shown in Fig. 11.5, could more grays scales be obtained? The answer, provided by *Lama* et al. [11.21], is that the total number of gray levels goes up very rapidly with additional gray levels in each cell. However, producing small dots with different gray levels on paper remains a challenging problem in electrophotography.

The next level of complication is to add pulse width modulation. Without changing the optics, the width of the pulse applied to the laser can be varied instead of allowing just the discrete values determined by the size of the laser beam. Consider the 3 × 3 matrix shown in Fig. 11.3. Across one line there appear to be 3 positions; by varying the pulse width, in principle, a continuous number of gray scales can be created. This manipulation of the gray scale information can be done external to the printer and is commercially available in cards which can be added to a personal computer. High "resolutions" such as "600 × 600" and even "1200" dots per inch equivalent quality printing are advertised. There are two prices to be paid for this technique. First, the amount of information per page has obviously been increased, increasing the cost of the electronics. Second, there is a minimum photoreceptor voltage associated with turning the laser beam on at one spot. If the laser beam is turned on for a time shorter than this, then the photoreceptor is not completely discharged, leading back to all the problems of analog gray scales. Choosing the number of gray levels, the laser on-times associated with these gray levels, and the patterns of dots to turn on are decisions built into the software that implements these techniques.

Of course, combining pulse width modulation and a small laser dot size is equivalent to actually increasing the resolution of the system. The price is electronics and optics costs. But the quality that can be achieved has the potential of being the best that electrophotography can produce. This is the route Canon chose in their CLC line of copiers. The laser spot size is elliptical in shape, 20 μm in the laser scan direction and 63.5 μm in the process direction. The 63.5 μm spot implies a screen of 160 cm^{-1} (400 dots per inch) in one direction. Each cell has 2 pixels in it, giving a line screen of 80 cm^{-1} in the other direction. The gray scale is obtained by pulse width modulation, effectively smearing out the 20 μm spot to 137 μm in the scan direction for full black. The pattern produced, seen under a microscope, looks like a series of vertical lines on the page, as the dots in adjacent rows merge (Fig. 11.6d). The line screen, 80 cm^{-1}, is better than most magazines. The number of electronic gray levels is 256, although the number of gray levels that can be distinguished in the final print is probably not that high. The resultant image quality is the best in the industry at this time and, for some images, compares favorably with the best magazine quality and with some offset printing. The recently announced Xerox 5775 appears to have a similar algorithm and equivalent quality.

11.2.2 Other Challenges

The desire to use digital halftone gray scales requires the printing of single dots in electrophotography. A critical test is producing uniform single dots on paper. Shown in Fig. 11.8c are single dots made in response to a very light gray input. Clearly the dots on paper are neither circular nor uniform, and are sig-

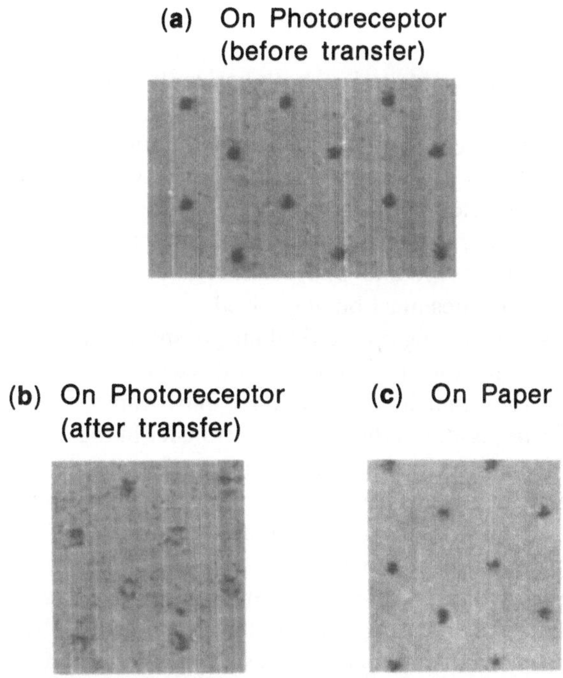

Fig. 11.8. Single dots made in response to a very light gray input. (**a**) The toner developed on the photoreceptor. (**b**) The dots on the photoreceptor after transfer. (**c**) The dots on paper

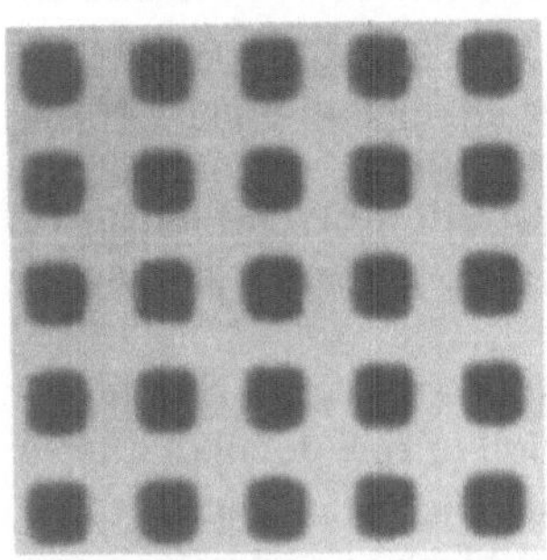

Fig. 11.9. Dots produced on a Linotronic Laser Imagesetter at 240 pixel/inch, 1 on, 1 off

nificantly inferior to dots produced in a Linotronic Laser Imagesetter (Fig. 11.9) in which a laser writes on films to create masters for an offset press. The reason for such poor quality dots can be found by inspecting the image on the photoreceptor before and after transfer. Before transfer, the developed toner dots are reasonably uniform (Fig. 11.8a); after transfer, doughnut shapes remain on the photoreceptor (Fig. 11.8b). Presumably, wrong sign toner develops onto the edges of the dots in response to the fringe electric fields; this wrong sign toner does not transfer. Clearly, the amount of wrong sign toner must be minimized in any development system that is expected to develop digital halftones.

In even the best color copiers available today, both horizontal and vertical lines can be seen in light colored areas. As mentioned earlier, lines in the process direction are usually due to corona wire contamination and lines perpendicular to the process directions are usually due to velocity variations or spacing variations between the development system and the photoreceptor.

To produce uniform colors requires improving the uniformity and stability of photoreceptor charging, dark decay, and photodischarge characteristics. Furthermore, exposure nonuniformities must be minimized.

With three or four toner layers making up the final image, special attention has to be given to the transfer and fuse subsystems. As discussed below, architectures which accumulate the toner images on paper must transfer toner onto already toned paper; this necessitated the use of roll transfer in the Xerox 6500. The use of an intermediate surface requires two transfers per image, emphasizing the importance of high transfer efficiency. Accumulation on the photoreceptor requires the transfer of at least three monolayers at one time.

Similar problems are encountered in the fuse subsystems. Much more toner must now be fused, as compared to black and white printing. This requires more energy. Furthermore, creating smooth layers (so light is not scattered) from the ≈ 10 μm diameter toner particles, especially important for transparencies, requires more thorough fusing than typically used for black and white printing.

Registration of the colored images is a concern, especially in those architectures in which the image is accumulated on paper (Sect. 11.3). Additional position sensing and feedback control systems can help in some subsystems. This problem is especially severe when text is copied on a color copier. If the copier builds up the black text from the three colored toners, misregistration can produce a rainbow of blurred images. Misregistration of 100 μm to as much as 250 μm can be seen on commercially available copiers which accumulate toner on paper.

One can also test characteristics such as the resolution and edge sharpness. Again, these criteria are important for the reproduction of fine detail or copies with both text and images. Generally, the resolution capabilities appear to be equal to, or at worst within a factor of two of, the best black and white copiers.

The market for black and white copying was driven partly by the desire to make copies of copies. Making a color copier that can faithfully copy its own copy requires the solution of two problems: maintaining color purity and copying the gray scale algorithm. As the reader can easily demonstrate, none of the commercially available copiers (to the author's knowledge) can faithfully copy its own copy. Most darken the cyan (which can be tested by copying a full color copy of cyan, produced by copying a solid black and requesting a single cyan color); some darken both cyan and magenta. Some cannot copy their own digital halftone algorithm.

11.3 Colored Toner Accumulation

Where the colored toners are accumulated has major architectural and subsystem implications. The choices are: on the paper, on an intermediate belt (or drum), and on the photoreceptor. One might expect that one of these choices would be preferable. However, all three have been implemented commercially.

The obvious first choice, and the one implemented initially by Xerox and later by Canon and Eastman Kodak, is accumulation on paper (Fig. 11.1). Such an architecture requires three or four transfers from the photoreceptor to the paper. The paper throughput is at least three times slower than the process speed. Another drawback to this choice is color registration. Paper is not a stable medium and alignment of each color image to within a fraction of the laser spot, typically 75 μm, is a major technical challenge. The resulting misalignment can be seen by inspecting the edge of a sharp line made up of two or more colors. Finally, each of the three development systems must be engaged one at a time against the photoreceptor; this can be done mechanically or magnetically (by rotating the magnets inside the roller of a magnetic brush development system).

An interesting variant of this architecture is to use four separate copiers each with its own exposure system, photoreceptor, and development system. The paper is carried from each copier to another. The advantage of this architecture is that it speeds up the throughput at the price of increased hardware complexity. Such an architecture has been implemented recently by Ricoh (Fig. 11.10) in their Artage 8000, which makes 15 ppm (pages per minute).

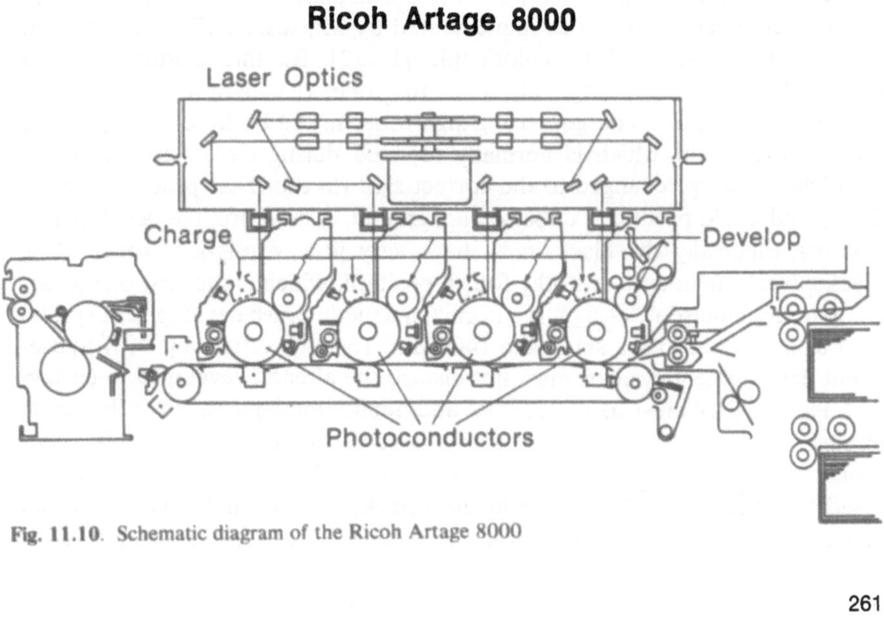

Fig. 11.10. Schematic diagram of the Ricoh Artage 8000

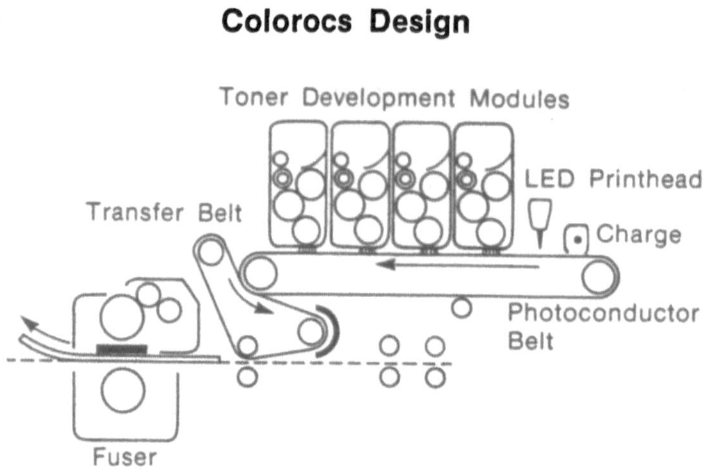

Fig. 11.11. Schematic diagram of the Colorocs color engine

To solve the misregistration problem, Colorocs Corporation suggested accumulating the toner images on an intermediate belt (Fig. 11.11). While belts are not necessarily the most stable surface, control systems were implemented to eliminate the misregistration problem. Note that the system throughput is again 1/3 or 1/4 (if one includes black development) of the process speed. Further, two transfer steps are now needed to move the toner image from the photoreceptor to the paper, requiring high efficiency and stable transfers to maintain image quality.

Perhaps a more elegant solution is to accumulate the toner images on the photoreceptor (Fig. 11.12) as implemented by Matsushita Electric Industrial Co. in their Panasonic FP-C1 color copier [11.22]. But this solution is also the most technically demanding: (1) Since the toner remaining on the photoreceptor is subject to recharge, wrong-sign toner in the background regions of the photoreceptor, which is normally rejected during electrostatic transfer, now has its charge changed to the correct sign (in the DAD process — Sect. 2.1.3) and could produce excessive background. Therefore, the development systems, especially the black, must have very little wrong-sign toner. The black development system in the fourth position, which is the first one to develop in the Panasonic design, is a normal magnetic brush development system with small carrier and a small gap (400 μm) for high efficiency. (2) Subsequent development systems must not disturb the already developed toner layers. Hence, they must be noncontact and noninteracting. As no such system existed, a new development system was invented (Sect. 11.4.3). (3) It was found that upon exposure to form the second (yellow) latent image, the unfused toner from the first toner image (black) on the photoreceptor would scatter into the latent yellow image in response to the fringe electric fields.

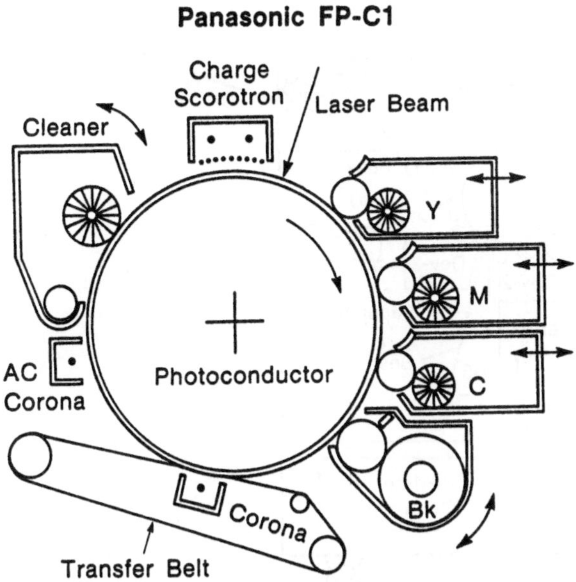

Fig. 11.12. Schematic diagram of the Panasonic FP-C1 color copier

This was corrected by reducing the initial potential for the yellow image and electronically separating the yellow image from the black image by one laser spot diameter [11.23]. (4) Because of the high dark decay of the Se-Te photoreceptor, the development system in the fourth position was subject to the greatest variability in the latent image. This was therefore chosen to be black. (5) With recharge, already developed toner layers could increase their charge, making it more difficult to achieve high transfer efficiency. To fix this and to decrease the toner scatter problem, an ac corona discharge is used in each cycle. (6) As *Chen* [11.24] and *Yamamoto* [11.25] have pointed out, developing toner on top of already developed toner layers is akin to placing an overcoat on a photoreceptor surface, which can lead to a reduction in developability and resolvability.

The Panasonic color copier still has a system throughput of 1/4 the process speed. *Kohyama* et al. of Toshiba have published [11.26] a solution to the throughput problem shown in Fig. 11.13. In this configuration, the three colored toners are accumulated on the photoreceptor in one pass requiring: (1) charge, (2) exposure, (3) develop (again noncontact, noninteracting), and (4) recharge (to bring the potential of the image back up to original charge level) for each color. Perhaps the most costly item in this architecture is the three (or four) laser systems and their associated electronics.

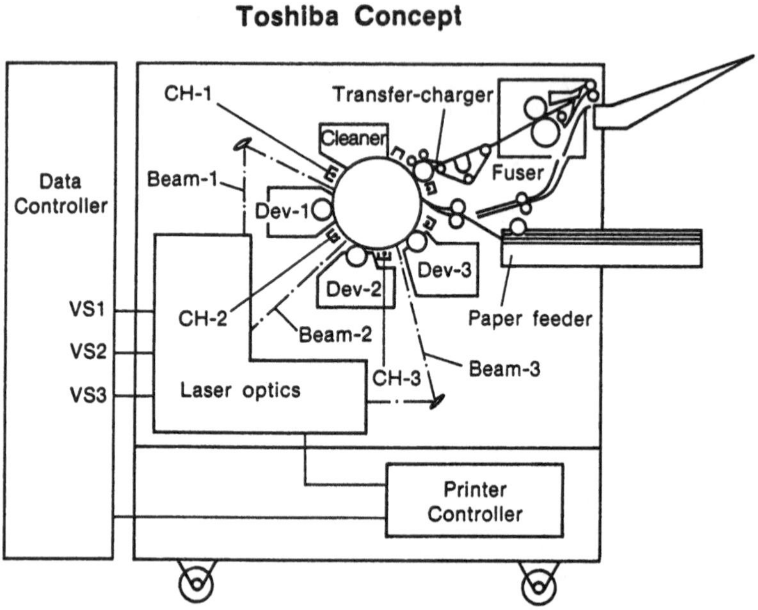

Fig. 11.13. Toshiba concept for a three color single pass copier

11.4 New Development Systems

In response to the need for improved image quality, more compactness, and toner accumulation on the photoreceptor in some architectures, several new development systems have been invented specifically for color electrophotography.

11.4.1 Image Quality

Consider the challenge of image quality. It determines some toner properties, which, in turn, affect the choice of development systems. First, the toner should be insulating, so that transfer nonuniformities at high relative humidity associated with conductive toner are avoided (Sect. 9.4). Second, the toner should be nonmagnetic, so that black magnetic material does not compromise the purity of the colors.

The development system, even with nonmagnetic insulative toner, is probably one of the more difficult subsystems to make uniform and stable. Uniformity is affected by low and high frequency image noise, some of which can be traced to the brush profile in the magnetic brush development system. History effects cause image defects at the leading or trailing edge of solids in some magnetic brush development systems and lines perpendicular to the process direction separated by one development roller circumference in some monocomponent systems. Lack of stability results from any change in the many parameters which affect development (see earlier chapters).

In principle, one would like very reproducible development, with no image defects. This suggests the need for a very efficient system which develops to neutralization, making the developed mass per unit area sensitive to the least number of hardware and material parameters (Sect. 6.3.1). Such a condition is approached by a development system invented by Eastman Kodak for the ColorEdge copier and also by liquid development.

Eastman Kodak, in designing the ColorEdge copier, which is the fastest available color copier today (23 full color pages per minute) actually faced two challenges. (1) If the efficiency of the development could be increased so that neutralization was approached, the image quality would be improved, as mentioned above. (2) Furthermore, they wanted to use one of their existing black and white copier engines (the Ektaprint 150, shown in Fig. 2.13) so that a new engine did not have to be designed. This required putting three development systems where one existed before. To accomplish these goals, they invented a new development system [11.27,28], which was mentioned in Sect. 6.7. In this system, which is a modified magnetic brush development system, small diameter (≈ 30 μm) permanently magnetized spherical carrier beads, with electrostatically attached toner, are attracted to a roller inside of which magnets rotate. The rotating magnetic field tends to move those carrier beads which have lost their toner particles, and are consequently charged, away from the latent image. Recall from Chap. 6 that these charged beads, if not moved away from the latent image, limit toner development because they reduce the electric field in the development zone to zero. The actual development characteristics of this system have not yet been discussed in the literature.

A liquid development system (Chap. 10) can, in principle, develop to neutralization and has been used in prepress proofing systems, discussed in Sect. 11.1. These operate at speeds of many minutes per copy in special contained environments due to use of noxious solvents.

Finally, improved image quality has been achieved in the standard magnetic brush development system. This has been achieved by reducing both the toner and carrier size. *Gruber* and *Dalal* [11.29] have suggested that reduced toner size leads to improved edge sharpness and reduced image graininess while reducing the overall pile height. *Chiba* and *Inoue* [11.30], following earlier work of *Goren* [11.31], point out that smaller toner particles produce better gray scales, improvements of dot shape and less noise in the halftone dot area. The black development system in the Panasonic color copier is a magnetic brush system with a smaller gap (400 μm) than usual (1250 μm) [11.22]. It is obvious from (6.57), which describes the factors controlling development, that a smaller gap gives higher M/A, simply because the electric fields are increased. A smaller gap requires smaller diameter carrier beads. This leads to

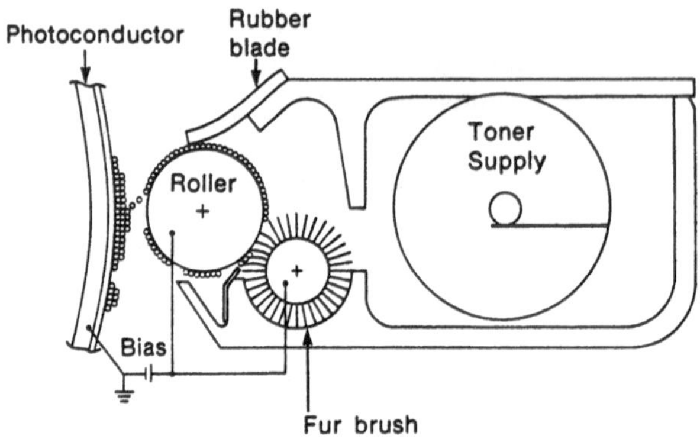

Fig. 11.14. Cross-sectional view of the monocomponent development system used in the Panasonic FP-C1 color copier

a secondary source of increased efficiency, as discussed in Sect. 12-6.2.2: the effective dielectric constant appears to increase as the carrier bead diameter is reduced. Furthermore, small carrier beads increase the spatial frequency of image noise associated with the bead motion through the developed toner layers, reducing its perception by the eye. One negative effect of smaller diameter carrier beads is that charged carrier beads can more readily adhere electrostatically to the photoreceptor and be carried out of the development system. These must be captured with magnets prior to reaching the paper. This use of smaller gaps and smaller diameter carrier beads for improved efficiency and reduced image noise can be found in both color (Canon) and some black and white copiers using magnetic brush development systems.

In principle, a monocomponent development system would seem to have an image quality advantage over most magnetic brush systems: development of all the toner on the roller would tend to produce more uniform development if the loading of the roller is uniform. However, the development efficiency of the color monocomponent development systems (Fig. 11.14) in the Panasonic color copier has not been published.

11.4.2 Compactness

As mentioned above, the development system in the ColorEdge copier was designed with compactness in mind: three development systems had to fit where one fit in the black and white machine.

The Panasonic monocomponent development systems are amazingly compact: They require only about 3.8 cm of surface length on the photoreceptor and are only 10 cm deep (Fig. 11.14).

The recent announcement of the Ricoh Artage 8000 [11.32,33] also describes the invention of a new more compact development system that com-

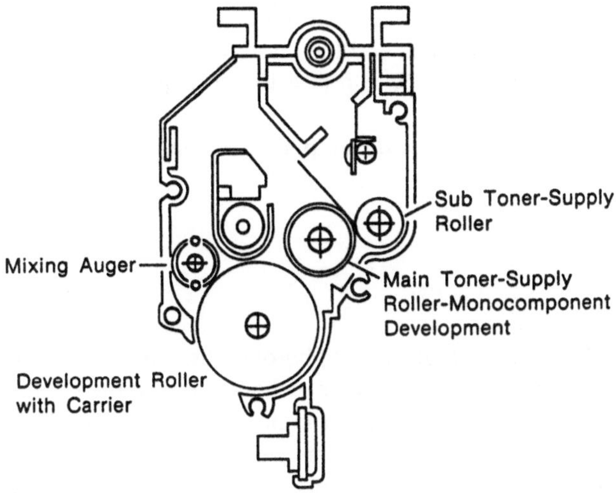

Fig. 11.15. Cross-sectional view of the Ricoh Artage 8000 development system

bines monocomponent and two component development. In this system, shown in Fig. 11.15, a toner is loaded onto a roller, as in a monocomponent development system. The roller then supplies toner to the beads on a magnetic brush development system. The toner Q/M on the monocomponent roller is $2 - 5\ \mu C/g$; by the time the toner particles are brought into the development zone the Q/M has increased to $\approx 20\ \mu C/g$. The claimed volume reduction of 1/5 is due to the elimination of the large volume usually used for mixing the toner and carrier beads.

11.4.3 Noncontact, Noninteracting Development System

The architecture in which the toner images are accumulated on the photoreceptor requires a noncontact, noninteracting development system since the development of the second and higher layers must not disturb the lower layers. Panasonic and Konica have commercialized and Toshiba has discussed such systems. The Panasonic system (Fig. 11.14) is based on a monocomponent system with nonmagnetic toner which jumps a gap to the photoreceptor. Only dc voltages due to the latent image are used to cause the toner to jump from the roller to the photoreceptor, thereby preventing the mixing of toner that occurs in the presence of ac voltages. (A review of monocomponent development can be found in Chaps. 8 and 9.) To dispense with an ac voltage required low average toner charge ($\approx 3\ \mu C/g$) and narrow charge distributions (to eliminate wrong-sign toner).

The Konica system is a magnetic brush development system with a 150 μm gap between the ends of the brush and the photoreceptor surface. To enhance development an ac bias of 1.4 kV p-p at 4 kHz is superimposed on the dc voltages [11.34].

A technical description of the Toshiba development system can be found in Sects. 8.3 and 9.6.

11.5 Color Market

The color market in 1990 indicates that electrophotographic copiers are already beginning to dominate all other color copier technologies. In the United States, according to Dataquest, of the 9000 color copiers sold, 7600 used the electrophotographic technology. 585M color copies were made using electrophotography, creating a revenue of $304M. For all color copying technologies (including thermal transfer, photographic, and Cycolor), 628M copies were made producing a total revenue of $378M.

Dataquest projections for 1995 show an even greater dominance of the electrophotographic technology. Of the 30 000 estimated 1995 placements, 28 000 will use electrophotography. 4944M color copies are projected to be made using electrophotography, producing a revenue of $1977M. For all technologies, 5000M color copies will be made for a total revenue of $2055M. This represents a compound annual growth rate (CAGR) of 45.4%, larger than any black and white printer's CAGR (Table 12.2).

11.6 Current Copier Products

Because so many new architectures and subsystems have evolved in response to the color requirements, most of the color copiers commercially available in the United States as of March 1992, listed in Table 11.1, will be described. In the table are listed the gray scale technique (analog or digital), the speed, price, toner accumulation method, and development system.

The Eastman Kodak ColorEdge (Fig. 11.16) introduced in 1988 is the fastest color copier on the market (23 ppm). This copier is based on the Ektaprint 150 architecture (Fig. 2.13) and accumulates the toner on paper, as done in the Xerox 6500 (Fig. 11.1). The paper is held on the bias transfer roller while the three toner images are accumulated. In place of the one black conductive magnetic brush development system (Chap. 7), three new development systems (Sect. 11.4.1) with the three subtractive colors, magenta, cyan, and yellow, were added. Maintaining the process speed used for black and white (70 ppm), gives a three-color print at a rate of 23 ppm, allowing for some dead time between each color. The gray scales were improved by intro-

Table 11.1 Color electrophotographic products available March 1992 in the U.S.

Company	Model (Introduction date)	Analog, Digital [A,D]	Speed [ppm]	Price [$]	Image accum.	Dev. system
Copier						
Kodak	ColorEdge (1988)	(Screened PC)	23	60K	Paper	New-Mag. Brush
Canon	CLC-1 (1987)	D	5		Paper	Mag. Brush
	CLC-200 (1989)			24K		
	CLC-500 (1989)			49K		
Panasonic	FP-C1 (1989)	D	5	16K	PC	Mag. Brush-Black DC Jump, Mono-Color
Konica	9028 (1991)	D	6.5		PC	New-Jump Mag. brush
Ricoh	Artage 5330 (1989)	A	4	17K	Paper	Mag. Brush
	Artage 8000 (1991)	D	15		Paper	Mono + Mag. Brush
Xerox	5775 (1991)	D	7.5	45K	Paper	Mag. Brush
Two-Color Printers						
Xerox	4850 (1991)	D	50	120K	PC	Con. Mag. Brush
AM Int.	ELECTROPRESS (1990)	D	300 ft/min	1500K	Paper	Liquid

ducing a screen behind the photoreceptor, which converts the analog gray levels reflected from the original into latent image dots on the photoreceptor.

The Canon CLC architecture, first introduced in 1987, is shown in Fig. 11.17. Again the toner images are accumulated on paper, which is held on a drum adjacent to the photoreceptor. Digital color was first introduced in this machine. The image is scanned by CCD arrays covered with appropriate filters to separate the three color images. With this information, a laser beam then sequentially exposes the photoreceptor. In the CLC-200, the laser spot is elliptical, 20×63.5 μm, and pulse width modulation is used to achieve the gray levels (Sect. 11.2.1, Fig. 11.6d). The introduction of digital color by Canon significantly improved the quality of the color copies, effectively achieving the goal of magazine quality. The development system remains magnetic brush development, although smaller toner is used in the CLC-500. Additional software refinements distinguish the CLC-500 from the CLC-200.

Kodak ColorEdge

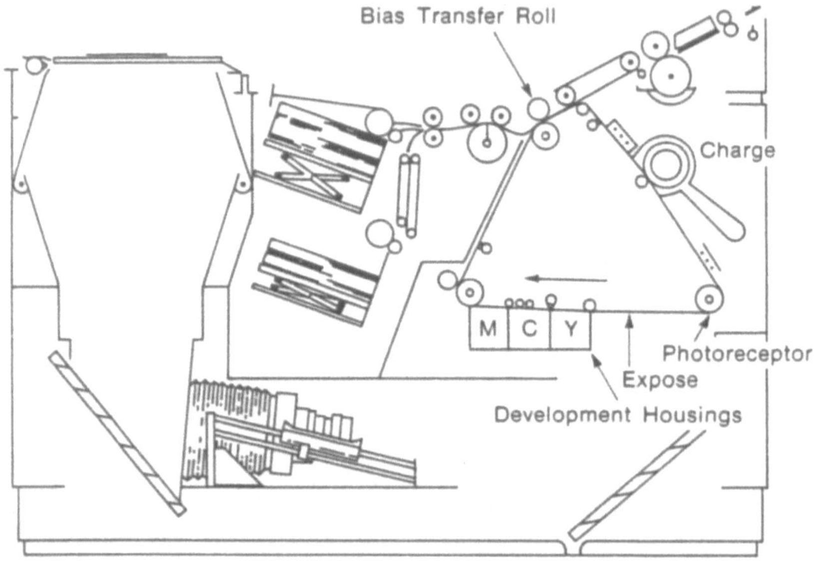

Fig. 11.16. Schematic diagram of the Eastman Kodak ColorEdge copier

Canon CLC

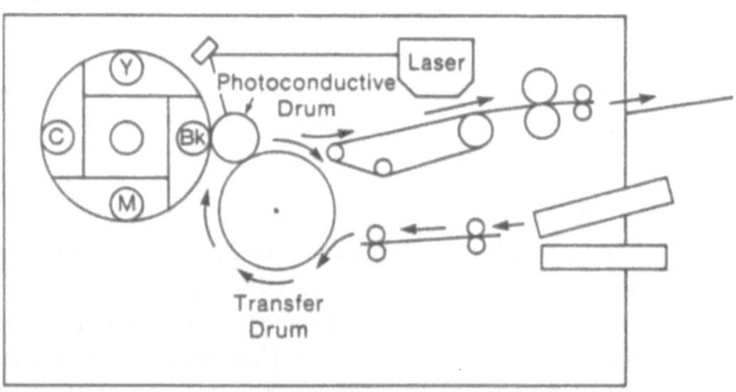

Fig. 11.17. Schematic diagram of the Canon CLC color copier

It makes five full color copies a minute at a resolution of 400 dots per inch. It uses a small diameter photoreceptor drum (8 cm) and a four-color rotating (turret-like) developing-station, which positions the colors as needed. Each color is sequentially transferred to paper, which is held on a transfer drum. In 1991 Canon introduced another member of the CLC family, the 300, for about half the price of the CLC-500; again software differences distinguish it from the other CLCs. Eastman Kodak also markets the Canon CLC-300 and CLC-500 as the ColorEdge 1525 and 1550, respectively.

The Panasonic FP-C1 digital color copier, introduced in 1989, departs significantly from the above architectures (Fig. 11.12). The toner images of each color are accumulated on the photoreceptor, after which the total toner image is transferred at one time to paper, which is carried on a transfer belt. The black image is developed first with a two component development system. The cyan, magenta, and yellow images are developed using a new development system, a nonmagnetic monocomponent jump gap system that uses only dc voltages. The technical problems associated with the architecture are discussed in Sect. 11.3. It makes five full color prints per minute and uses digital halftoning to achieve its gray scales. In the photo-mode, the resolution is 400 dots per inch, and the line screen is 28 cm^{-1} at 45°, giving 33 gray levels.

Colorocs (Fig. 11.11), whose color engines are made by Sharp and marketed by Sharp (CX-7500) and Savin (Prism 1), accumulates the toner images on an intermediate belt. The transfer processes are electrostatic. The machine uses a modular structure for improved extendibility, lower manufacturing cost, and easier maintenance. Further, it physically uncouples all the subsystems by going to a software controlled asynchronous operation – as opposed to the conventional synchronous system with one main drive motor for all functions [11.1]. It makes 7.5 ppm and uses analog gray scales. The four development systems are conventional magnetic brush. Sharp, who manufactures these engines, is no longer an active participant in the color market. Moreover, Colorocs has filed for bankruptcy.

Konica (Fig. 11.18) chose a similar strategy as Panasonic – accumulation of the toner images on the photoreceptor. The development system is a magnetic brush system, but with a gap between the magnetic bead chains and the photoreceptor surface. The model 9028 delivers 6.5 ppm. Only highlight color quality (no gray scale) is claimed.

Ricoh recently announced (October 1991) yet another architecture in their digital Artage 8000 (Fig. 11.10). The toner images are accumulated on paper. Four complete electrophotographic engines run simultaneously, allowing for faster throughput, 15 ppm.

Xerox announced a new color copier, Xerox 5775, in October 1991 (Fig. 11.19). The architecture is based on accumulating the toner images on paper, as done in the earlier Xerox color copiers. The gray scale algorithm appears to be similar to the Canon algorithm: 400 dots per inch input scanner and 400 × 1600 dots per inch addressability output. It is not known whether the laser beam is elliptical in shape. The copy quality is excellent and matches the Canon copy quality.

Konica 9028

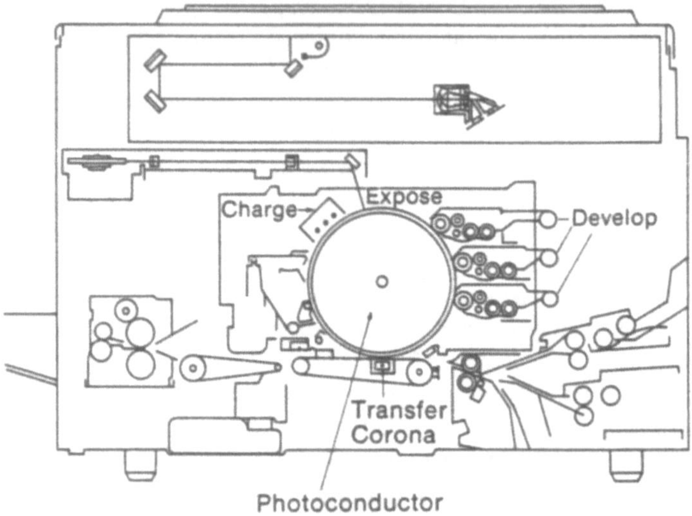

Fig. 11.18. Schematic diagram of the Konica 9028 color copier

Xerox 5775

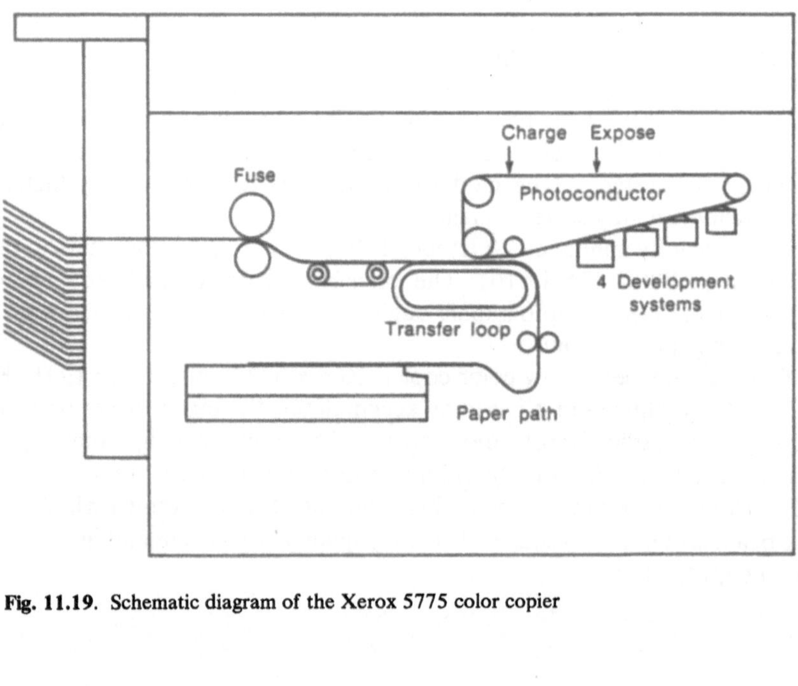

Fig. 11.19. Schematic diagram of the Xerox 5775 color copier

11.7 Current Printer Products

While it is relatively straightforward conceptually to convert a digital color copier into a printer by adding a controller (which are being offered commercially for some of the copiers listed in Table 11.1), as yet, color printers are not generally available (see discussion at beginning of this chapter). An exception is the Xerox 4850, which is based on a black and white printer. The Xerox 4850, introduced in October 1991, is a two-color printer, which operates single pass at full speed, 50 ppm. It is based on the Xerox 4050 printer, which was based on the Xerox 1075 copier (Fig. 2.14). On the bottom segment of the photoreceptor belt, a second conductive magnetic brush color development system is added (Fig. 11.20). Both colors are developed in a single

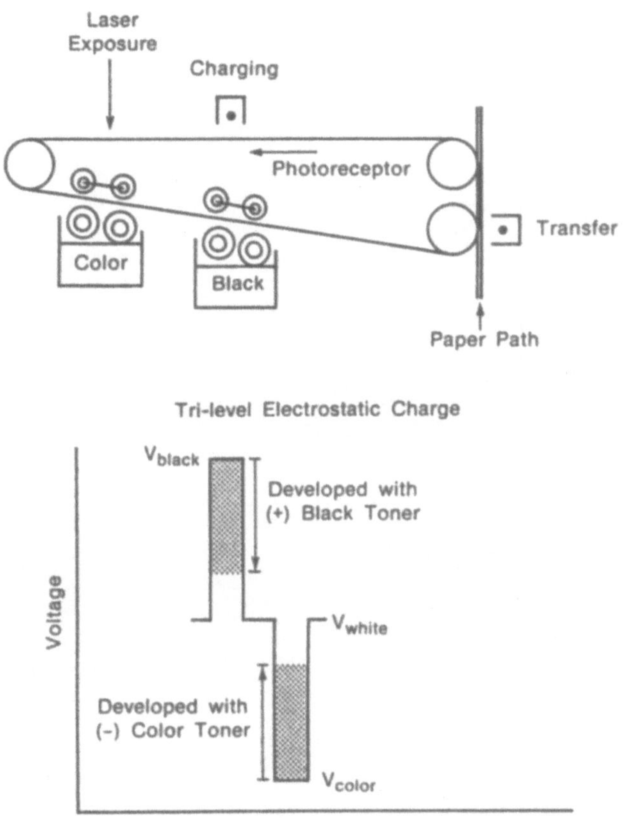

Fig. 11.20. The Xerox 4850 color printer was built by adding a second development system to the Xerox 4050 black and white printer. The two colors are developed in a single pass by using toners of opposite polarity

pass by using toners of opposite charge polarity. Various concepts [11.35–38] have been suggested before for two-color single pass systems, starting in 1981 by Hitachi at the first International Non-Impact Printing Conference [11.38]. The scheme adapted by Xerox [11.39] uses three electrostatic levels on the photoreceptor. The areas to be printed in highlight color are fully exposed and discharged (DAD, Sect. 2.1.3) while black areas are unexposed and remain fully charged (CAD, Sect. 2.1.3). The intermediate charge level, which received the middle exposure level, does not attract either toner and remains white. To electrostatically transfer the toner, a pretransfer corona changes the polarity of all the toner to one sign.

Finally, AM International is selling a two-color printing press, operating at 300 feet/min, based on liquid electrophotography. It uses corona transfer and the usual reverse roll metering. Exposure is accomplished with 300 LEDs per inch and the image is accumulated on "plain" paper. It is apparently being used for mass mailings.

12. Update of Chapters 1–10

This chapter updates the first edition of the book. It has been designed to allow the reader to quickly identify and learn about those areas in which progress has been made: each section number after 12- corresponds to a section earlier in the book. For almost all the sections in this chapter, it is assumed the reader is familiar with the material contained in the first ten chapters of the book.

It is amazing to realize, as one goes through the literature, how much progress has been made since the first edition of this book was written (October 1987) until now (March 1992). Particularly interesting are the advances in toner adhesion (Sect. 12-2.1.4), the theory of insulator charging (Sect. 12-4.3), the analysis of toner charge measuring tools (Sect. 12-4.4.4), the resolution of the effective dielectric constant problem (Sect. 12-6.2.2), and the work on background development (Sect. 12-6.6). Only one completely new section had to be added, Sect. 12-4.4.5, in which recent work on the life characteristics of insulating magnetic brush development systems is discussed, and Sect. 12-2.1.4 had its title broadened to Transfer and Toner Adhesion.

12-1.2 Copier Market

Data on the U.S. copier market provided by Dataquest are shown in Table 12.1; this represents Dataquest's latest forecast. The table shows the 1990 market and projections for 1995. The 1987 definitions of the market segments are given in Sect. 1.2; the 1991 definitions differ from the earlier definitions mostly in price.

The U.S. copier market was \$19 billion dollars in 1990 and is projected to increase to \$21 billion in 1995. This represents a compound average growth rate of 2.0%. The segment which is expected to grow the most is Segment 4, which includes highly featured machines in the speed range of 45–69 cpm, making 24 000 copies per month. Segment PC, Segment 3 and Segment 6 are all anticipated to grow with a compound annual growth rate of between 3.1% and 3.8%.

Comparing the 1990 actual revenues (Table 12.1) with the 1991 prediction given in Table 1.2 (p. 15) shows that Segment PC did not grow as fast as predicted, partly due to the larger than expected growth of Segment 1 and

Table 12.1. U.S. copier market (Dataquest, September 1991)

Segment definition		Market[a]		
	Speed [cpm]	1990 [Millions of U.S. dollars]	1995	Compound annual growth rate [%]
PC	up to 12	355	413	3.1
1	up to 20	2 675	2 280	–3.1
2	21–30	3 144	3 077	–0.4
3	31–44	2 932	3 418	3.1
4	45–69	3 558	5 078	7.4
5	70–90	2 830	2 517	–2.3
6	91+	3 600	4 338	3.8
Total		19 094	21 121	2.0

[a]Includes hardware, service, and supplies.

partly due to saturation of distribution channels such as office supply chains. Segments 5 and 6 both grew significantly more than predicted, indicating the continued desire for high speed copiers/duplicators.

12-1.3 Printer Market

Data on the U.S. printer market provided by Dataquest are shown in Table 12.2; this represents Dataquest's latest forecast. The table shows the 1990 market and projections for 1995. The definitions of the market segments are given in Sect. 1.3. The numbers represent hardware and supplies only; Table 1.5 also only includes hardware and supplies (not service, as indicated in Table 1.5).

The U.S. printer market is expected to grow significantly from 1990 to 1995, from $9 billion in 1990 to $16 billion in 1995, which represents a compound annual growth rate of 12.2%. All segments will grow at rates greater than 13%, except Segment 1, printers up to 10 ppm, and Segment 7, printers that operate at 151+ ppm. The largest growth rate is in Segment 2, the 11–20 ppm printers, which represent the personal page printers now available from Hewlett Packard (the LaserJet series which use Canon engines), IBM (made by Lexmark, a company spun off in 1991 from an IBM division in Lexington, Kentucky) and others.

Table 12.2. U.S. printer market (Dataquest, September 1991)

Segment definition		Market[a]		
	Speed [ppm]	1990 [Millions of U.S. dollars]	1995	Compound annual growth rate [%]
1	up to 10	5 573	8 008	7.5
2	11– 20	785	3 239	32.8
3	21– 30	256	695	22.1
4	31– 50	240	447	13.3
5	51– 80	290	709	19.6
6	81–150	1 068	2 119	14.7
7	151+	851	870	0.4
Total		9 063	16 087	12.2

[a]Includes hardware and supplies only.

Comparing Table 12.2 and Table 1.5 (p. 19) indicates that the growth of 30.6% predicted for Segment 1 was exceeded (!) while the growth of Segments 2–5 was not achieved; Segments 4 and 5 appear to have even decreased from the 1986 levels. Segment 6 grew faster than expected while Segment 7 decreased faster than expected. The difficulty of making accurate predictions for the printer market probably reflects normal behavior for an evolving business.

12-1.4 Alternative Powder Marking Technologies

Alternative powder marking technologies continue to be explored.

A method of combining clean, charge, expose, and develop into one step was suggested [12.1]. In this concept, the photoconductor is charged and exposed while it is in the development zone. Conductive magnetic toner is used. The conductive toner brush charges the photoconductor at the nip entrance of the development system, and image exposure at the nip exit through the back of the photoreceptor establishes the latent image to which the conductive magnetic toner adheres. Similar ideas using conductive magnetic brush development were suggested by *Kimura* et al. [12.2].

Schmidlin [12.3] suggested a one step "direct marking" process: toner on a donor roll could be attracted to paper with a dc electric field. The stream of toner particles can be shut off by causing the toner to pass through apertures which are opened and closed electronically with fringe electric field.

Another, simpler, version of "Magnestylus" (Sect. 1.4) was reported by *Nishigaito* et al. [12.4]. The need for large conductivity anisotropy (along versus between toner chains) was clarified and pressure transfer was incorporated into the system.

As many have realized, the cleaner can be eliminated if the development system scavenges off untransferred toner. An implementation is discussed by *Hosoya* et al. [12.5].

By uniformly charging a polymer film and using a thermal printhead to image-wise discharge the film, the photoconductor and laser scanning system can be eliminated [12.6].

A printer using only two steps can be built by image-wise exposing a uniformly charged toner layer to charged ions from an ion print head, thereby changing the sign of the toner which will form the real image. The toner is then transferred to paper by electrostatic transfer [12.7].

A thorough review of electrographic printing was recently published by *Bugner* [12.8]. In electrographic printing a master is first formed which is used to make multiple copies. Methods of master formation are divided into physical and chemical methods. An example of a physical method is to produce a toner image by normal electrophotographic methods and to fuse the toner to the photoreceptor. The fused toner serves as a mask for subsequent printing cycles: using uniform charge and blanket exposure creates a charge image on top of the fused toner. Another example: workers at Xerox [12.9,10] have described using a photoreceptor with small (0.35 μm) diameter Se particles embedded in the top surface as a master. After a uniform charge and image-wise exposure, the Se particles under the image are charged. Subsequent heat drives the Se particles into the bulk of the photoreceptor, creating a master. After uniform charge and blanket exposure of the master, only those areas with Se particles on top retain their full charge. An example of a chemical method of making a master is to incorporate into the photoreceptor a halogenated hydrocarbon that decomposes (into HBr or HI) upon exposure, creating a region on the photoreceptor surface that can no longer hold a charge. An example favored by the author is to incorporate onium salts into a polymer film; exposure creates strong acids, rendering the area more conductive.

12-1.4.1 Magnetography

The Bull Peripheriques group continues to explore the limits of their version of the magnetography technology. They have extended their earlier work [12.11] on high speed printing and report new results on increased resolution.

Building on earlier successful high speed printing results, up to 180 m/min, a full function prototype was built [12.12]. An interesting aspect of this prototype is that it operates at variable speed, to mimic a requirement of printing presses. Speeds can be varied from 18 to 90 m/min without noticeable effect on print quality.

In their high resolution work [12.13], up to 200 dots per cm printing are reported, twice the resolution in current products (95 dots per cm). A new printhead design was chosen based on a single needle inside a conductor circuit which has many turns. A theoretical analysis suggested a pole diameter of 50 μm was necessary for the 200 dots per cm printing. Prototypes were built with 50 adjacent poles and run in their M6090 printer, which has a print speed of 33 cm/s. Without sacrificing optical density, improved resolution and line straightness were achieved.

Meanwhile, a considerably simplified printhead was suggested by *Keefe* et al. [12.14]. In their design, the magnetic structure is one piece and consists of a two-dimensional array of elements formed upon a common ground plane. Each print element is enclosed by two turns of conductor and the array of elements is addressed using a multiplexed method for economy of electronic drives. Good print quality was achieved at 100 dots per cm at 30 cm/s paper speed.

12-1.4.2 Ionography

Work by *Caley* et al. [12.15] on the Delphax ionographic printhead has produced a surprising result. The negative charges, which form the latent image, are due to electrons, not ions. The evidence presented includes measurements of the transit time for the charges to travel from the printhead to the latent image surface. The values are only consistent with electron motion. It must be that there is insufficient time for negative ions to form in their system (Sect. 2.1.1).

Omodani et al. [12.16] suggest that ionography systems similar to the Delphax configuration are ideal for producing continuum gray scales since time modulation of the aperature electrode controls the charge in each dot. Using a liquid development system, 8 dots per mm monochrome and color printing with 256 gray levels was demonstrated. (See the discussion of gray scale printing in Sect. 11.2.1.)

12-2.1.4 Transfer and Toner Adhesion

Several discussions of toner adhesion, critical for understanding both transfer and cleaning, have appeared. While most people continue to emphasize the role of electrostatistics in adhesion, recent results from Eastman Kodak have rekindled the debate on the role of nonelectrostatic forces.

(1) Centrifuge Studies

Takeuchi and *Onose* [12.17] recently reported centrifuge studies, comparing the adhesion distribution of essentially uncharged and charged toner. They found, consistent with expectations, that charged toner had higher adhesion than uncharged toner; the range of adhesion forces found covered two orders of magnitude.

(2) The Role of Nonuniform Charge Distributions in Adhesion

Evidence has been accumulating that the toner adhesion problem requires an understanding of the nonuniform charge distribution on the toner particles (Sect. 2.1.4).

Schmidlin [12.18] pointed out, following earlier work, that the exchange of one charge separated by 1Å is equivalent in force to 5×10^4 charges sitting at the center of a 10 μm diameter toner particle. Therefore, the exchange of a few charges could appear as though a nonelectrostatic, i.e., van der Waals, force is operative.

Another advance in understanding toner adhesion has been made by *Hays* and *Wayman* [12.19]. They succeeded in measuring the nonuniform charge distribution on a toner particle and, using this distribution, they quantitatively predicted the electric field necessary to detach the toner particle from a surface. The experimental technique involved capturing an individual toner particle between two electrodes supplied with a charge sensitive amplifier. As the charged toner particle bounces between the electrodes, charge is induced in the electrodes, which is detected by the electronics. If the center of charge does not correspond to the center of mass (as measured by the dipole moment), the toner particle will rotate in response to the torque due to the electric field. The induced signal due to the rotation corresponds to the center of charge rotating through π radians. As the particle comes to rest on an electrode, a rocking motion due to the nonuniform distribution of charges on the surfaces is observed. This provides a measure of the quadrupole moment. Figure 12.1 shows a signal for a 99 μm diameter particle; the induced charge due to translation between the electrodes, Q_T, the induced charge due to rotation Q_R, and the rocking period T_R are identified. After analysis, it was found that much of the charge is centered at the poles (Fig. 12.2), which fraction increases as the particle is bounced between the electrodes and the toner acquires further charge.

To what extent does this represent the charge distribution on actual toner particles, which are a factor of 10 smaller in diameter? The charging of these

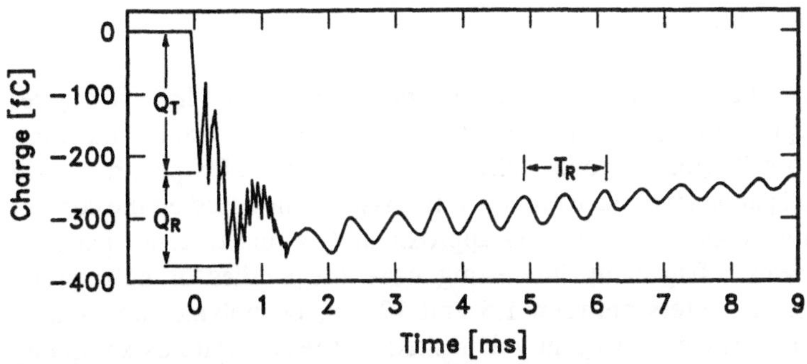

Fig. 12.1. The charge induced in the charge-sensing electrode for a 99 μm bouncing and rocking particle detached by 580 V across a 137 μm gap [12.19]

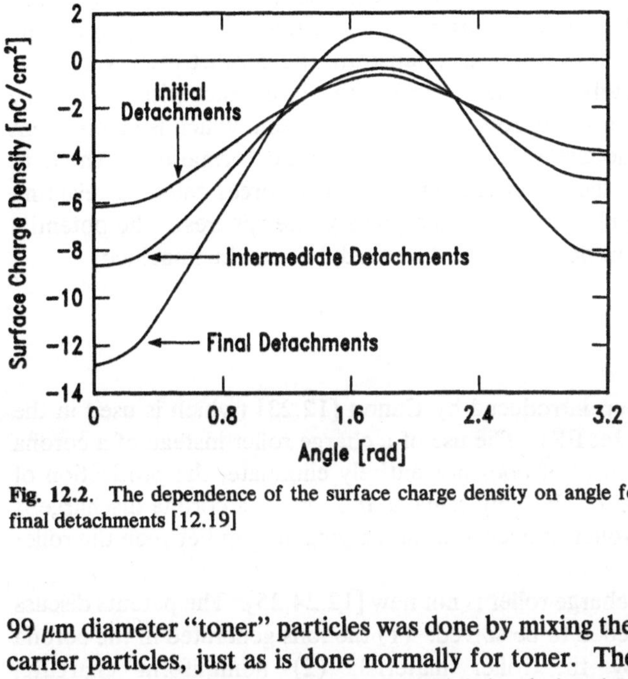

Fig. 12.2. The dependence of the surface charge density on angle for initial, intermediate and final detachments [12.19]

99 μm diameter "toner" particles was done by mixing them with large diameter carrier particles, just as is done normally for toner. The experiment was carried out in a vacuum environment, which is obviously different than normal but would not appear to change its distribution. Finally, van der Waals forces become increasingly important as the diameter is reduced, but the authors argue that this force is probably unimportant due to the natural roughness of the toner surface, which reduces the actual area of contact.

(3) Nonelectrostatic Adhesion

One might have thought that the work described above would have finally resolved the dispute on the relative importance of the electrostatic and nonelectrostatic contributions to toner adhesion. But a series of papers by *Rimai* et al. [12.20–22] appears to have reopened this issue. They studied the contact radii between spheres with diameters approximately equal to toner particles on flat substrates. Especially interesting were the studies of polystyrene spheres, having diameters between 1.5 and 12 μm, on polymer and silicon substrates. The particles were gently deposited on the substrate by sprinkling them from a height of 2 cm. They were presumably initially uncharged or low charged, since they were not mixed with a second powder, although some charge exchange probably occurred when contact was made with the substrate. The particles were left for two weeks before being prepared (with metal evaporation) for a scanning electron microscopic evaluation of the contact radii. Rimai et al. found the surprising result that when the polystyrene spheres were placed on a polyurethane substrate the contact radii were relatively large, approximately half the particle radii! In contact with silicon, the contact radii vary as the square root of the particle radii, which is inconsistent with elastic response models which assume Hertzian deformations. The data appear to indicate that the stress caused by surface forces causes a resulting strain that exceeds the elastic limit of the polystyrene spheres. The potential implications for toner particles are not discussed, but can be imagined.

12-2.3.2 Charge

A charge roller has been introduced by Canon [12.23] (which is used in the Hewlett Packard LaserJet IIP). The use of a charge roller instead of a corona wire significantly reduces, but does not entirely eliminate, the production of ozone. The mechanism of charging remains the same: a corona discharge is formed. For a charge roller, it occurs in the varying air gap between the roller and the photoreceptor.

The idea of using a charge roller is not new [12.24,25]. The patents discuss the problems which needed to be solved: (1) the ions generated in the corona discharge can destroy the roller material; (2) nonuniform charging; (3) pinholes in the photoreceptor will draw large currents, decreasing the potential of the charge roller, causing decreased charging over the entire length of the roller; (4) the roller itself will charge up with counterions and needs to be discharged.

The solution proposed by Canon is to use a two-layer roller built on a metal core. The inner layer is a conductive elastic layer and the outer layer is resistive. A dc voltage plus an ac voltage are applied to the metal core, with the ac peak-to-peak voltage greater than twice the dc voltage. One example given is a 6 mm diameter metal core, coated with 3 mm thick nitrile butyl rubber in which carbon is dispersed (resistivity 10^3 Ω cm), further coated with 50 μm of nylon (resistivity 10^{10} Ω cm). The applied voltage is 700 V dc plus 1300 V p-p 1 kHz ac.

12-2.3.5 Transfer

The electrostatics of the roll transfer system in the Eastman Kodak ColorEdge copier has been described in detail by *Zaretsky* [12.26]. *Kimura* et al. [12.27] also discuss roll transfer. A discussion of the conditions which give rise to air breakdown during the transfer step and some of the resultant copy quality defects has been given by *Fletcher* [12.28]. It was pointed out that pre-nip breakdown, often a concern for roll transfer, tends to reverse the polarity of the toner, leading to nonuniform transfer deletions. Even during "intimate" paper contact the roughness of the paper and high toner piles at the edge of images produce finite air gaps which limit the electric field that can be used for transfer. Above these limiting fields wrong sign toner can be generated in the toner pile, leading to a loss of transfer efficiency. Air breakdown during paper separation from the photoreceptor is a normal occurrence which tends to reverse the polarity of any residual toner on the photoreceptor and to drive the transferred toner on paper in the direction of the original toner polarity.

12-2.3.7 Clean

A comparison between magnetic brush cleaning and magnetic brush development with the same hardware has shown that magnetic brush cleaning can be more "efficient" than development [12.29]. "Efficiency" is determined by comparing the toner charge per unit area Q/A that can be cleaned or developed at the same voltage and speed ratio. It is shown experimentally that a conductive magnetic brush development can clean up to 6 times more Q/A than it can develop under the same conditions (with the polarity of the voltage reversed). The suggested explanation of this effect is based on an analysis of the magnitude of the electric field in the nip under the conditions of cleaning and development.

12-4 Toner Charging for Two Component Development Systems

Considerable progress has been made in our understanding of insulator charging since the first edition of this book. These advances have come from a comparison of toner charging characteristics and the surface state model.

Application of charge measuring tools (Sect. 12-4.4.4) to toner charging measurements has further verified the predictions of the surface state theory (Sect. 12-4.4.1). Significantly, a means of experimentally distinguishing between the high and low density limits of the surface state theory has been identified in the toner-carrier geometry (Sect. 12-4.3); only the high density limit is consistent with experiments. This represents a major advance in the theory of insulator charging. This advance suggests that the contact charging of insulators occurs to neutralize an electric field created in the interface during the contact event.

12-4.3 Insulator-Insulator Contact Charging

Cognizant of the experimental and theoretical difficulties in the field of contact charging, *Schein* et al. [12.30] chose to study insulator charging using a geometry in which reproducible results have been obtained: mixtures of insulating powders used in electrophotography. This system has empirically been made electrostatically reproducible in order to make copying and printing reproducible. With this system, progress has been made: an experiment was identified which can be explained by the high density limit of the surface state theory and is inconsistent with the low density limit of this theory.

Recall from Sects. 4.2.4, 4.4.1, and 12-4.4.1 (below) and [12.31] the discussions of the surface state theory. In this theory, charge is exchanged between surface states of the two materials, driven by the "surface work function" difference between the materials. The theory has two limits, schematically indicated for metal-insulator contacts in Fig. 4.7 and for insulator-insulator contacts in Fig. 12.3. In the high surface state density limit (Fig. 12.3b), charge exchange is large enough to raise the insulator with the larger work function (before charging) to the energy level of the insulator with the smaller work function (on the left in Fig. 12.3b). This limit requires the charge exchanged per unit area to be orders of magnitude larger than is experimentally observed and is therefore usually dismissed. In the low surface state density limit, charge is exchanged to fill the states between the two work functions (Fig. 12.3a), from the surface of the material with the lower work function to the surface of the material with the higher work function. Such a theory can explain the observed linear dependence of insulator charging on metal work function (Sect. 4.2.4) only if the density of surface states per unit energy is constant. It appears to describe toner-carrier charging experiments (Sects. 4.4.1, 12-4.4.1). But it also implies electron exchange, which is difficult to reconcile with the evidence for ion transfer (Sect. 4.2.3). It assumes a "surface work function" can be defined for an insulator, which requires thermodynamic equilibrium. However, these materials have virtually zero conductivity, making it difficult to understand how charges can move to es-

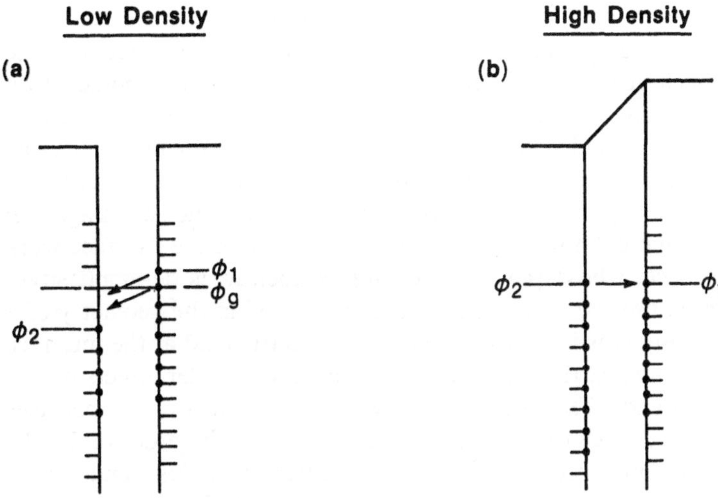

Fig. 12.3. Schematic diagram of the insulator-insulator contact in the plane-plane geometry for the low (**a**) and high density (**b**) limits of the surface state theory. ϕ_1 and ϕ_2 are the "surface work functions" (energy difference from highest occupied level to vacuum) of the insulators and ϕ_g is the final common work function after charge is exchanged. A dash at the interface represents a surface state; a dot on the dash indicates a filled surface state before contact and an arrow indicates the movement of charges during charging.

tablish an equilibrium condition. Further, the nature of the "surface states," which corresponds to 1 molecule in $10^4 - 10^5$ being charged, has never been identified.

What Schein et al. were able to do was to suggest an experiment which distinguishes the low from the high density limit of the theory in the geometry of a mixture of two insulating powders. Only the high density limit is consistent with the data. Further, it was shown that this limit of the theory can explain almost all available insulator charging experiments.

The argument is as follows: Consider the predictions of the charge-to-mass ratio Q/M versus the toner concentration C_t in the two limits of the theory. The low surface state density theory, as applied to the toner-carrier geometry, is discussed in Sect. 4.4.1; the prediction of M_t/Q_t versus C_t is given by

$$\frac{M_t}{Q_t} = RC_t\left(\frac{\rho_c}{3\Delta\phi e N_c}\right) + r\left(\frac{\rho_t}{3\Delta\phi e N_t}\right), \tag{4.14}$$

where N_c (N_t) is the carrier (toner) surface state density per unit area per unit energy, ρ_c (ρ_t) is the density of the carrier (toner) and $\Delta\phi$ is the difference of the carrier and toner "surface work function."

In the high density limit, in the single contact plane-sphere geometry, charge is exchanged so that the electrostatic energy $\sigma_s ze/\varepsilon_0$ caused by the charge exchange offsets the change in work functions at the interface, where σ_s is the charge exchanged per unit area and z is the distance between the insulators at which charge exchange ceases. Alternatively, one could physically interpret the high density limit to indicate that charge is exchanged to neutralize an electric field created at the interface between the contacting surfaces determined by the difference in work functions $(\phi_c - \phi_t)$ and the gap z. The field is simply $(\phi_c - \phi_t)/ez$. This interpretation has the advantage that it combines all of the unknown parameters, and the role of the "surface work function" difference, whose physical meaning is unclear, is de-emphasized. This model can be used to derive the charge exchanged in the geometry of a mixture of insulating powders by calculating the electric field at the interface between the toner and carrier particles. Ignoring second order effects due to the image charges and effects due to the dielectric constants of the particles (which would only change the values by small factors on the order of 2), the electric field at the interface between a toner particle of radius r and charge Q and a carrier particle of radius R and charge Q_c can be written by inspection

$$E = \frac{1}{4\pi\varepsilon_0}\left(\frac{Q_c}{R^2} + \frac{Q}{r^2}\right) . \tag{12.1}$$

Other toner particles on the carrier will contribute to E, but their effect is a small correction to the smaller Q/r^2 term due to the strong dependence on the distance between these charges and the interface. Using charge neutrality $-Q_c = nQ$, and assuming charging ceases when E reaches a material-dependent value E_e, an expression for M_t/Q_t is easily found

$$\frac{M_t}{Q_t} = RC_t\left[\frac{\rho_c}{3\varepsilon_0 E_e}\right] + r\left[\frac{\rho_t}{3\varepsilon_0 E_e}\right] . \tag{12.2}$$

Note that (12.2) (the high density limit) is identical to (4.14) (the low density limit) with $\Delta\phi e N_c$ and $\Delta\phi e N_t$ replaced with $\varepsilon_0 E_e$. There is an important difference between (4.14) and (12.2): the slope-to-intercept ratio of M_t/Q_t versus C_t is determined entirely by known parameters $R\rho_c/(r\rho_t)$ in (12.2); it is determined by the product of this parameter and N_t/N_c in (4.14). This result has obvious experimental implications that are not present in the single contact plane-sphere geometry. (Prior workers [12.32–34] have suggested (12.2) is another explanation for observation of linearity in M/Q versus C_t data.)

These predictions can be compared with experimental measurements of M/Q versus C_t by *Lee* [12.35] and *Anderson* [12.36] and data shown in Fig. 12.4 (in which the concentration of charge control agent, Sect. 4.4.3, was also varied), as summarized in Table 12.3. The interesting result is that the slope-to-intercept ratio agrees with the prediction of the high density limit of the theory within a factor of two for almost all the experiments with no adjustable parameters. One must therefore argue that $N_c = N_t$, within a factor of two, for all of the different toner-carrier systems characterized by these sets of data, taken with different toners and carriers at different laboratories, which seems unlikely, or that the high density limit is the correct description of the data.

The analysis of Fig. 12.4 in Table 12.3 is also independent evidence for the validity of the high density limit. In this experiment, only the percentage of charge control agent relative to the toner was changed; the carrier remained unchanged. The increased charging due to the addition of charge control agent in the toner can be ascribed to a change in N_t in the low density limit of the theory. This predicts only a change in intercept. Nonetheless, experimentally both the slope and intercept changed, maintaining the ratio constant, inconsistent with the low density limit, but consistent with the high density limit of the theory. Clearly E_e, the field being neutralized in the high density version of the theory, changes as the amount of charge control agent is changed ($E_e = 8.8\ V/\mu m$ at 0% CCA and $21.3\ V/\mu m$ at 2.5% CCA), indicating that E_e is determined by some material properties.

Further evidence for this result can be found in other studies by *Anderson* and *Bugner* [12.37] and life studies of toner-carrier systems published by *Nash* and *Bickmore* [12.38]. Anderson and Bugner showed that the intercept changed as the CCA concentration changed at low toner concentrations, consistent with these results. Nash and Bickmore showed that the Q/M of toner in a two-component mixture degrades with time as

$$\frac{Q}{M} = \frac{A_c(t)}{C_t + k_0} \tag{12.3}$$

where $A_c(t)$ is the reduction in fresh carrier surface area with time as toner is fused permanently to the carrier surface (Sect. 12-4.4.5); k_0 is a constant "unaffected by developer aging." Note that the slope-to-intercept ratio of M/Q versus C_t is just k_0^{-1}, a constant. This is further evidence that toner-carrier charging is determined by the neutralization of an effective electric field, as predicted by the high density limit. The low density limit of the theory predicts a change in slope only [in (4.14), a decrease in A_c affects only the slope], which is inconsistent with the experimental result. The physical picture that emerges is that the area of the carrier not covered with permanently stuck toner continues to exchange charge with the toner to neutralize the electric field E_e created during the contact. The reason why toner charging decreases is that the area of fresh carrier surface is decreasing with time.

The difficulty with the high density limit of the surface state theory is that the charge exchanged per unit area predicted using reasonable estimates of the parameter is orders of magnitude higher ($6 \times 10^{12}\ cm^{-2}$) than observed ($10^{10}\ cm^{-2}$). This results from assuming $\Delta\phi = 1$ eV and $z = 10$ Å. It is also not physically realistic that the insulators used, toner and carrier particles, have associated fields before mixing. If an electric field existed at the surface of an

Table 12.3. Analysis of M/Q versus C_t data. The figure numbers in parentheses refer to this book; the other figure numbers refer to the original papers

Data	Material parameters R [μm]	r [μm]	ρ_c [g/cm^3]	ρ_t [g/cm^3]	Slope/Intercept Observed	$\frac{R\rho_c}{r\rho_t}$
Lee [12.35]						
Fig. 2 (Fig. 4.12b)	50	7.1	5.5	1	50	38
Fig. 3	50	7.1	5.5	1	29	38
Fig. 4	50	7.1	5.5	1	18	38
Fig. 5 (Fig. 4.12c)	125	7.1	5.5	1	236	96
Anderson [12.36] Fig. 2	50	5	4	1	32	40
	15	6	5.5	1	3.7	14
(Figure 12.4)						
0% CCA	100	5	7.7	1	185	154
0.5%	100	5	7.7	1	142	154
2.5%	100	5	7.7	1	141	154

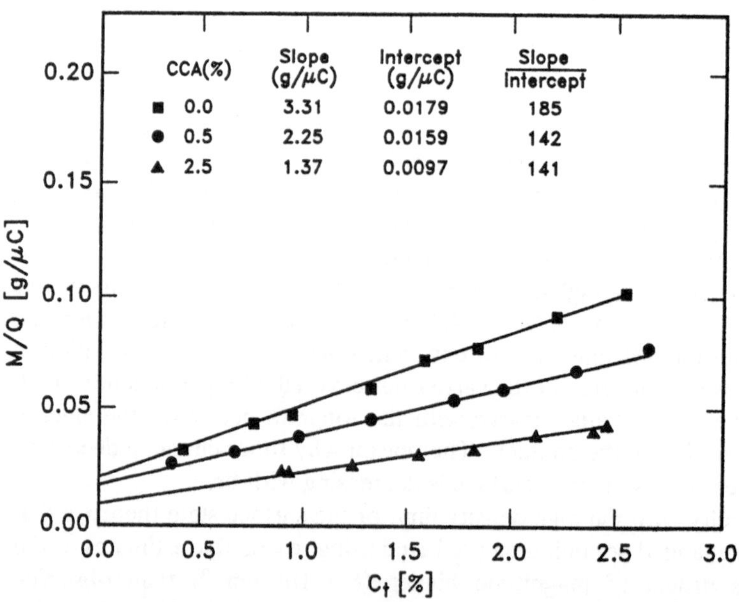

Fig. 12.4. Measured mass-to-charge ratios M/Q versus the toner concentration C_t for the same carrier and three toners with charge control agent percentages of 0, 0.5 and 2.5%. The least squares determined slope, intercept, and their ratio are given in the figure [12.30]

insulator, the insulator would have a net charge; yet these powder insulators are neutral before mixing (as can be demonstrated by putting them into a Faraday cage). Even if they were charged, stray charges from the atmosphere would neutralize them with time.

This is the first experiment, to our knowledge, which can distinguish between the low and high density limit of the surface state theory. The experimental results indicate that only the high density limit describes the experimental data. This theory explains not only data on mixtures of insulating powder, but also experiments in the single contact plane-sphere geometry [12.30]. The numerical difficulties associated with this theory remain unresolved, which implies that the association of the electric field E_e with material properties, such as the difference in surface work functions and the gap between the materials when charging ceases, remains unclear. Nonetheless, the agreement between experiments and theory clearly indicates that charge is exchanged to neutralize an electric field created during the contact event. That material properties determine E_e is clear since E_e changes for each toner shown in Fig. 12.4 and in fact for every toner-carrier mixture. Therefore, this theory focuses attention on where work is needed to complete the theory of insulator charging: finding the physical significance of E_e.

12-4.4 Toner-Carrier Charging

As is well known, insulator-insulator or metal-insulator charging requires many contacts before the charge exchange stabilizes (Fig. 4.4). The same is true for the toner-carrier system (Fig. 4.11). *Anderson* [12.39] showed that this equilibrium value is independent of mixing rate, as shown in Fig. 12.5.

12-4.4.1 Surface State Theory

Several advances were made in the last four years which continue to indicate that the surface state theory describes toner charging. (1) Further evidence was given that toner-carrier charging is a surface phenomena. (2) The surface state theory was generalized to a nonuniform density of states and correlated with molecular properties. (3) The theory was generalized to describe the distribution of charge among the various size toner particles within one mix.

(1) Toner-Carrier Charging Is a Surface Phenomenon

Further evidence that toner-carrier charging (and therefore insulator charging) is a surface phenomena, as assumed in Lee's surface state model (Sect. 4.4.1), has been presented in two publications. *Hutcheson* and *Sukovich* [12.40] showed that various surface treatments of carrier made of ferrite, such as acid

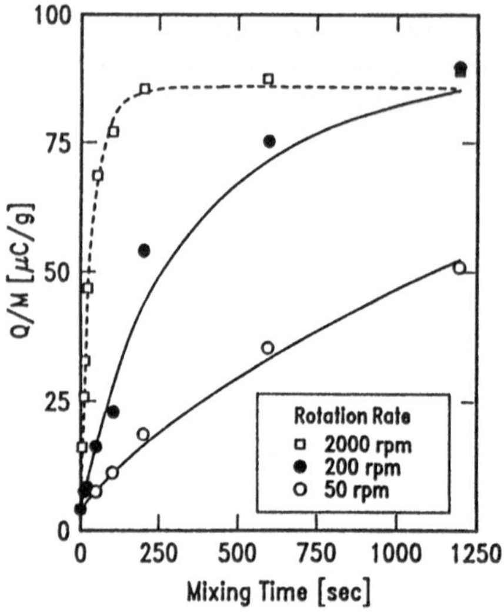

Fig. 12.5. Charging rate as a function of magnet rotation rate (unpigmented toners on uncoated ferrite) [12.39]

washing, chemical treatments or mechanical processing (rubbing the carrier beads against themselves), can dramatically change the amount of charge that the carrier exchanges with toner particles. *Nash* and *Bickmore* [12.38] systematically studied the changed toner charging characteristics as the carrier ages by accumulating permanently stuck (fused, sometimes called impacted) toner on its surface. This occurs when toner particles are trapped between carrier-carrier impacts during the mixing process within the development system housing. This contamination of the carrier surface reduces the area for charging fresh toner particles, also demonstrating that charging is a surface phenomena.

(2) Surface State Model — Further Evidence

In a series of papers *Anderson* et al. [12.36,37] generalized the surface state theory to a nonuniform density of states (a Gaussian distribution was assumed), taking into account the probability that the states were occupied, allowing for both donor and acceptor sites on both surfaces, while retaining the ability to move the energy levels consistent with knowledge of the molecular energy levels of the charge control agents.

Anderson's generalization of the theory required a computer to accomplish the calculations. The fundamental predictions of the surface state theory remained: M/Q is still predicted to be linear in C_t. Experimental data which were fit with the theory include M/Q versus C_t (at only very low C_t's) with the amount of charge control agents as a parameter and Q/M versus the energy of the highest occupied molecular orbital (HOMO) of the charge control agent, which had been modified chemically to change its electronic structure

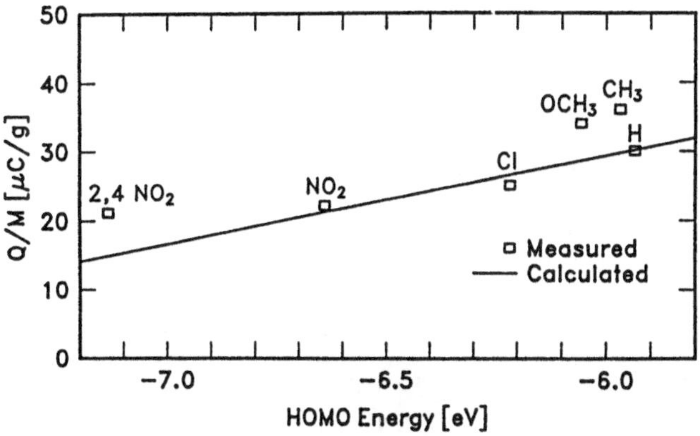

Fig. 12.6. Q/M as a function of HOMO energy for anion-substituted charge agents [12.37]

systematically (Fig. 12.6). *Suzuki* and *Isoda* [12.41] observed the sign of the toner charge to invert under certain conditions, and were able to account for such data by invoking the surface state model and contributions to charging by the various components of toner.

(3) Surface State Model — r Distribution

Schein et al. [12.42] inquired whether the surface state theory can describe the distribution of charge among the various size toner particles within one mix. Obviously, a tool which measures charge distributions is necessary for such a study. Using the charge spectrometer shown in Fig. 4.15, it was shown that, indeed, the surface state theory remains consistent with the data.

Theoretically, (4.11–14) need to be generalized. Assuming two different radii toner particles, labelled with subscripts 1 and 2, allows for simplicity in deriving the equations; generalization to a distribution of radii is trivial. The generalization of (4.11) and (4.12) proposed is

$$Q_c = eA_cN_c(\phi_g - \phi_c) \quad \text{unchanged,} \tag{12.4}$$

$$Q_1 = eA_1N_t(\phi_g - \phi_t), \tag{12.5}$$

$$Q_2 = eA_2N_t(\phi_g - \phi_t), \tag{12.6}$$

i.e., the charge on toner particle with charge Q_i is proportional to its area A_i. Charge neutrality demands

$$-Q_c = n_1Q_1 + n_2Q_2. \tag{12.7}$$

Again, ϕ_g is eliminated from the equations. If

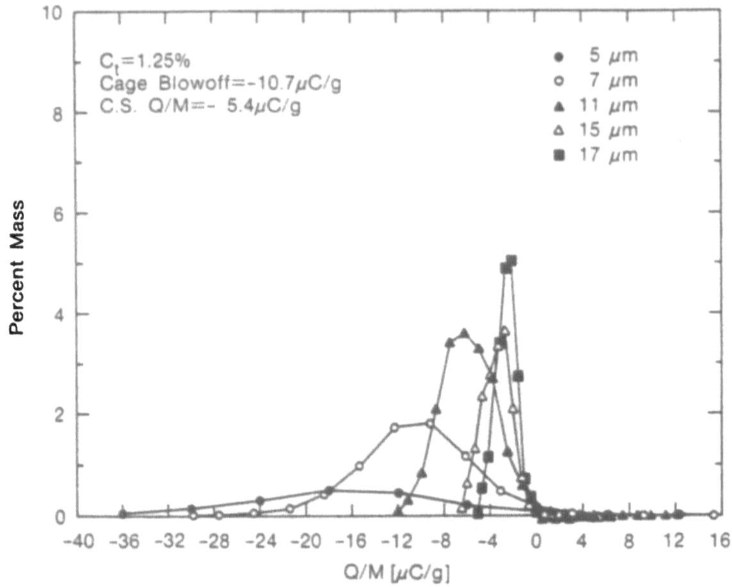

Fig. 12.7. Charge spectrometer measurements at $C_t = 1.25\%$ [12.42]

$$< r > \equiv (n_1 r_1^3 + n_2 r_2^3)/(n_1 r_1^2 + n_2 r_2^2), \tag{12.8}$$

$$C_t \equiv (n_1 M_1 + n_2 M_2)/M_c, \tag{12.9}$$

are defined, then the M_1/Q_1 for the toner of radius r_1 is

$$\frac{M_1}{Q_1} = RC_t\left(\frac{\rho_c}{\Delta\phi e N_c}\right)\frac{r_1}{< r >} + r_1\frac{\rho_t}{3\Delta\phi e N_t} \tag{12.10}$$

and the total M/Q for all the toner is

$$\frac{M}{Q} = RC_t\left(\frac{\rho_c}{\Delta\phi e N_c}\right) + < r > \left(\frac{\rho_t}{3\Delta\phi e N_t}\right). \tag{12.11}$$

In comparing (12.10) and (12.11) with (4.14), notice that r is replaced by $< r >$ in the expression for M/Q, and an extra factor $r_1/ < r >$ appears in the expression for the slope M_1/Q_1.

In order to experimentally test (12.10), measurements were made of the peak Q_i/M_i as a function of r_i in a charge spectrometer for a range of toner concentration, 1%–2.1% for an experimental toner and a 200 μm diameter uncoated steel carrier. Results at 1.25% and 2.1% toner concentrations are shown in Figs. 12.7 and 12.8. Note that at 2.1%, which slightly exceeds a

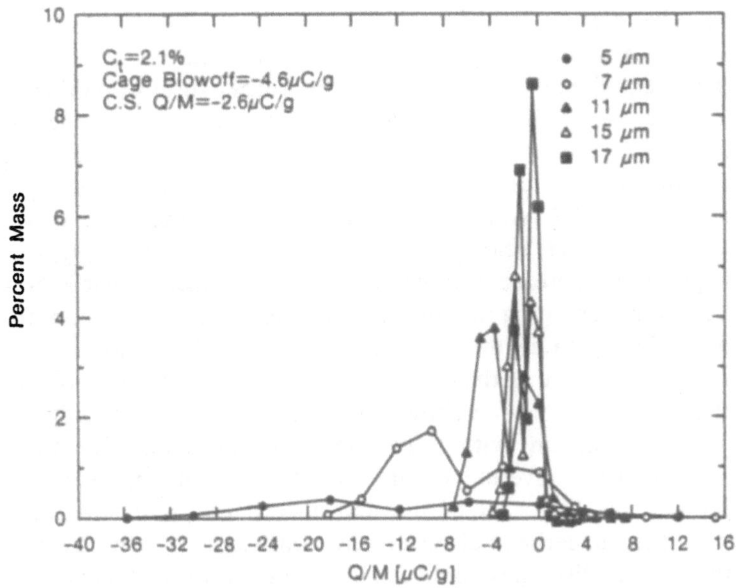

Fig. 12.8. Charge spectrometer measurements at $C_t = 2.1\%$ [12.42]

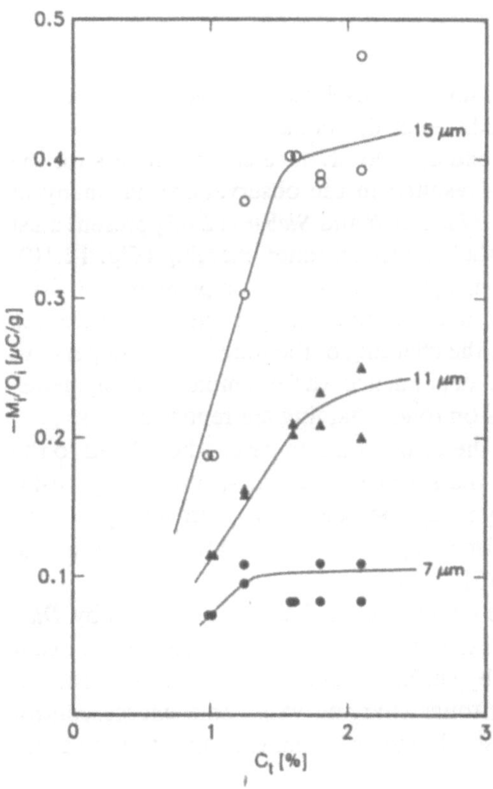

Fig. 12.9. Peak M_i/Q_i versus C_t for various toner diameters [12.42]

monolayer coverage, some of the distributions have two peaks, including a lower one near zero charge. It appears that above a monolayer some of the toner particles make limited or no contact with the carrier surface, producing low or uncharged toner, which is an experimental observation with some important technological implications. Plotting the peaks of these distributions for the 7, 11 and 15 μm diameter toner particles (ignoring the peak at zero charge) against C_t results in Fig. 12.9. Above a monolayer ($\approx$ 1.7%) the M_i/Q_i versus C_t curves level off. However, below a monolayer, the slope of the M_i/Q_i versus C_t curves appears to increase as r_i increases. This is exactly what is predicted by (12.10), demonstrating that the surface state theory can account for the distribution of charge among the different size toner particles within a mix.

Equations (12.4–11) were derived assuming the low density limit of the surface state theory. In Sect. 12-4.3 above it was shown that, in fact, only the high density limit is consistent with (other) experimental data. Equations (12.10) and (12.11) are converted to the high density limit by replacing $\Delta\phi e N_c$ and $\Delta\phi e N_t$ with $\varepsilon_0 E_e$. All of the above conclusions remain valid. Therefore, while these experimental results are consistent with the surface state theory, they do not distinguish between the two limits of the theory.

12-4.4.3 Charge Control Agents

As expected (Sect. 4.4.5), significant new results on charge control agents have been reported since the first edition of this book.

The increased use of pigments and dyes to produce colored toners for the new color copiers (Chap. 11) has resulted in the observation that many of these act as charge control agents. *Mackholdt* and *Sieber* [12.43] present a list of pigments ordered according to their effect on toner charging (Fig. 12.10). The effect is not simple: the charging appears to depend on pigment resistivity over certain ranges of resistivity and on pigment particle size once the resistivity effect is saturated. Also, the charging of the toner is time dependent (as is well known) and this time dependence varies among the pigments. Substituent and crystallinity effects on toner charging are reported.

It has been shown [12.44] that the charge on a toner can be related to the amount of polyolefin additives on the surface of styrene-acrylic or polyester resin based toner, as determined by x-ray photoelectron spectroscopy.

The importance of properly dispersing the charge control agents in the toner is discussed by *Hayashi* [12.45].

The use of isonomers as charge control agents has been reported by *Diaz* et al. [12.46–48]. Observations of the ionic part of the charge control agent on the carrier surface after mixing by surface science tools [12.46–49] has led to the suggestion [12.46–48] that toner charging with some charge control agents is due to mobile anions physically moving from the toner to the carrier

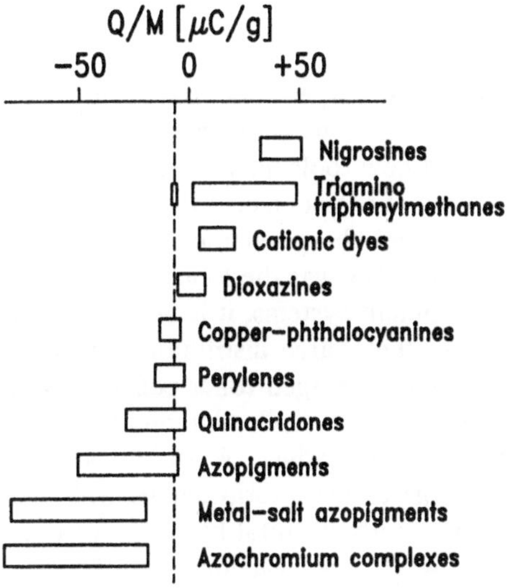

Fig. 12.10. The typical range of triboelectric effects after 30 min of mixing for various pigment classes [12.43]

surface. This concept has the difficulty that it cannot account for the fact that a toner particle's charge depends on the number of toner particles on the carrier surface, i.e., M_t/Q_t is linear in C_t (Sect. 12-4.3).

It was pointed out in Sect. 4.2.3 that water layers on polymer, i.e., toner, surfaces should be considered in any model of toner charging (p. 73). *Matsui* et al. [12.50] have provided a detailed picture of how water may be playing a central role in toner charging together with partial experimental verification. Their picture assumes that the charge control agent is oriented on a toner surface such that a hydrophobic resin causes the charge control agent to orient with its hydrophobic part facing inward and its hydrophilic part facing outwards. Surface water therefore forms a water bridge preferentially from the hydrophilic end of the charge control agent to the adjacent carrier surface. The presence of the charge control agent causes the net charge on the oxygen atom of the water molecule to be more negative. Subsequent dissociation of an end water molecule is then responsible for the toner charging: OH^- goes to the toner; H^+ goes to the carrier. Correlation of the toner Q/M with net calculated charge on the oxygen atom and the lack of correlation with any other electronic property of the charge control agent tends to support this model. However, it is not clear that such a model can explain why a toner particle's charge is experimentally observed to depend on the presence of other toner particles on the carrier (Sect. 12-4.3).

12-4.4.4 Charge Measuring Tools

As it becomes clearer that wrong sign and zero charged toner have important negative effects on image quality, efforts to measure and understand toner charge distributions have increased. Several new tools have been introduced and the results have been analyzed in an attempt to understand the physics and chemistry of toner charging. For each new toner charge characterization tool, it is useful to keep in mind several questions. How was the toner charged? If it is not by methods used in normal development systems, it is not clear that the charge distribution obtained will reflect the charge distribution of toner during the development process. How was the charged toner collected and injected into the charge characterization apparatus? Collection and injection of the toner can preferentially select by charge or radius. A useful test is to determine whether the observed total charge-to-mass ratio agrees with the total charge-to-mass ratio obtained independently by total blowoff, for dual component development, or plate blowoff (or vacuum liftoff), for monocomponent development.

(1) Millikan's Oil Drop Experiment

The classic experiment of Millikan, which determined the magnitude of the electron's charge by levitating particles using opposing gravity and Coulomb forces, can be used to measure the charge on individual toner particles.

Folan et al. [12.51] focused their experiment on a fundamental issue: what is the role of surface water layers in insulator charging? They built an apparatus that could inject a single spherical particle (made of a polystyrene-divinyl benzene copolymer), levitate it (and hence measure its charge), contact it onto a Ni surface, and re-levitate it. The charge was observed to increase in approximately equal amounts ($-0.045\ \mu C/g$ or almost 1500 charges per contact) each time it touched the Ni surface. Then the inside of the apparatus, including the levitated particle, was evacuated to about 10^{-3} Torr for 4 h to evaporate any water vapor before being back filled with purified N_2. Contacts with the Ni plate resumed. The same charging, within experimental error, was observed. As a further check, the system was again evacuated for 24 h; again, the same amount of charging occurred at each contact. One can conclude either: (1) that aqueous layers (i.e., H^+ and OH^-) do not play a role in the charging of the polymer particles tested or (2) the evacuation was insufficient to remove a monolayer of aqueous water on the surface. (Direct observation of the removal of almost all the water from the particle was made by using the levitator as a pico-balance and by elastic light scattering.)

Measurements on actual toner particles made on the Millikan oil drop apparatus have shown that the charge appears to increase as the third power of the toner radius [12.52]. This is a surprising result because it suggests that the bulk and not the surface of an insulator is involved in charging, despite evidence to the contrary (Sect. 12-4.4.1). This result may reflect the toner charging mechanisms used: it appears that the particles were blown through tubes, charging by contacting the walls ("cloud developing method"), or by using inertial forces to strip toner off the carrier by dropping the carrier on a mesh.

(2) Charge Spectrometer

The charge spectrometer, which uses laminar air flow and crossed electric fields (Fig. 4.15), was first introduced by *Lewis* et al. [12.53] and later by *Terris* and *Jaffe* [12.54]. A more detailed description of the instrument has been presented and a thorough analysis of its capabilities has been carried out.

(a) Description of Charge Spectrometer. *Terris* and *Fowler* [12.55] presented a description of their charge spectrometer (which was also used by Schein et al., see below and Sect. 12-4.4.1). Of particular interest is a modified injection system that allows the characterization of toners from a monocomponent system, shown in Fig. 12.11. It consists of two concentric cylinders which are brought into contact with a toner-bearing surface. The air flow from the outer cylinders dislodges the toner from the surface and transports the particles into the inner cylinder, which then directs the toner into the funnel in the charge spectrometer (Fig. 4.15).

Other interesting aspects of the device are described [12.55]. For example, (1) the filter paper that captures the toner must be grounded to prevent the field of the landed toner from distorting the trajectory of incoming toner. But grounding the filter paper distorts the electric field at the bottom of the charge spectrometer. A series expansion of the electric field and appropriate correction factors are given. (2) As shown below, the average Q/M calculated from charge spectrometer data is lower than the Q/M obtained by total blowoff (for dual component) or vacuum liftoff (for monocomponent). For

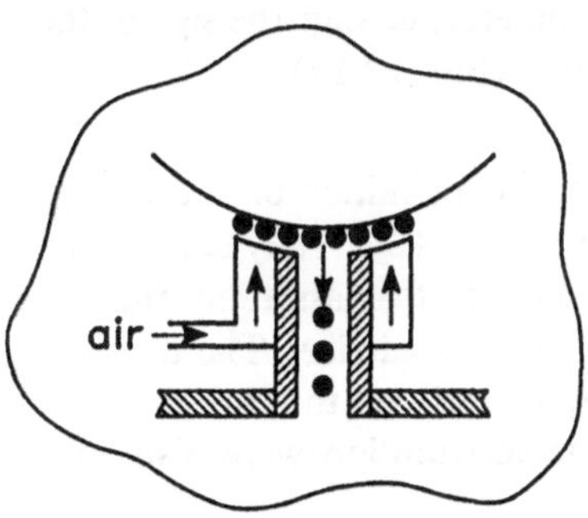

Fig. 12.11. Injector for monocomponent toner [12.55]

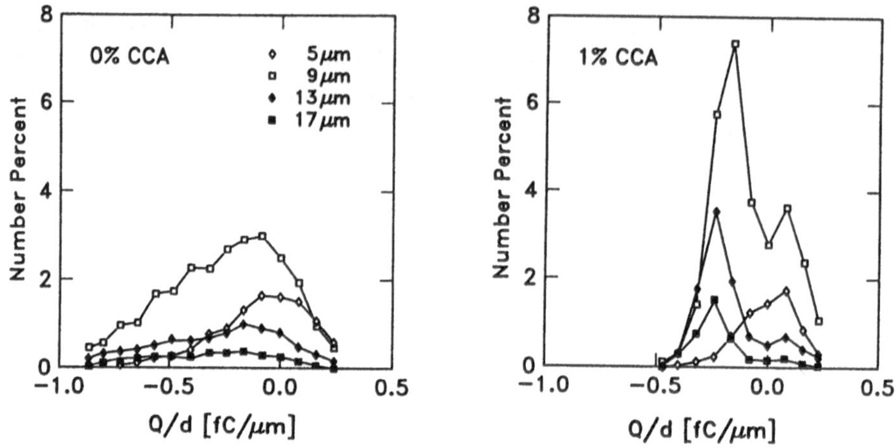

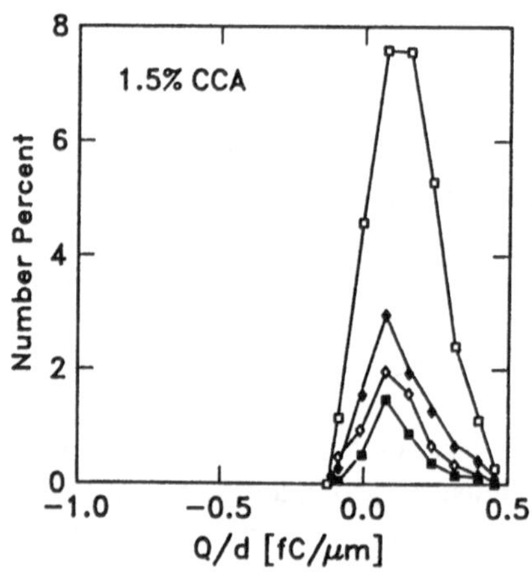

Fig. 12.12. The charge distributions for toner with three levels of charge control agents, 0%, 1.0% and 1.5%. As the charge control agent is added, the distribution shifts from negative charge to positive charge. For clarity, data from only four particle sizes are shown [12.55]

the dual component system, part of the problem is toner left on the carrier; for the monocomponent system, the origin of the discrepancy was not determined. (3) It is verified that used mixes have more wrong sign toner (Fig. 4.15c) than fresh mixes and that background development on the photoreceptor has more wrong sign toner than image development (Sect. 12-6.6). (4) It is also shown how the addition of charge control agent can completely change the sign of the toner charge in one particular dual component mix (Fig. 12.12).

(b) Analysis of Capabilities. An analysis of the capabilities of the charge spectrometer has been carried out by *Schein* et al. [12.42]. They studied the effect on the final result of the average toner charge-to-mass ratio and the needle air pressure used to separate the toner from the carrier. The dependence of the peak charge and the half-width to peak ratio on toner radius, and the effect of toner concentration on the charge distribution were also analyzed.

In this study the carrier and toner particles were charged by the usual technique: rolling them together. Then a small amount was put into a blowoff cage built into the top of the charge spectrometer (see Fig. 4.10 and description in [12.55]). A needle was used to direct an air stream at the mixture, driving the toner off the carrier and through the bottom screen into the charge spectrometer. The maximum air velocity out of the needle is determined by the requirement that the air velocity out of the glass tube which injects toner into the charge spectrometer must match the air velocity in the charge spectrometer (about 30 m/s).

It was found that for toner-carrier mixtures with $Q/M \approx 9\mu C/g$ almost all the toner is removed from the carrier. With $Q/M > 20\mu C/g$ not enough toner is stripped from carrier to give a representative sampling. Even when all of the toner particles are removed from the carrier, only approximately 10% of the toner ends up on the filter paper. The rest are caught on the walls of the glass tubes and in the glass frit. To attempt to determine the extent to which the data are affected by this loss of toner, experiments were carried out with varying needle air pressure. It was found that the peak and relative width remained unchanged, but the relative heights of the peaks changed. Higher pressures are required to carry the smaller toner particles to the filter paper. The cumulative size distribution of the original toner and the captured toner confirmed this. Calculations of the (total Q)/(total M) were always lower than blowoff Q/M by about a factor of 2. This could be correlated with the loss of small diameter toner particles (which have large Q/M's).

Using a double log plot of peak Q/M versus d, n in $Q \propto d^n$ was found to be between 1 and 2 with n observed to be lower at higher toner concentrations, consistent with earlier results [12.56]. This was rationalized in terms of an area effect; it was hypothesized that when n is less than 2, only a fraction of the toner area is able to contact the carrier surface.

The relative width of the peaks was observed to be ≈ 0.4 for one toner carrier system (with some tendency to decrease as r increased) and ≈ 1 for another system (in which case the relative width was independent of r). It was pointed out that the size of the bins into which the toner diameters are sorted by the instrument itself will produce a relative width on the order of 0.2, close to one of these results.

Finally, the effect of toner concentration was studied. One interesting result, shown in Fig. 12.8, is that above a monolayer of toner on the carrier surface the distributions have two peaks, one near zero. It appears that once a monolayer is achieved, some of the toner can no longer contact the carrier surface and obtain its "share" of charge. The study of the effect of toner concentration on charging has implications for the physics of insulator charging; this is discussed in Sect. 12-4.4.1.

(3) Gravity and Crossed Electric Fields

Takahashi and *Nakabayashi* [12.57] continued work (p. 90) on their method in which the toner size and charge are measured simultaneously by letting toner particles fall under gravity in a crossed electric field and capturing the fall on video tape. The toner was injected by using an insulative fur brush to mechanically separate toner which had been developed on a photoreceptor. While one might be concerned that the fur brush could affect the toner charge, a comparison of Q/M calculated and measured by the blowoff method agreed. A large amount of wrong sign toner can be seen in the data. Also, they found $Q \propto d^3$.

In another paper, the same authors [12.58] report using this tool to study the charge distribution of toner on carrier. Here the toner and carrier were put into a blowoff cage. The cylinder was vibrated at 40 Hz to release toner. The Q/M calculated was not given, making it difficult to judge the reliability of the sampling. Clear evidence for a broader distribution for aged developer was shown, consistent with earlier results (see Figs. 4.15 and 4.16 for two examples). They listed the problems with this tool: difficulty of getting complete separation of toner and carrier, need for reduction of measuring time, and need for simplification of operation.

(4) Laminar Air Flow, Crossed Electric Fields; Measure Optical Density

In order to reduce the measurement time of the charge spectrometer [Sect. 12-4.4.4, (2) above], *Kimura* et al. [12.59] suggested measuring the optical density on the filter paper. This, of course, provides no information about the d dependence, but it does produce the Q/d distribution integrated over all d. The method was shown to be reproducible, although a comparison of Q/M calculated and directly measured was not given. The injection method used was to flow nitrogen gas at a toner-carrier mixture captured magnetically on a roller. Efficiency of toner release was not discussed.

(5) Gravity, Crossed Electric Field; Measure Toner Velocity

Tabata [12.60] has suggested a method which differs from the above techniques in that it measures the toner velocity, captured on video film, before it lands on a substrate. The toner is released from carrier beads by a weak air jet and then falls under gravity in a crossed electric field. The velocity, obtained from (4.16), is proportional to Q_t/r. Images are recorded on video tape and computer analyzed. The toner diameter distribution was verified, and the Q/d distribution was checked against the E-SPART method [12.61]. However, the values of Q/M obtained, $\approx 1\ \mu C/g$, appear low for a dual component system.

(6) Laminar Air Flow, Crossed Electric Fields; Capture Toner on Walls

Epping [12.62] has developed a commercially available tool in which the toner charge distribution is determined by analyzing the position along the walls at which the toner particles land after experiencing laminar air flow and crossed electric fields, somewhat similar to the instrument shown in Fig. 4.14. A description of this work [12.62] and a suggested optimization of its parameters have been published [12.63].

(7) Three-Dimensional Method

Yet another technique, suggested by *Kutsuwada* and *Nakamura* [12.64], is to use air flow, electric fields and gravity, as shown in Fig. 4.17. The toner, as it lands on the walls, is automatically separated by charge (contour lines shown) and diameter (horizontally). Additional measurements are reported in [12.64].

12-4.4.5 Life Characteristics

A subject not discussed in the first edition of this book is life characteristics of the toner-carrier mixture. It is observed that the Q/M of the toner decreases in a copier as many copies are made. This degradation is due to toner sticking to the carrier surface, which is sometimes called toner impaction [12.27]. Fresh toner then has less carrier surface to be charged against.

The cause of impacted toner is toner particles fused at carrier-carrier impact points during the mixing process within a developer housing. *Nash* and *Bickmore* [12.38] presented a theory of impaction in two parts. First they assumed that first order kinetics governed the impaction of toner on carrier surface, giving an exponential decrease in the area $A_c(t)$ of fresh carrier surface with time (t). (They ignored the possibility of impacted toner-toner charging.) Then they empirically expressed the Q_t/M_t in terms of this $A_c(t)$,

$$\frac{Q_t}{M_t} = \frac{A_c(t)}{C_t + k_0} , \tag{12.12}$$

where k_0 is a constant. They state k_0 "is a characteristic offset term for any given carrier/toner system" which is "unaffected by developer aging." The significance of this observation was pointed out in Sect. 12-4.3; the slope-to-intercept ratio of M/Q versus C_t is just k_0^{-1}, giving additional evidence that toner-carrier charging is determined by the neutralization of an electric field.

The authors show that $A_c(t)$ has the intuitively expected dependence on hardware and material parameters, such as sump size [$A_c(t)$ increases as less developer is placed in the sump since each carrier bead is subjected to more impacts per unit time], toner diameter (smaller diameter toner particles impact faster), and addition of nanometer-diameter silica additive (reduces impaction rates). Such results provide guidelines for extending the life of the developer.

Another idea for extending the life of the developer was proposed by *Hart* et al. [12.65]. Generally, when the toner charge falls below a certain level, image quality degrades and the developer (the carrier and toner) is replaced. Instead, Hart et al. suggest slowly adding fresh developer all the time. The fresh developer mixes with the "aged" developer, providing more carrier area for toner charging. Excess carrier "trickles" out of the developer housing and is captured and presumably disposed of.

12-6.2.2 Effective Dielectric Constant

The physical significance of the effective dielectric constant has been identified and both theoretical and experimental results are in agreement. The effective dielectric constant, K_E, which measures the enhancement of the electric field over the air gap value in the insulative magnetic brush development system, is simply equal to the dielectric constant K of the toner-carrier mixture as claimed in [12.66] and by *Hays* [12.67]. An interesting and surprising result is that the dielectric constant, and therefore the effective dielectric constant (and therefore M/A), increases as the carrier diameter decreases, because higher carrier packing fractions are achieved in the nip.

The experimental demonstration of this result was achieved [12.68] by varying both K_{max} and K and measuring K_E. K_{max} is the ratio of the maximum and the air gap value of the electric field (which occurs directly under a carrier chain). By varying the carrier diameter and carrier coating thickness, K_{max} was varied (6.27). The mass flow rate F of the carrier beads per unit length of developer was measured and is related to the carrier packing fraction p by

$$F = \rho_c v_r p / L \,, \tag{12.13}$$

where ρ_c is the carrier density, v_r is the roller surface velocity and L is the gap. Given p, the dielectric constant can be determined using Fig. 6.8. The effective dielectric constant was determined experimentally by using the equation of development, (6.57), and ignoring the small term d_s/K_s, giving

$$K_E = \frac{M}{A}\frac{Q}{M}\frac{L}{V\varepsilon_0 v} \,. \tag{12.14}$$

The slope of M/A versus V curves and measurements of Q/M were used to calculate K_E.

The results are shown in Table 12.4. Note that the experimental values of K_E vary from 7 to 11, with the larger values associated with the small diameter

Table 12.4. Measurements of the effective dielectric constant.

Carrier diameter, $2R[\mu m]$	Coating thickness, $1[\mu m]$	K_E experiment	F [g/cm s]	p	Theory: K Disorder	Theory: K BCC	Theory: K_{max}
100	2.3	11	34.74	0.626	∞	13	18
200	2.2	7.6	33.70	0.608	10.5	9	36
300	2.3	7.0	31.95	0.576	7	7	54
300	3.9	6.4	32.36	0.584	8	8	45
300	5.4	7.0	31.85	0.574	7	7	39

carrier beads. From F, p is calculated, and from Fig. 6.8, K, the dielectric constant, is determined for the disordered and base-centered cubic BCC packing. The experimentally determined K_E appears to agree with K. K_{max} is also calculated in the last column; K_E does not appear to be described by K_{max}.

The theoretical explanation for this result was given first by *Hays* [12.67] and later by *Schein* et al. [12.68]. It follows from a careful consideration of the nonuniformities in the system. Consider the carrier charge σ_c required to neutralize the electric field due to the applied voltage V. This is given by (6.62). For simplicity, ignore the small term associated with the toner charge σ_t, and assume Λ is given by L/K,

$$\sigma_c = \frac{V\varepsilon_0}{L} K . \tag{12.15}$$

This equation describes the charge per unit area σ_c in the carrier bead layer adjacent to the photoreceptor which neutralizes the electric field due to the applied voltage V. Consider two cases. First, consider the situation directly under a carrier chain. Here the electric field is maximum, VK_{max}/L, σ_c is maximum, and the K in (12.15) is therefore K_{max}. Second, consider the situation over a region large compared to a carrier bead. Now σ_c is the average value, implying K in (12.15) is the usual dielectric constant. Hence the correct K in (12.15) depends on the size scale of interest. The calculation of M/A (toner development) requires the calculation of the toner charge per unit area over a large distance. Therefore

$$\frac{M}{A}\frac{Q}{M} = \sigma_t = \nu\sigma_c^{\text{large}} \tag{12.16}$$

[using (6.63)], where σ_c^{large} is σ_c over large distances, i.e., (12.15) with K the dielectric constant. Hence

$$\frac{M}{A} = \frac{1}{Q/M} \frac{V \varepsilon_0 \nu}{L} K, \tag{12.17}$$

where the K in (12.17) is the dielectric constant, consistent with the experimental results.

12-6.4 Solid Area Development Experiments

In the past few years, several solid area development experiments have been published which challenge the equilibrium theory, as discussed in Sect. 6.3.4. The most puzzling results are the observed effect of varying the toner diameter. Indirect evidence was given that adding an ac field can move the carrier charge away from the beads adjacent to the photoreceptor. Others suggested that, even in the usual insulative magnetic brush (without ac bias), the carrier charge may be near the center of the brush and that toner adhesion to carrier (ignored in the equilibrium theory by assuming that toner develops only during three-body contact events) may be significantly contributing to development.

(1) Effect of Toner Diameter

In recent years, there has been increased interest in using smaller diameter toner particles to increase the quality of images. A study by *Yamazaki* et al. [12.69] using these smaller diameter toner particles appears to have resulted in an experiment which is inconsistent with the equilibrium theory. Yamazaki et al. measured the slope of M/A versus V curves for toner diameters of 6.8, 10.1 and 13.5 μm. They found that plotting $(M/A)/V$ against Q/M did *not* result in a universal curve. Instead, three separate curves resulted, with the smaller diameter toner having larger values of $(M/A)/V$. This is qualitatively predicted by (6.57) with $\Lambda = L/K_E$, since K_E increases as the toner diameter decreases. However, in the analysis presented, K_E was not set equal to the dielectric constant of the mix (Sect. 12-6.2.2) nor was the dielectric constant determined. Results of *Hays* [12.70] indicate that the dielectric constant does increase as the toner diameter decreases but, at least in Hays' mixtures, the effect is probably not large enough to account for Yamazaki et al.'s observations. On the other hand, plots of the dielectric constant (Fig. 6.8) clearly indicate that data is being taken near a divergence in the dielectric constant. Clearly it would be useful to repeat these measurements and simultaneously measure the dielectric constant of the mix.

(2) Addition of an ac Field

Kurita [12.71] electrically isolated a magnetic brush roller and a photoreceptor and was able, under certain conditions, to measure the currents due to toner development. In the presence of an ac field, these currents significantly increased. He suggested that the reason for the increased current is the migration of the carrier charge away from the beads adjacent to the photoreceptor, presumably due to toner migration in response to the large ac electric field between beads.

(3) Carrier Charge Migration in Insulative Magnetic Brush Development

The possibility that the carrier charge migrates beyond the first layer of carrier beads adjacent to the photoreceptor during the development process in a normal magnetic brush development system has been considered by *Hays*, in two studies [12.67,72] and *Schmidlin* [12.73]. This is of obvious importance because the electric field due to charged carriers is then reduced, enhancing M/A. In terms of the equilibrium theory, (6.57), the idea suggests Λ is smaller than L/K_E. Schmidlin postulates that the mechanism of development is primarily field stripping, to which the electric field (or forces) due to the build up of carrier charge must be added. The experimental verification of carrier charge migration in an insulative magnetic brush development system beyond the layers adjacent to the photoreceptor remains unclear in both studies. In Hays' work: (1) the data were not taken on an actual development system but on a microscopic flat plate capacitor structure, (2) the observed magnitudes of M/A were always in quantitative agreement with (6.57) (see also Sect. 12-6.2.2), and (3) the enhancement of M/A due to carrier migration was always coincidentally approximately cancelled by a decrease due to a development probability function. The probability function takes into account toner adhesion to carrier, a force assumed to be cancelled in the equilibrium theory due to toner adhesion to the photoreceptor in three-body contact events.

Similarly, Schmidlin argues that the carrier charge has migrated significantly away from the layer adjacent to the photoreceptor and calculates M/A and Q/M from first principles, making assumptions about the adhesion distribution functions. The evidence Schmidlin presents is a slight departure from linearity of M/A versus V curves and a Q/M versus V curve exhibiting lower Q/M at low V, exactly what one would qualitatively expect for a field stripping theory (Fig. 6.2). However, neither of these effects has been observed before in the insulative magnetic brush development system. Figure 6.23 shows M/A versus V not only linear down to the lowest field, but practically unchanged when the toner size distribution, i.e., the adhesion distribution, is changed. Figure 6.26 and data in [12.74] show that Q/M is independent of V. Of course, in reverse bias Q/M changes sign as the development mechanism changes and wrong sign toner develops (Sect. 12-6.6.1); the transition of Q/M from forward to reverse bias is expected to take place over a finite voltage range. One would also expect that stronger departures from linearity of M/A versus V curves would be predicted (Fig. 6.2) over some reasonable ranges of the parameters determining the adhesion distribution. It would be interesting to see these predictions and to compare them with experimental data. While the dynamics of development is probably a complicated set of events involving the adhesion distributions, the success of the equilibrium theory indicates that the final state (the final M/A observed) is determined by the simple physical concept of field collapse.

12-6.6 Background Development

The mechanism(s) responsible for background development was identified in the first edition of this book as one of the fundamental unsolved problems in electrophotography. Two significant studies of this problem have been published since then.

Hays [12.75] has studied background development on the photoreceptor in a simulated development system. He has found, in agreement with previous work [12.74], that background development either increases or remains constant as reverse bias is increased. He also found that the average charge of the toner decreases and then becomes wrong sign as reverse bias increases, also in agreement with prior work. Using a charge spectrometer, it was shown that, within experimental error, wrong sign toner did not exist in the developer mixer he used. The logical conclusion is that wrong sign toner is being generated during the development process. It was hypothesized that the electric field due to the reverse bias drives the charge exchange between the toner and carrier to cause the toner to become wrong sign, an idea which is reasonable given the observation of such effects in the plane-plane geometry. It was observed that developer mixes whose average Q/M had decayed after resting were more susceptible to this electric field induced generation of wrong sign toner.

The second study looked at background development on both the photoreceptor and paper and again implicated wrong sign toner as the primary, albeit indirect, cause of background development [12.76,77].

On the photoreceptor both M/A and Q/M were measured as a function of reverse bias, as shown in Fig. 12.13 for a premix (fresh from the factory), a broken-in mix, and a failed mix. As can be seen, even in reverse bias, toner develops onto the photoreceptor. The reason for this is obvious from the bottom of the figure: the toner is wrong sign and therefore is naturally driven to the photoreceptor.

Pictures of the toner on the photoreceptor (PC), on the PC after electrostatic transfer, on paper unfused and fused are shown in Fig. 12.14. Here it

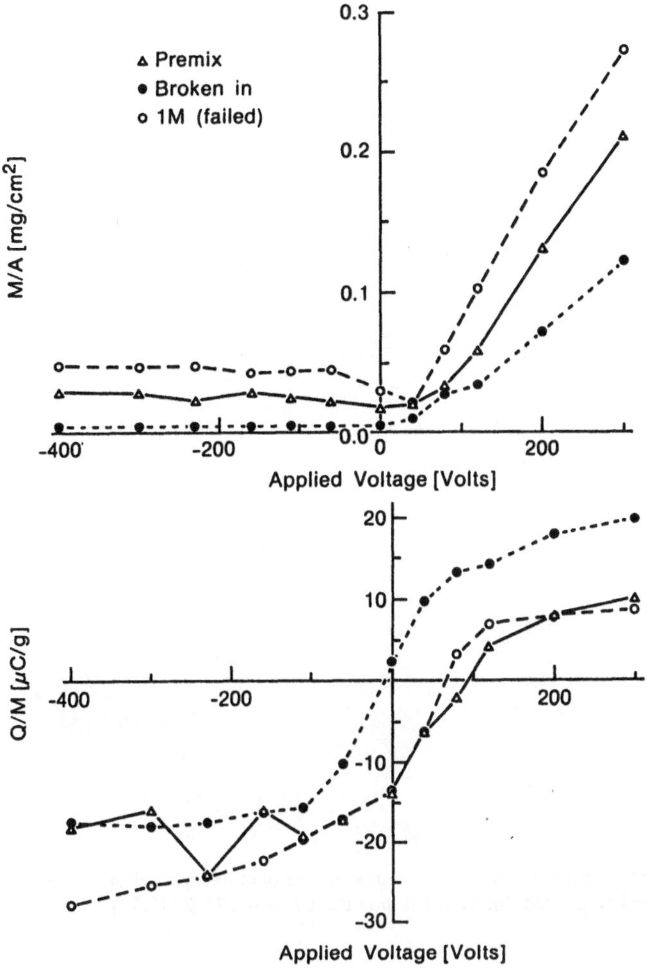

Fig. 12.13. Measurements of M/A and Q/M versus bias for premix, "broken-in" mix, and "failed" mix (run for 10^6 prints). Note that toner developed in reverse bias is wrong sign toner [12.77]

is seen that even though a substantial amount of toner develops on the photoreceptor in reverse bias (Fig. 12.14a), only a very small fraction transfers to the paper (Fig. 12.14b), as one might expect, since electrostatic transfer tends to reject wrong sign toner. So what is the nature of the toner that does transfer to the paper?

The nature of the toner which transfers to paper was determined by obtaining a special mix, whose development characteristics are shown in Fig. 12.15. For this mix, which was purposely designed to have little or no wrong sign toner, the development characteristics differed from normal mixes. In this

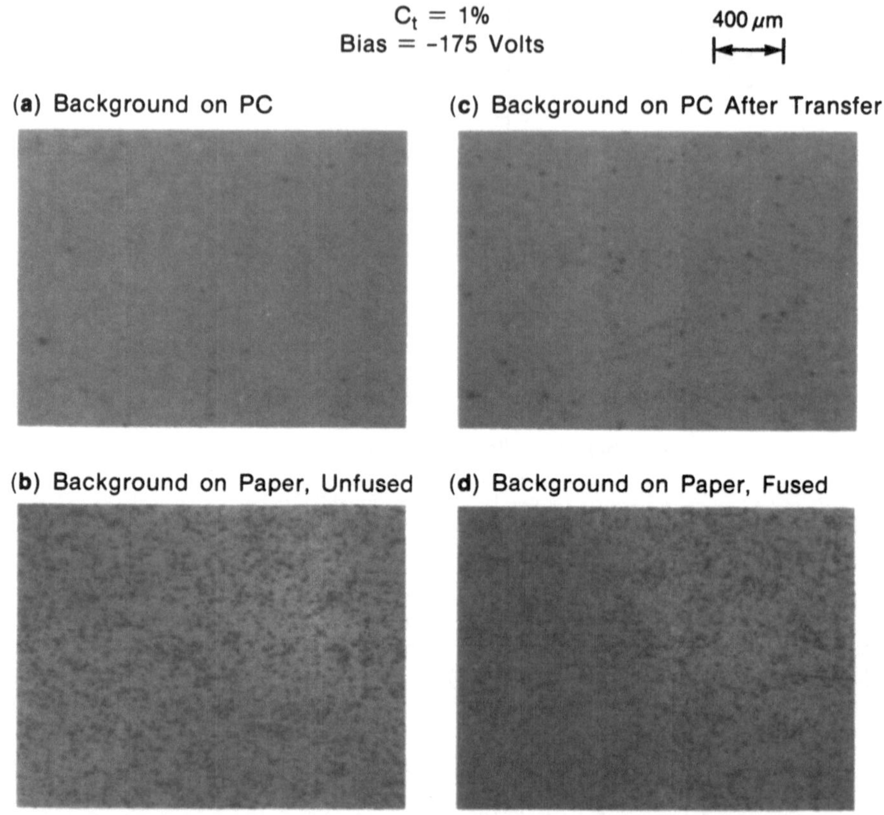

Fig. 12.14. Photomicrographs of toner on the photoreceptor, on the photoreceptor after transfer, on paper, and on paper after being fused, for $C_t = 1\%$ premix at $V = -175$ V [12.77]

case M/A is much smaller and approaches zero as reverse bias increases and Q/M is always right sign as far as can be determined; beyond −100 V the amount of charge and mass were too small to make reliable measurements. Pictures of background development using this mix are shown in Fig. 12.16. Note in this case the number of toner particles on the photoreceptor approximately equals the number on the paper.

These results can be accounted for by a generalization of Schein's 1975 theory of background development (Sect. 6.6, [12.74]). Consider, as shown in Fig. 12.17, that $F_p - F_c$, the difference in toner-photoreceptor and toner-carrier adhesions, is a distributed property, an idea discussed by *Schmidlin* [12.78]. Then, increasing reverse bias adds a third force to the toner particles, decreasing the probability that positive toner will stay on the photoreceptor. Since $F_p - F_c$ is distributed, increasing reverse bias decreases background, but

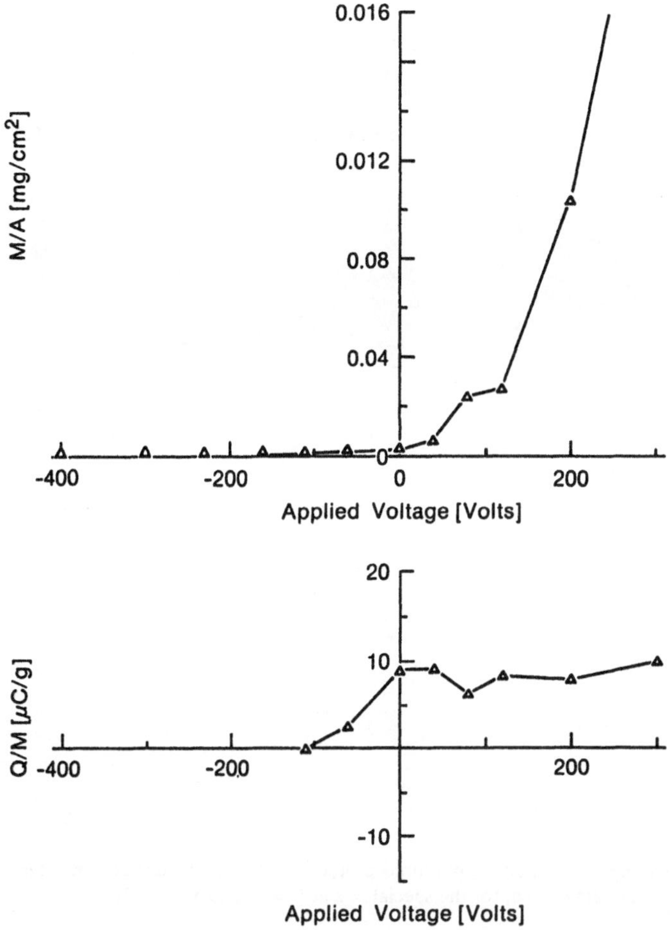

Fig. 12.15. Measurements of M/A and Q/M versus bias for the special mix. M/A goes to zero as reverse bias is increased. The toner on average is always right sign [12.77]

this only slowly approaches zero depending on the details of the $F_p - F_c$ distribution.

This theory appears to describe the behavior of the special mix. It obviously can also describe what happens to right sign toner even when wrong sign toner is also present. Some of the background development observed with all developer mixes must therefore be due to this mechanism.

However, note that background on paper, which occurs with normal mixes that obviously have some wrong sign toner (compare Fig. 12.13 with Fig. 12.15) appears to be higher. The data suggest that the presence of wrong sign toner enhances background development. Since electrostatic transfer tends to reject wrong sign toner, it must be that the presence of wrong sign toner enhances the development of right sign toner onto paper.

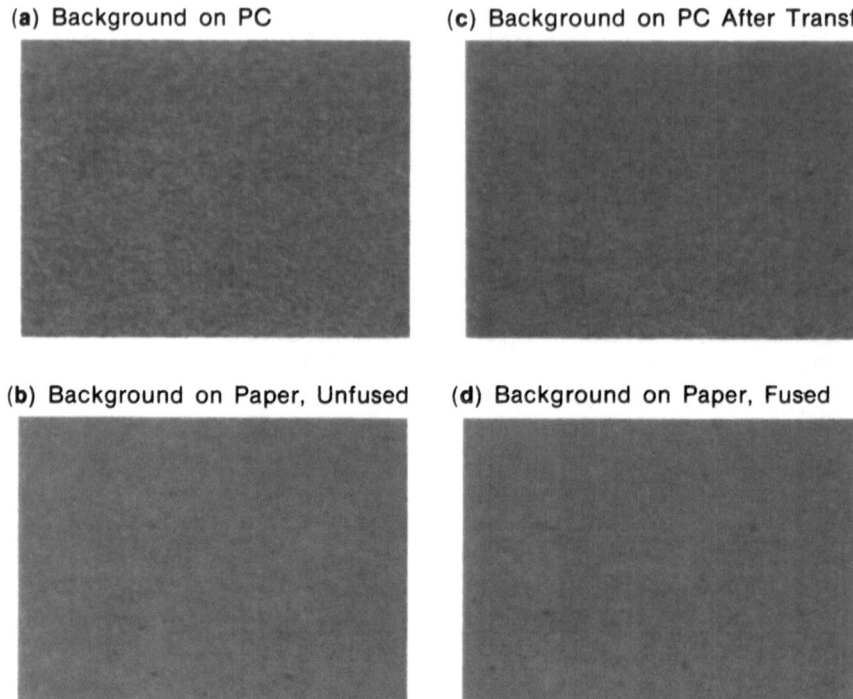

Fig. 12.16. Photomicrographs of toner on the photoreceptor, on the photoreceptor after transfer, on paper, and on paper after being fused, for the special mix at $V = -180$ V [12.77]

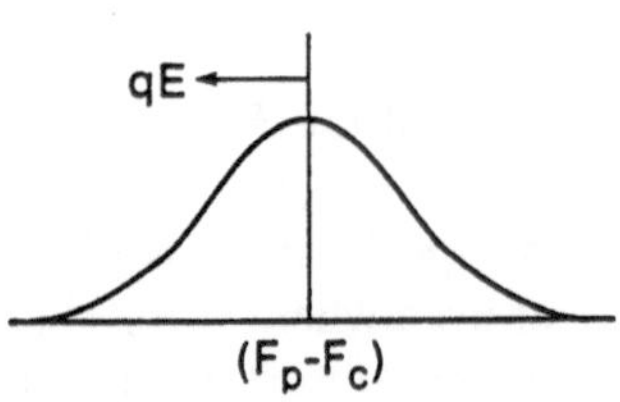

Fig. 12.17. If the toner-photoreceptor adhesion F_p exceeds the toner-carrier adhesion F_c, then the toner particle will develop. If $F_p - F_c$ is a distributed property (shown), then the application of reverse bias will reduce the probability that toner will develop on the PC. The Coulomb force on the toner particle is qE in the presence of an electric field E [12.77]

Three mechanisms by which wrong sign toner enhances the development of right sign toner onto paper were suggested. First, when wrong sign toner develops onto the photoreceptor, it leaves behind a positive charge in the carrier particles adjacent to the photoreceptor. This charge produces a field, and therefore a Coulomb force, forcing right sign toner to the photoreceptor. Second, it is possible that net wrong sign doublets are developed onto the photoreceptor which are broken apart during transfer. No direct evidence for this effect was given. Third, it is possible that the $F_p - F_c$ distribution is altered by the presence of wrong sign toner. Again, no direct evidence for this is provided by the data.

12-7.3 Infinitely Conductive Theory

Several advances were made in the theory of conductive magnetic brush development, all related to the conductivity of the brush. It was shown how to account for developer degradation, i.e., loss of conductivity, empirically. The effect of additives on conductivity was studied. And finally, a microscopic theory of development, with conductivity as a parameter, was proposed.

(1) Developer Degradation

Equation (7.23) describes the development characteristics of the ideal, infinitely conducting bead chains in the conductive magnetic brush development system (far from neutralization). One mode of degradation of this system is for the chains to decrease their conductivity, perhaps because toner particles stick to the carrier at the exposed points of the sponge carrier that normally make conductive contacts. How can this degradation be detected and measured?

A suggestion made by *Schein* et al. [12.68] is to add a factor K_c to the right side of (7.23) that is equal to 1 in the ideal case but decreases when the conductivity decreases. Solving for K_c gives

$$K_c = \frac{M}{A}\frac{Q}{M}\frac{1}{C_t v V}\frac{\rho_t}{p\rho_c}\frac{r}{8\varepsilon_0} . \qquad (12.18)$$

Measurements of the parameter on the right side of (12.18) gives K_c. Figure 12.18 shows measurements of K_c for an experimental sponge carrier mix. Note clear evidence of a decrease in K_c with usage, suggesting developer degradation. Also note that the extrapolated value of K_c to time zero is very close to 1, another verification of this theory.

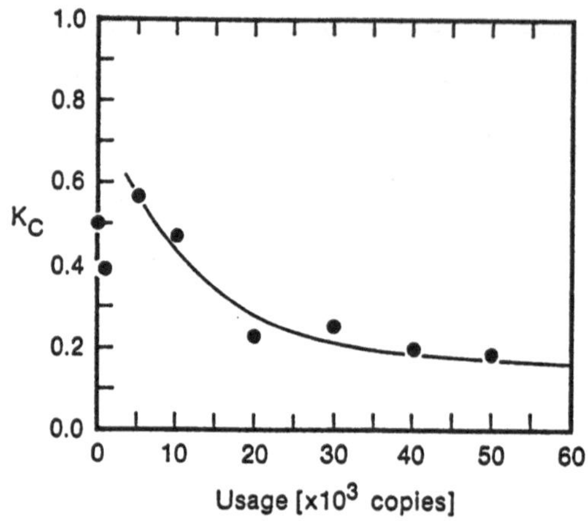

Fig. 12.18. Measurement of K_c, the efficiency of a conductive magnetic brush system, as given by (12.18). Note the decrease with usage: by 40k copies this mix has effectively failed [12.68]

(2) Effect of Additives

Nash [12.79] studied the effect on the conductivity of a conductive magnetic brush system of adding various external flow additives such as Aerosil, Kynar, and zinc stearate. Such additives, which are commonly used for flow and cleaning enhancement, might detrimentally decrease the brush conductivity. The data clearly indicate that zinc stearate maintains the conductivity better than the other additives because it maintains the close packing of the carrier beads necessary for maintaining the conduction pathways.

(3) Theory with σ as a Parameter

An extension of the theories of insulative ($\sigma = 0$) and conductive ($\sigma = \infty$) magnetic brush development is to formulate a theory of development in which the conductivity σ is a variable, giving a comprehensive model of magnetic brush development. This would provide a link between prior theories and also a model for intermediate conductivities. Such a theory has now been presented by *Folkins* [12.80] which, in the appropriate limiting cases, gives the results derived above (Sect. 6.3.4, Sect. 7.3).

The theory assumes the usual plane parallel geometry of a photoreceptor, developed toner layer, air gap, and magnetic brush, all of which are characterized by their thickness and dielectric constants. Differential equations are written to describe the behavior of four variables, the charge per unit area on the magnetic brush tips, the developed toner charge per unit area, the available toner mass per unit area on the magnetic brush tips, and the developed toner mass per unit area. The equations are written as a function of distance along the tip. Toner resupply is not allowed but charge flow down the magnetic brush is determined by the product of the brush conductivity and the electric field within the brush. Finally, it is assumed that the amount of toner which develops is proportional to the toner mass per unit area in the brush tips times the electric field in the air gap. The solutions are coupled simultaneous non-linear differential equations, for which no general analytic solution is apparent. However, in the various limits, such as zero and infinite conductivity, analytic solutions are obtained which agree with prior results.

12-8.3 Contact Charging

A critical requirement of a development system is charging of the toner particles. Several discussions of toner charging for the nonmagnetic, contact, monocomponent development system have appeared.

A system similar to the Ricoh system (Fig. 8.12) has been commercialized in the IBM 4019 LaserPrinter. A thorough discussion of the hardware has been given by *Thompson* [12.81]. Toner charging occurs both at the supply roller interface and at the blade-roller interface. A bias of 100 V between the conductive supply roller and the roller improves both the charging and reloading of toner on the roller; the bias also assists in discharging residual charge on the surface of the roller following the removal of toner during development. The bias on the metering blade (325 V) significantly reduces the amount of wrong sign toner passing through the metering blade-roller gap.

Shinozaki et al. [12.82] discuss the use of a doctor roller instead of a doctor blade in the same development system. The rationale for considering a doctor roller is that the variables which control toner charging and metering, such as speed of rotation, pressure, and electric fields, are more easily controlled for a roller. They chose to rotate the doctor roller against (anti-parallel to) the developer roller and at approximately 1/7 the surface speed. The Q/M of the negative toner on a conductive roller was observed to be independent of negative bias of the doctor roller, but M/A increased linearly with voltage to $\approx 0.7\ \mathrm{mg/cm^2}$ at 600 V. This suggests that the doctor roller is not participating in the toner charging process but is effective in the metering process. By studying the charge flow and electrostatic charging of the roller (by blowing off the toner immediately after the doctor roller) Shinozaki et al. were able to establish that during the doctoring process there appears to be charge flow to the roller surface. Therefore, if the electrical resistance of the roller is high, charging of the roller surface will influence the electric field in the toner layer at the interface and hence its final Q/M and M/A characteristics.

Hirose et al. [12.83] and *Kamaji* et al. [12.84] discuss the use of a porous elastic roller and a biased metallic doctor blade in a configuration very similar to Fig. 8.12. They report that this configuration extends the operational environmental range from 10°C, 20% RH–32°C, 80% RH to 0°C, 20%

RH–40°C, 80% RH. The development roller is made of a porous elastic polyurethane and the doctor blade is made of stainless steel, 0.1 mm thick, with a rounded working edge, applied at contact pressures of 50 g/cm. The doctor blade is placed tangent to the roller surface.

12-9.3 Theory of Monocomponent Development

Further work on the theory of monocomponent development has been directed towards understanding the effect of the ac bias and the underlying physics determining the voltage width (Fig. 9.6).

(1) ac Bias

In most monocomponent jump gap development systems, an ac bias is applied to enhance development (see Sect. 11.4.2 for an exception). In Sect. 9.3 it was suggested that the effect of the ac bias is to cause the toner to "project" back and forth from the photoreceptor to the roller, essentially freeing the toner particles of any electrostatic attraction to the roller.

Yanagida et al. [12.85] have pointed out that, in a nonmagnetic system, part of the mechanism by which toner is caused to leave the roller surface is by collisions with toner returning from the photoreceptor. Their conclusion results from the observation that solid area development is sometimes reduced at high contrast potentials, producing a condition in which the toner may not be returning to the development roller. One experiment carried out to verify their model was to rectify the ac voltage so that the toner would have no forces returning it to the roller, while the electrostatic forces remain unchanged. As their model would predict, much less development occurred.

(2) Voltage Width

A theory of monocomponent development must account for the threshold voltage (Fig. 9.6), the voltage width, and the maximum M/A. In Sect. 9.3 it was suggested that the voltage width V_w is due to the distribution of adhesion forces of toner to carrier. New results include computer simulations and considerations of the effect of the toner space charge.

Computer simulations have been carried out which predict the voltage width [12.86,87]. It was noted that the adhesion force on a toner particle includes the force of attraction to its image charge as well as to the image charge of its neighbors. Therefore, one might expect that as particles develop from the roller and their image charge disappears, the attraction of other particles decreases. In a single diameter, single charge approximation, the whole layer

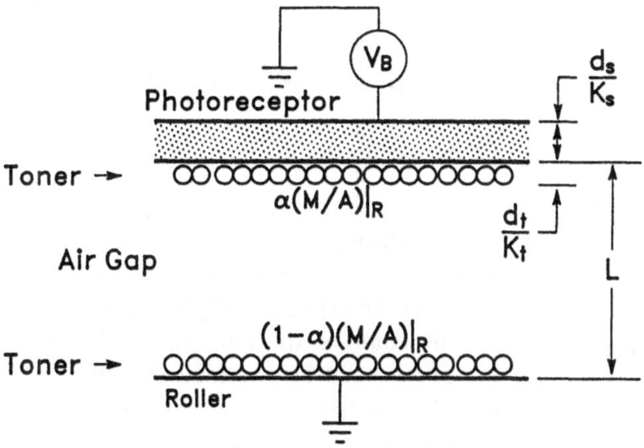

Fig. 12.19. Idealized electrostatic model of toner layers on the roller and photoreceptor [12.91]

"unzips" at one voltage as suggested earlier by *Goal* and *Spencer* [12.88] and in [12.89]; in the simulation with a more realistic diameter and charge distribution, voids are predicted in the layer on the roller. The computer simulation was carried out assuming a log normal distribution of toner radii and a normal distribution of charge, with the average charge assumed to be proportional to the diameter to the 1.1 power. Some experimental evidence based on photomicrographics appears to support this conclusion.

Studies of the effect of toner space charge on the voltage width were published by *Tachibana* et al. [12.90], *Thompson* [12.81], *Castle* et al. [12.91], and *Schein* et al. [12.92].

To understand the effect of the toner space charge on V_w, assume a delta function adhesion distribution and a single value of Q/M for all the toner particles. Consider the sub-monolayer of toner shown in Fig. 12.19, some of which remains on the conductive development roller $(1-\alpha)(M/A)\,|_R$ and some of which has developed on the photoreceptor $\alpha(M/A)\,|_R$, where $(M/A)\,|_R$ is mass per unit area initially on the roller. The toner itself creates electric fields in the development zone. For a toner with only one value of Q/M, these values can be converted to the charge per unit area by multiplying by M/A. By using Gauss's Law, it is easily shown that the electric field in the air gap E_{air} due to the toner charge and the applied bias V_B is

$$E_{\text{air}} = \frac{1}{(L+d_s/K_s)}\left[\left(\alpha\frac{M}{A}\frac{Q}{M}\right)\frac{1}{\varepsilon_0}\left(\frac{d_s}{K_s}+\frac{d_t}{2K_t}\right) - \left((1-\alpha)\frac{M}{A}\frac{Q}{M}\right)\frac{1}{\varepsilon_0}\left(\frac{d_t}{2K_t}\right)+V_B\right] \qquad (12.19)$$

or

$$E_{\text{air}} = \frac{1}{(L + d_s/K_s)}\left[V_B + \frac{1}{\varepsilon_0}\frac{M}{A}\frac{Q}{M}\left(\alpha\frac{d_s}{K_s} + \frac{d_t}{2K_t}(2\alpha - 1)\right)\right]. \tag{12.20}$$

This represents the dc electric field governing the strength of the force F_{dc} on a given toner particle, i.e., $F_{dc} = QE_{air}$. To clarify the implications of (12.20), assume $d_s/K_s = 0$. This is useful because the experimental verification described below was done with the photoreceptor exposed to light; under working conditions, where d_s/K_s is finite, only some of the numerical values will change. Toner will develop onto the photoreceptor provided it is liberated from the bonding forces (F_{es}, F_M and F_{sc}, see Sect. 9.3) holding it onto the roller and provided it experiences a positive field E_{air} directed towards the photoreceptor. This leads to two possible idealized models for development, which can be classified as follows:

(a) Field limited toner development. This assumes that all toner is liberated in the gap and that development continues until $E_{air} = 0$.

(b) Force limited toner development. This assumes that the toner requires a threshold dc field (E_{th}) to liberate it into the gap and that when $E_{air} > E_{th}$ toner is liberated into the gap. Once this occurs, development continues until $E_{air} = 0$.

(a) Field Limited Toner Development. Toner can be liberated if an ac field of sufficient magnitude is applied in the gap. Once toner is liberated, forces will act to deposit it on both the photoreceptor and roller until the condition $E_{air} = 0$ is reached. Setting $E_{air} = 0$ in (12.20) and solving for α, the fractional amount of toner developed on the photoreceptor, gives

$$\alpha = \frac{1}{2} - \frac{V_B}{\frac{1}{\varepsilon_0}\frac{Q}{M}\left.\frac{M}{A}\right|_R \frac{d_t}{K_t}}. \tag{12.21}$$

The voltage at which development starts, V_{th}, is obtained by setting $\alpha = 0$ in (12.21), giving

$$V_{th} = \frac{1}{2}\frac{1}{\varepsilon_0}\frac{Q}{M}\left.\frac{M}{A}\right|_R \frac{d_t}{K_t}. \tag{12.22}$$

The development curve predicted by this equation is shown in Fig. 12.20. From this it follows that the width of the development curve is

$$V_w = 2V_{th} = \frac{1}{\varepsilon_0}\frac{Q}{M}\left.\frac{M}{A}\right|_R \frac{d_t}{2K_t}. \tag{12.23}$$

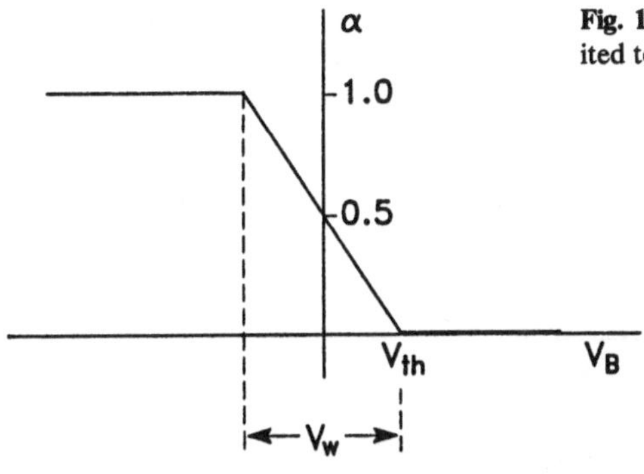

Fig. 12.20. Development curve for the field limited toner development case [12.91]

Several remarks can be made based on these results. First, the predicted development curve is found to be independent of L. This occurs because the condition $E_{air} = 0$ requires the surface potentials to be equal no matter what the spacing. Secondly, it can be seen that 50% of the toner will develop at zero bias. This is a result of the space charge fields of the toner, since equal layers of toner on the photoreceptor and the roller result in $E_{air} = 0$. It should be noted that this observation is based upon the assumption of $d_s/K_s = 0$ (with finite d_s, α at $V_B = 0$ is $[2 + (d_s/K_s)(2K_t/d_t)]^{-1}$). A third observation is that the slope of the toner development curve is inversely proportional to Q/M and V_{th} is directly proportional to Q/M. This means that V_w increases as the magnitude of the toner Q/M increases.

(b) Force Limited Toner Development. If ac fields are not present or are of insufficient strength, toner will be liberated once $F_{dc} > F_{es} + F_M + F_{sc}$. Thus, the threshold development field is given as

$$E_{th} = \frac{F_{es} + F_M + F_{sc}}{Q} \; . \tag{12.24}$$

Setting $\alpha = 0$ in (12.20) and solving for the bias voltage required to achieve toner liberation, V_{th}, gives

$$V_{th} = E_{th}\left(L + \frac{d_s}{K_s}\right) + \frac{1}{2\varepsilon_0} \frac{M}{A}\bigg|_R \frac{Q}{M}\frac{d_t}{K_t} \; . \tag{12.25}$$

This predicts a development curve as shown in Fig. 12.21. It is identical in shape to that of Fig. 12.20, but offset by $E_{th}(L + d_s/K_s)$.

Equation (12.25) shows that V_{th} is linearly dependent on L and is directly dependent upon the magnitudes of F_M, F_{es} and F_{sc} (as already derived in Sect. 9.3).

Thus, if the ac field is of sufficient strength to completely liberate the toner from the roller, the field limited development prediction should apply. If the

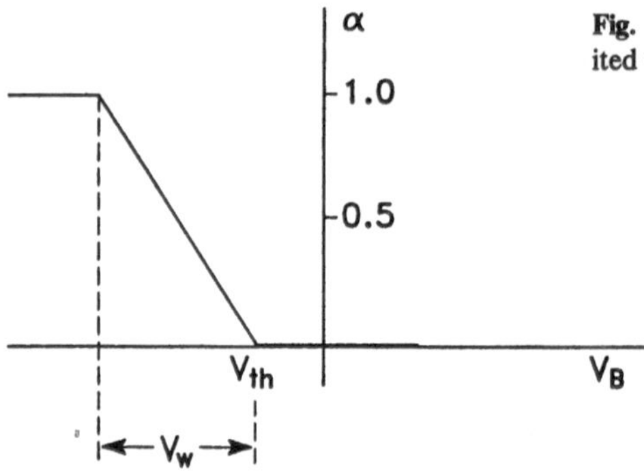

Fig. 12.21. Development curve for the force limited toner development case [12.91]

ac field is not present or of insufficient strength, the restraining forces present will determine the voltage offset of the development curve.

The above predictions for both development models are based upon the assumption of a constant Q/M in the toner. As discussed in Sect. 9.3, if a distribution of Q/M exists, then there will be a range of values of V_{th} for which the toner will develop. Clearly this will result in a broadening of the development curve (i.e., an increase in V_w).

*(c) **Experimental Results.*** Figures 12.22 and 23 show the development curves of a monocomponent magnetic toner system in which the following parameters were varied: V_B (–300 to +1200 V), gap (contact, 70 μm, 350 μm), magnetic field, Q/M (2.5, 5.3 μC/g), and a superimposed ac bias (chosen to maintain the same electric field for the different gaps) of $E_{max} = 1.5$ V/μm, 1.7 kHz. The ac bias is reported only for gaps of 350 μm and 70 μm since no effect of ac was found for the contact case. Figures 12.22a-d show the results with dc and ac bias applied; Figs. 12.23a-d show the results with only dc bias applied.

Consider the effect of the gap. First, in each of graphs in Fig. 12.22, the threshold voltage for each pair of results, i.e., the two gaps tested, is approximately the same. This shows that V_{th} is independent of L. Secondly, with the exception of some variation in Fig. 12.22a, the curves for each gap overlap, showing that the development is also independent of L. Figure 12.22c shows a development curve most closely in agreement with that predicted by the field limited theory. The development at $V_B = 0$ is approximately 25% (lower than the 50% predicted) and the value of V_w is significantly greater than predicted. As discussed above, this is believed to be due to the distribution of Q/M values in the toner.

Consider the effect of Q/M. Figure 12.22c shows results for $Q/M = 2.5$ μC/g; Fig. 12.22d shows results for $Q/M = 5.3$ μC/g; the higher Q/M toner gives a slope of the development curve that is much less steep. This is as predicted by (12.20). The fact that only 50% of the toner is developed at maximum bias is because the ac liberation field was not high enough

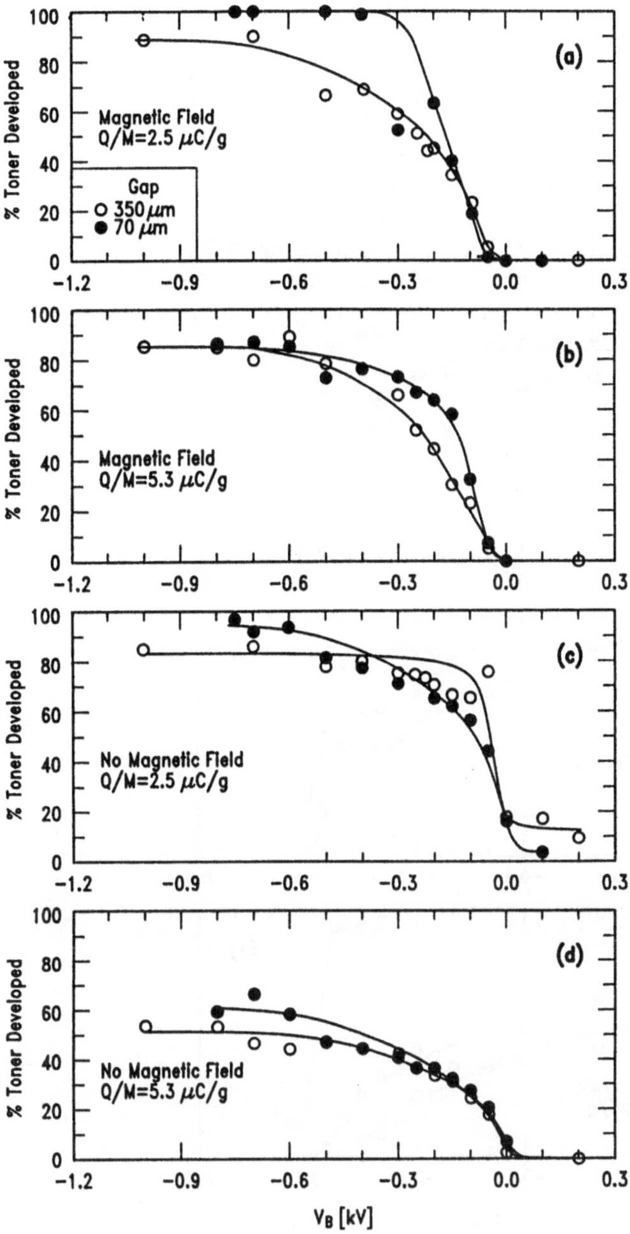

Fig. 12.22. Monocomponent development characteristics with dc and ac fields [12.91]

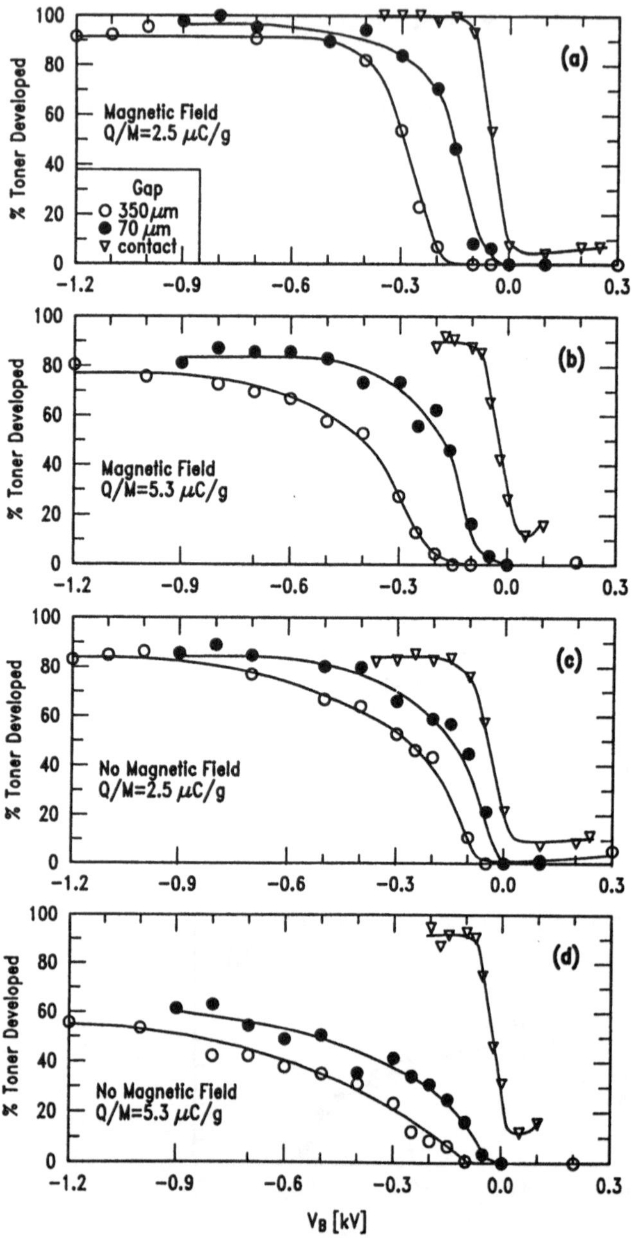

Fig. 12.23. Monocomponent development characteristics *without* ac field [12.91]

to liberate the upper range of the highly charged toner. Consider the effect of the magnetic field. Figures 12.22a and b, with the magnetic field present, show values of V_{th} higher than the nonmagnetic cases, Figs. 12.22c and d. This shows that F_{ac} is not sufficient to completely overcome the effect of F_M. However, since F_M depends only on the magnetic field gradient and not on charge, once liberated, the field limited model is satisfied to a first approximation.

Figure 12.23 shows the development curves for the same range of variables as in Fig. 12.22, but in the absence of ac bias.

Looking first at the 350 μm and 70 μm cases, it can be seen that, as predicted by the force limited development model, V_{th} is dependent on L. Also, as expected, the threshold voltage is higher in the magnetic case than the nonmagnetic case and for the higher charged toner. This shows the relative importance of the effects of F_M and F_{es} in each case.

Secondly, it can be observed that the pair of development curves for the cases of the 70 μm and 350 μm gap in each figure are of similar shape but displaced in voltage. This confirms the prediction that, once liberated, toner development is independent of L. In all cases the slopes of the comparable curves are smaller than those in Fig. 12.22 with the ac present, showing the influence of the distribution of Q/M on the force limited development.

The case of contact development shows a different behavior than the case with finite gaps. All cases, particularly the nonmagnetic examples, are consistent with the field limited development model. This can be explained by the fact that in contact development the image forces are cancelled by similar forces to both surfaces, liberating the toner.

Therefore, the theory and experimental results have identified and confirmed the existence of two regimes of development, which have been classified as field limited toner development and force limited toner development. Field limited toner development exists when the toner particles are liberated either by the presence of ac fields of sufficient magnitude or by the cancellation of electrostatic adhesion forces in contact development. In this situation the development is determined by the applied bias and the space charge due to charge in the toner layers. This results in threshold voltages and development widths that are independent of L. Force limited toner development exists when the development force must first overcome adhesion forces and liberate the toner from the roller. In this case the value of the threshold voltage is dependent upon L whereas the development width is independent of L. In all cases the values of V_{th} and V_w depend upon the magnitude of the toner Q/M, and thus the uniformity of Q/M distribution within the toner material is very important in determining the shape of the development characteristics.

12-9.4 Conductive Toner

The Océ system, which uses conductive toner, was discussed by *Huijben* at the Seventh IS&T Non-Impact Printing Conference [12.93]. This system uses a ZnO-resin photoreceptor. The photoreceptor life is claimed to be 10^5 copies, which results from (1) gentle magnetic cleaning (after both the toner and photoreceptor are discharged by illumination) and (2) a gentle indirect transfer-transfuse system based on a silicon rubber intermediate belt. The intermediate belt transfer system also solves one of the problems associated with conductive toner (Sect. 9.4)—it cannot be reliably transferred electrostatically.

12-9.6 Nonmagnetic, Insulative Toner

The greatest interest in monocomponent development appears to be in the nonmagnetic, insulative toner type, as judged by the number of technical papers which discuss this system. In addition to the work by Ricoh and Toshiba reviewed in the first edition of this book, new results have been published by two groups from Fiji Xerox, three groups at (or associated with) IBM, and one group at Matsushita. Four deal with a jump development system; one (an IBM system) deals with a contact system. (Ref. [8.15] from the Ricoh group has been published in a journal and is now [12.94].)

The jump system discussed by *Yamamoto* et al. [12.95] was designed specifically for their color copier, which requires a noncontact, noninteracting development system (Sect. 11.4.2). The system is shown in Fig. 11.14. A fur brush places the toner on a roller which is doctored by a rubber blade. The toner layer is about 30 μm thick and must jump a gap of approximately 150 μm to the photoreceptor surface *without* the use of an ac bias (so that toner back contamination does not occur as the four-color images are built up on the photoreceptor). In order to accomplish this with the normal 800 V potentials, very low charged toner is required, $\approx 3\ \mu C/g$. Measurements of optical density versus bias voltage indicate a threshold voltage of about 200 V and a voltage width of about another 200 V.

One of the IBM discussions [12.81] focused on all aspects of the contact development system in the IBM LaserPrinter family. Results relating to the toner charging mechanism are discussed in Sect. 12-8.3 and aspects that relate to the voltage width are discussed in Sect. 12-9.3. The importance of the roller resistivity is pointed out. A long electrical time constant, i.e., high resistivity, does not allow the potential on the surface of the roller to reach its full value in the development zone, thereby decreasing the electric field in the gap, reducing M/A. Normalized development characteristics are given in Fig. 12.24.

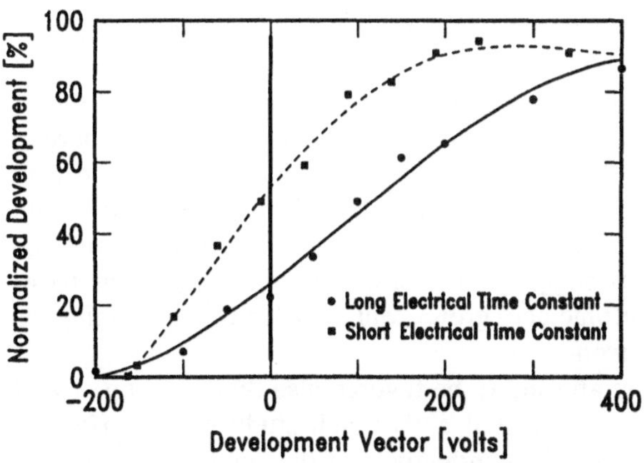

Fig. 12.24. Normalized development versus development vector for a roller with a long and a short electrical time constant [12.81]

A similar point was made by *Tachibana* et al. [12.90]: roller resistivity below a certain value (10^{11} Ω cm) strongly affects the development curve, and similar data are published. In fact, they go one step further, suggesting using this fact to vary the tone reproduction curve (in a presumably analog gray scale system).

The other Fuji Xerox paper [12.85] focused on the mechanism by which an ac voltage improves development, and is also discussed in Sect. 12-9.3.

The work by *Castle* et al. [12.91] is focused on the physics of development and is discussed in Sect. 12-9.3.

An extensive characterization of the toner charge distribution, as measured by a charge spectrometer, of toner in a jump development system is presented by *Lee* et al. [12.89]. They measured the average toner charge-to-mass ratio Q/M and the charge distribution of toner on the roller before development, on the roller after development, and on the photoreceptor. They found that the toner which jumped has $Q/M \approx 5\ \mu C/g$ almost irrespective of the Q/M on the roller or the dc voltage used to cause the toner to jump the gap. Even the charge distributions appeared to be reasonably similar under a variety of conditions, although it appears that smaller diameter particles do not jump as well at low voltages. It is argued that these results may be understood if interactions among the toner particles are taken into account.

12-10.1.1 Toner Charging

Several papers have appeared describing toner charging in liquid development systems and its mechanism(s). For example, *Larson* [12.96] suggests that acid-base considerations probably predominate. The idea is that the formation of a negatively charged toner particle is caused by proton transfer to the charge director molecules, as originally suggested by *Fowkes* et al. [12.97] and *Croucher* et al. [12.98]. Similar ideas were recently discussed by *Pearlstine* et al. [12.99]. These charge director molecules such as lecithin, barium arylsufonates, etc., form inverse micelles with a highly polar interior and with the hydrocarbon-like end of each molecule extended outward, compatible with the solvent (Isopar). Proton exchange occurs between the micelle and the toner particle. Such concepts suggest that increasing the acidity of the toner resin, e.g., by attaching a sulfonic acid group, will enhance toner charging, as was done successfully by Larson.

This acid-base mechanism appears to be favored also by the group at AM International [12.100]. They presented a thorough study of the effect of varying lecithin concentration on the conductivity of the liquid with and without toner. They found the surprising result that the conductivity was the same for the liquid with toner and for the liquid with the toner centrifuged out (the centrate). This was explained by a re-equilibration of the neutral and ionic species as one ionic species is removed. By measuring the centrate more quickly after centrifuging, the expected lower conductivity was observed. These results suggest that attempting to obtain conducting values by centrifuging can be in error due to the time dependent re-equilibration of the ionic and neutral species.

Larson et al. [12.101] went on to show that aminoalcohols can enhance toner charging, but not by the acid-base mechanism. In this case, he suggested that both proton exchange and solubility of the aminoalcohol in the polar interior of the sulfactant micelles need to be considered. The protonated amine inside the micelle serves to increase the charge separation in the double layer surrounding the particle, thus increasing the electrophoretic mobility.

Gibson and *Luebbe* [12.102] noted a transient suppression of the conductivity when a lecithin/isopar liquid system is exposed to electric fields. Since recovery could take minutes to days, this could affect the stability of the development characteristics. A mechanism is proposed (a disturbance of the equilibrium between the neutral and charged species) and a solution suggested (addition of alkylated N-vinyl pyrrolidone copolymers to the toner).

An issue related to toner charging is the existance of excess ions. Excess ions are defined as ions in addition to the toner particles and the counter charges required for charge neutralization. *Novotny* [12.103–105] knew that excess ions exist in liquid developers and used a combination of techniques to identify their properties. Excess ions result from dissociation of the charge director molecules (among other sources), as an equilibrium is established between the charged and uncharged molecules in the liquid developer, as discussed by *Davis* et al. [12.100] and others. The evidence for these excess ions can be seen in Davis's paper and in papers by *Buxton* (Fig. 6 in [12.106]), and *Almag* et al. (Fig. 1 in [12.107]), etc., in which developed mass is plotted versus conductivity, which is used to measure the concentration of the charge director molecules. As the concentration of charge director molecules increases, M/A increases to a peak and then decreases. The decrease in M/A is explained by the competition between the toner and other like-charged excess ions to develop the latent image. As the concentration of excess ions increases, the fraction of the latent image developed by toner decreases.

12-10.4 Recent Developments

A qualitative discussion of ElectroInk has been given by *Landa* et al. [12.108], who pointed out its benefits (see also [12.109]). Computer modeling of ElectroInk was presented by *Nathaniel* et al. [12.110]. They included in their theory excess negative charge director molecules (improving on earlier work [12.111]) and predicted the development process taking into account the space charge fields. Experimental verification was not given. *Almag* et al. [12.107] presented experimental results of M/A versus conductivity, which is an indirect measure of the charge director concentration. A qualitative discussion was given, arguing that the M/A peaks fall rapidly and then more slowly. They argued that the initial decrease is due to an increase in the toner charge to mass ratio; the slower decrease is due to competition with excess ions.

References

Chapter 1

1.1 R. Schaffert: *Electrophotography* (Focal, New York 1965, revised several times up to 1980)
1.2 C. Carlson: "History of Electrostatic Recording", in *Xerography and Related Processes*, ed. by J. Dessauer, H. Clark (Focal, New York 1965) p. 15
1.3 G. Jackson, J. Hillkirk: *Xerox, an American Samurai* (Macmillan, New York 1986)
1.4 R. M. Schaffert: Photogr. Sci. Eng. **22**, 149 (1978)
1.5 C. F. Carlson: U.S. Patent 2297691 (1942)
1.6 G. C. Lichtenberg: Novi Comment. Göttingen **8**, 168 (1777)
1.7 P. Selenyi: J. Appl. Phys. **9**, 638 (1938)
1.8 C. F. Carlson: U.S. Patent 2357809 (1944)
1.9 J. Dessauer, H. Clark (eds.): *Xerography and Related Processes* (Focal, New York 1965)
1.10 Information supplied by Monica David, Senior VP, Director Office Equipment Group, Dataquest, San Jose, CA 95131
1.11 Information supplied by Bhanu Bhattasali, Director, Electronic Printer Industry Service, Dataquest, San Jose, CA 95131
1.12 S. M. Pytka: Comput. Software News, April 27, 15 (1987)
1.13 J. Weigl: Angew. Chem., Int. Ed. Engl. **16**, 374 (1977)
1.14 C. J. Young, H. G. Greig: RCA Rev. **15**, 469 (1954)
1.15 J. A. Amick: RCA Rev. **20**, 753 (1959)
1.16 H. P. Kallmann, J.Rennert, M. Sidran: Photogr. Sci. Eng. **4**, 345 (1960)
1.17 H. P. Kallmann, J. R. Freeman: Phys. Rev. **109**, 1506 (1958)
1.18 J. R. Freeman, H. P. Kallmann, M. Silver: Rev. Mod. Phys. **33**, 553 (1961)
1.19 H. P. Kallmann, J. Rennert, J. Burgos: Photogr. Sci. Eng. **6**, 65 (1962)
1.20 V. M. Fridkin, I. S. Zheludev: *Photoelectrets and the Electrophotographic Process* (Consultants Bureau, New York 1961)
1.21 V. M. Fridkin: J. Opt. Soc. Am. **50**, 545 (1960)
1.22 B. L. Shely: U.S. Patent 3563734 (1971); U.S. Patent 3764313 (1973)
1.23 L. E. Walkup: U.S. Patent 2825814 (1958)

1.24 C. F. Carlson, H. Bogdonoff: U.S. Patent 2982647 (1961)
1.25 L. E. Walkup: U.S. Patent 2833648 (1958); U.S. Patent 2937943 (1960)
1.26 R. M. Schaffert: IBM J. Res. Dev. **6**, 192 (1962)
1.27 I. Brodie, J. A. Dahlquist: J. Appl. Phys. **39**, 1618 (1968)
1.28 A. H. Sporer: Photogr. Sci. Eng. **12**, 213 (1968)
1.29 R. W. Gundlach, C. J. Claus: Photogr. Sci. Eng. **7**, 14 (1963)
1.30 P. Cressman: J. Appl. Phys. **34**, 2327 (1967)
1.31 H. J. Budd: J. Appl. Phys. **36**, 1613 (1965)
1.32 C. Snelling: U.S. Patent 3220324 (1965)
1.33 G. Pressman: In *Electrophotography, Second International Conference*, ed. by D. R. White (SPSE, Washington, DC, 1974) p. 37
1.34 E. G. Johnson, B. W. Neher: U.S. Patents 3010883, 3010884 (1961); U.S. Patents 3257304, 3285837 (1966)
1.35 V. Tulagin, R. F. Coles, R. A. Miller: U.S. Patent 3172827 (1964)
1.36 D. K. Meyer, A. G. Ostrem, G. J. Pollman: U.S. Patent 3130655 (1964)
1.37 N. R. Nail: U.S. Patent 3096260 (1963)
1.38 D. R. Eastman: U.S. Patent 3095808 (1963)
1.39 S. Tokumoto, E. Tanaka, C. Hara, O. Ogasawara, S. Murata: Photogr. Sci. Eng. **7**, 218 (1963)
1.40 M. C. Zerner, J. F. Sobieski, H. A. Hodes: Photogr. Sci. Eng. **13**, 184 (1969)
1.41 J. E. Kassner: J. Imaging Technol. **12**, 325 (1986)
1.42 F. W. Schmidlin: IEEE Trans. ED-**19**, 448 (1972); F. W. Schmidlin: In *Photoconductivity and Related Phenomena*, ed. by J. Mort, D. M. Pai (Elsevier, New York 1976) Chap. 11
1.43 M. E. Scharfe, F. W. Schmidlin: "Charged Pigment Xerography", in *Advances in Electronics and Electron Physics*, Vol. 38, (Academic, New York 1975) p. 83
1.44 G. C. Hartmann, L. M. Marks, C. C. Yang: J. Appl. Phys. **47**, 5409 (1976)
1.45 Y. C. Cheng, G. C. Hartmann: J. Appl. Phys. **51**, 2332 (1980)
1.46 R. W. Gundlach: Jpn. Patent 43-2242 (1967)
1.47 W. L. Goffe: Photogr. Sci. Eng. **15**, 304 (1971)
1.48 P. S. Vincett, G. J. Kovacs, M. C. Tam, A. L. Pundsack, P. H. Soden: J. Imaging Technol. **30**, 183 (1986)
1.49 V. Tulagin: J. Opt. Soc. Am. **59**, 328 (1960)
1.50 L. Carreira, V. Tulagin: U.S. Patent 3477934 (1969)
1.51 V. Tulagin, L. Carreira: U.S. Patent 3881920 (1975); V. Tulagin: U.S. Patent 3535221 (1970)
1.52 G. Hartmann, F. Schmidlin: J. Appl. Phys. **46**, 266 (1975)
1.53 P. Cressman, G. C. Hartmann: J. Chem. Phys. **61**, 2740 (1974)
1.54 P. Warter, V. Tulagin: L. Carreira; H. Hermanson, P. Warter; R.

Gruber L. Carreira; J. Grover; L. Cass; V. Tulagin: In *Third International Congress on Non-Impact Printing Technologies*, ed. by J. Gaynor (SPSE, Springfield, VA 1987) pp. 419-494
1.55 C. Snelling: U.S. Patent 3741760 (1973)
1.56 W. L. Little Jr., R. H. Townsend: U.S. Patent 3703459 (1972); U. S Patent 3952700 (1976)
1.57 J. B. Wells: U.S. Patent 3645874 (1972); U.S. Patent 3784294 (1974)
1.58 V. M. Marquart, R. H. Townsend: U.S. Patent 3427242 (1969)
1.59 J. B. Wells: U.S. Patent 3772013 (1973)
1.60 J. B. Wells, P. C. Swanton, J. W. Weigl, E. Forest: U.S. Patent 3850627 (1974); U.S. Patent 3920330 (1975); U.S. Patent 3954465 (1976)
1.61 R. H. Luebbe, M. S. Maltz, G. Reinis, W. G. VanDorn: In *Electrophotography, Second International Conference*, ed. by D. R. White (SPSE, Washington, DC, 1974) p. 48
1.62 C. C. Yang, G. C. Hartmann: IEEE Trans. ED-23, 308 (1976)
1.63 V. Tulagin: U.S. Patent 3512968 (1970)
1.64 A. R. Kotz, O. L. Nelson: In *Advances in Non-Impact Printing Technologies for Computer and Office Applications*, ed. by J. Gaynor (Van Nostrand Reinhold, New York 1982) p. 704
1.65 A. E. Berkowitz, J. A. Lahut, W. H. Meiklehjohn, R. E. Skoda, J. J. Wang: IEEE Trans. Magn. **18**, 1976 (1982)
1.66 K. Kokaji, K. Kinoshita, T. Urano, K. Saitoh: In *Advances in Non-Impact Printing Technologies for Computer and Office Applications*, ed. by J. Gaynor (Van Nostrand Reinhold, New York 1982) p. 769
1.67 J. J. Eltgen, J. G. Magnenet, J. P. Bresson: In *Third International Congress on Advances in Non-Impact Printing Technologies*, ed.by J. Gaynor (SPSE, Springfield, VA 1987) p. 547
1.68 H. A. Hermanson, R. E. Drews; D. G. Parker, F. Tomak, S. Swackhamer: In *Electrophotography, Fourth International Conference*, ed. by S. W. Ing, M. D. Tabak, W. E. Haas (SPSE, Springfield, VA 1983), pp. 541-570
1.69 J. J. Eltgen: In *Electrophotography, Fourth International Conference* ed. by S. W. Ing, M. D. Tabak, W. E. Haas (SPSE, Springfield, VA 1983) p. 519
1.70 G. D. Springer: In *Second International Congress on Advances in Non-Impact Printing Technologies*, (SPSE, Springfield, VA 1984) p. 73
1.71 D. G. Parker, F. Tomek, S. Swackhamer: In *Electrophotography, Fourth International Conference*, ed. by S. W. Ing., M. D. Tabak, W. E. Haas (SPSE, Springfield, VA 1983) p. 561
1.72 E. Schlömann: IEEE Trans. MAG-**10**, 60 (1974)
1.73 G. Bottlik, G. Cann; B. Ai, J. P. Maume, C. Mayoux, G. Sauret;

R. Miida, M. Ohnishi, K. Tomura, K. Samejima; R. Schayes, P. Gustin; M. Kimura, I. Kondo, M. Horie, H. Takahashi; T. Toyooshima, T. Todo, T. Kimoto, K. Nakano, S. Tomiyama: In *Advances in Non-Impact Printing Technologies for Computer and Office Applications*, ed. by J. Gaynor (Van Nostrand Reinhold, New York 1982), pp. 531-703

1.74 J. R. Rumsey, D. Bennewitz: J. Imaging Technol. **12**, 144 (1986)

1.75 M. Omodani, Y. Hoshino, T. Tanaka: J. Phys. D **18**, 153 (1985)

1.76 M. Omodani, T. Tanaka and Y. Hoshino: In *Third International Congress on Advances in Non-Impact Printing Technologies*, ed. by J. Gaynor (SPSE, Springfield, VA 1987) p. 295

Chapter 2

The interested reader will find books and review articles on electrophotography in [2.1–18]. In addition, many important papers have been presented and published in the proceedings of the IEEE-IAS Annual Conferences, the SPSE International Electrophotography Conferences, and the SPSE International Congresses on Advances in Non-Impact Printing Technologies.

2.1 J. Dessauer, H. Clark (eds.): *Xerography and Related Processes* (Focal, New York 1965)

2.2 R. M. Schaffert: *Electrophotography* (Focal, New York 1980)

2.3 M. Scharfe: *Electrophotography Principles and Optimization* (Research Studies Press., Letchworth, England 1984)

2.4 E. M. Williams: *The Physics and Technology of Xerographic Processes* (Wiley, New York 1984)

2.5 H. Keiss: RCA Rev. **40**, 59 (1978)

2.6 J. W. Weigl: Angew. Chem., Int. Ed. Engl. **16**, 374 (1977)

2.7 F. W. Schmidlin: In *Photoconductivity and Related Phenomena*, ed. by J. Mort, D. M. Pai (Elsevier, New York 1976) Chap. 11.

2.8 D. Winkelmann: J. Electrost. **4**, 193 (1977)

2.9 D. M. Burland, L. B. Schein: Phys. Today **39** (5), 46 (1986)

2.10 M. H. Lee, J. Ayala, B. D. Grant, W. Imaino, A. Jaffe, M. R. Latta, S. L. Rice: IBM J. Res. Dev. **28**, 241, (1984)

2.11 M. M. Shahin: In *Advances in Non-Impact Printing Technologies for Computer and Office Applications*, ed. by J. Gaynor (Van Nostrand Reinhold, New York 1982) p. 1350

2.12 E. S. Baltazzi: J. Appl. Photogr. Eng. **8**, 224 (1980)

2.13 E. S. Baltazzi: J. Appl. Photogr. Eng. **6**, 147 (1980)

2.14 D. Winkelmann: J. Appl. Photogr. Eng. **4**, 187 (1978)

2.15 R. M. Schaffert: Photogr. Sci. Eng. **22**, 149 (1978)

2.16 R. B. Comizzoli, G. S. Lozier, D. A. Ross: Proc. IEEE **60**, 348 (1972)

2.17 W. F. Berg, K. Hauffe (eds.): *Current Problems in Electrophography* (de Gruyter, Berlin 1972)

2.18 A. B. Jaffe, D. M. Burland: In *Hard Copy Output Devices*, ed. by R. C. Durbeck, S. Sherr (Academic, New York 1988) p. 221

2.19 R. G. Vyverberg: In *Xerography and Related Processes*, ed. by J. Dessauer, H. Clark (Focal, New York 1965) Chap. 7

2.20 A. J. Rushing: IEEE Trans. **AC-25**, 1078 (1980);
D. G. Parker: IEEE-IAS Annu. Conf. Proc., 363 (1974);
T. F. Hayne: IEEE-IAS Annu. Conf. Proc., 345 (1974);
D. G. Parker: IEEE-IAS Annu. Conf. Proc., 367 (1973)

2.21 J. D. Cobine: *Gaseous Conductor* (Dover, New York 1958)

2.22 J. Meek, J. D. Craggs: *Electrical Breakdown in Gases* (Wiley, New York 1978)

2.23 E. Nasser: *Fundamentals of Gaseous Ionization and Plasma Electronics* (Wiley, New York 1971)

2.24 R. S. Sigmond: J. Appl. Phys. **53**, 891 (1982)

2.25 M. M. Shahin: J. Chem. Phys. **45**, 2600 (1966)

2.26 M. M. Shahin: Appl. Opt., Suppl. No. 3 Electrophotography 106 (1969)

2.27 M. M. Shahin: Photogr. Sci. Eng. **15**, 322 (1971)

2.28 C. F. Gallo: IEEE Trans. **IA-11**, 739 (1975);
C. Gallo: IEEE Trans. **IA-13**, 550 (1977);
P. Walsh, C. Gallo, W. Lama: Photogr. Sci. Eng. **28**, 109 (1984)

2.29 T. G. Davis, G. J. Safford: U.S. Patent 4086650 (1978)

2.30 F. W. Hudson, J. E. Cranch: IEEE Trans. IA-**13**, 366 (1977)

2.31 B. E. Springett, F. M. Tesche, A. R. Davies, J. A. L. Thompson: Photogr. Sci. Eng. **22**, 200 (1978)

2.32 The literature is extensive; one might start with R. G. Enck, G. Pfister: In *Photoconductivity and Related Phenomena*, ed. by J. Mort, D. M. Pai (Elsevier, Amsterdam 1976) Chap. 7, and D. M. Pai, R. C. Enck: Phys. Rev. **B11**, 5163 (1975)

2.33 W. D. Gill: J. Appl. Phys. **43**, 5033 (1972)

2.34 M. D. Shattuck and U. Vahtra: U.S. Patent 3484237 (1969)

2.35 W. J. Dullmage, W. A. Light, S. J. Marino, C. D. Salzberg, D. L. Smith, W. J. Staudenmayer: J. Appl. Phys. **49**, 5543 (1978)

2.36 P. M. Borsenberger, A. Chowdry, D. C. Hoestery, W. Mey: J. Appl. Phys. **49**, 5555 (1978)

2.37 D. M. Pai, J. Yanus: Photogr. Sci. Eng. **27**, 14 (1983)

2.38 D. M. Pai: J. Non-Cryst. Solids **60**, 1255 (1983)

2.39 J. Mort, G. Pfister: Polym. Plast. Technol. Eng. **12**, 89 (1979)

2.40 P. J. Melz, R. B. Champ, L. S. Chang, G. S. Keller, L. C. Liclican, R. R. Neiman, M. D. Shattuck, W. J. Weiche: Photogr. Sci. Eng. **21**, 73 (1977)

2.41 J. Mort, D. M. Pai (eds.): *Photoconductivity and Related Phenomena* (Elsevier, Amsterdam 1976)

2.42 J. Mort, G. Pfister: In *Electronic Properties of Polymers*, ed. by J. Mort, G. Pfister (Wiley, New York 1982) Chap. 6

2.43 M. Stolka, J. F. Yanus, D. M. Pai: J. Phys. Chem. **88**, 4707 (1984)

2.44 L. S. J. Santos, J. Hirsh: Philos. Mag. B **53**, 4707 (1984)

2.45 L. B. Schein, A. Rosenberg, S. L. Rice: J. Appl. Phys. **60**, 4287 (1986)

2.46 S. K. Ghosh, W. E. Bixby: J. Appl. Photogr. Eng. **6**, 109 (1980)

2.47 T. Kawamura, N. Yamamoto, Y. Nakayama: "Electrophotographic Applications of Amorphous Semiconductors" In *Japanese Annual Reviews in Electronics, Computers, and Telecommunications*, Vol. 6, ed. by Y. Hamakawa (Ohmsha, Tokyo, and North-Holland, Amsterdam 1983)

2.48 C. J. Claus, E. F. Mayer: In *Xerography and Related Processes*, ed. by J. Dessauer, H. Clark (Focal, New York 1965) Chap. 12

2.49 V. Tulagin: J. Opt. Soc. Am. **59**, 328 (1969)

2.50 P. S. Vincett, G. J. Kovacs, M. C. Tam, A. L. Pundsack, P. H. Soden: J. Imaging Technol. **30**, 1983 (1986)

2.51 I. Chen: Photogr. Sci. Eng. **26**, 153 (1982)

2.52 J. Bares: Photogr. Sci., Eng. **28**, 111 (1984)

2.53 N. Kawamura, M. Itoh: In *Third International Congress on Advances in Non-impact Printing Technologies*, ed. by J. Gaynor (SPSE, Springfield, VA 1986) p. 154

2.54 P. G. Andrus, F. W. Hudson: In *Xerography and Related Processes*, ed. by J. Dessauer, H. Clark (Focal, New York 1965) Chap. 14

2.55 H. Krupp, G. Sperling: J. Appl. Phys. **37**, 4176 (1966)

2.56 H. Krupp: Adv. Colloid Interface Sci. **1**, 111 (1967)

2.57 D. A. Hays: Photogr. Sci. Eng. **22**, 232 (1978)

2.58 D. A. Hays: Photogr. Sci. Eng. **22**, 965 (1978)

2.59 D. K. Donald: J. Appl. Phys. **40**, 3013 (1969)

2.60 D. K. Donald: J. Adhes. **4**, 233 (1972)

2.61 C. J. Mastrangelo: Photogr. Sci. Eng. **26**, 194 (1982)

2.62 M. H. Lee: SID Proc. **27**, 9 (1986)

2.63 M. H. Lee, J. Ayala: J. Imaging Technol. **11**, 279 (1985)

2.64 N. Goal, P. Spencer: Polym. Sci. Technol. **9B**, 76 (1975)

2.65 C. C. Yang, G. C. Hartmann: IEEE Trans. **ED-23** 308 (1976)

2.66 M. Hida, J. Nakajima, and H. Takahashi: IEEE-IAS Annu. Conf. Proc. 1225 (1982)

2.67 L. H. Lee: Adhes. Sci. and Technol. **98**, 831 (1975)

2.68 P. E. Castro, W. C. Lu: Photogr. Sci. Eng. **22**, 154 (1978)

2.69 L. Nebenzahl, J. Borgioli, V. De Palma, K. Gung, C. Mastrangelo, F. Pourroy: Photogr. Sci. Eng. **24**, 293 (1980)

2.70 V. M. De Palma: Photogr. Sci. Eng. **26**, 198 (1982)

2.71 G. Harpavat: IEEE-IAS Annu. Conf. Proc. 569 (1977); IEEE Trans. **IA-15**, 681 (1979)

2.72 B. V. Deryagin, N. A. Krotova, V. P. Smilga: *Adhesion of Solids* (Consultants Bureau, New York 1978)
2.73 G. Abowitz: IEEE-IAS Annu. Conf. Proc. 153 (1974)
2.74 T. B. McMillen, D. P. Salamida: IEEE-IAS Annu. Conf. Proc. 161 (1974)
2.75 N. R. Lindblad, I. Rezanka: U.S. Patent 4279499 (1981);
I Rezanka: U.S. Patent 4272184 (1981)
2.76 W. S. Jewett: IEEE-IAS Annu. Conf. Proc. 557 (1977)
2.77 M. S. Doery: U.S. Patent 4508447 (1985)
2.78 W. D. Hope, M. Levy: In *Xerography and Related Processes*, ed. by J. Dessauer and H. Clark (Focal, New York, 1965) Chap. 4
2.79 I. Shimizu, T. Komatsu, K. Saito, E. Imaine: J. Non-Cryst. Solids **35**, 773 (1980);
K. Wakita, Y. Nakayama, T. Kawamura: Photogr. Sci. Eng. **26**, 183 (1982)
2.80 Y. Nakayama, A. Sugimura, M. Nakano and T. Kawamura: Photogr. Sci. Eng. **26**, 188 (1982);
E. Inoue and I. Shimizu: Photogr. Sci. Eng. **26**, 148 (1982)
2.81 I. Chen, J. Mort, F. Jansen, S. Grammatica, M. Morgan: J. Imaging Sci. **29**, 73 (1985)
2.82 L. Cheung, G. M. Foley, P. Fournia, B. Springett: Photogr. Sci. Eng. **26**, 245 (1982)
2.83 A. R. Melnyk, J. S. Berkes, L. B. Schein: In *Advances in Non-Impact Printing Technologies for Computer and Office Applications*, ed. by J. Gaynor (Van Nostrand Reinhold, New York 1981) p. 508
2.84 S. Faria: U.S. Patent 4374917 (1983)
2.85 M. Lutz, B. Reimer: SPSE Annual Meeting, Rochester, April 1982
2.86 S. Arora, W. Murphy: SPSE Annual Meeting, Rochester, April 1982
2.87 R. B. Champ, M. D. Shattuck: U.S. Patent 3824099 (1974);
R. Wingard: IEEE-IAS Annu. Conf. Proc. 1251 (1982)
2.88 R. O. Loutfy, C. K. Hsiao, P. M. Kazmaier: Photogr. Sci. Eng. **27**, 5 (1983)
2.89 K. Y. Lee: J. Imaging Technol. **31**, 83 (1987)
2.90 M. Chang, P. Edelman: U.S. Patent 4353971 (1982)
2.91 S. P. Clark, G. A. Reynolds, J. H. Perlstein: U.S. Patent 4327169 (1982)
2.92 K. Arishima, H. Hiratsuka, A. Tate, T. Okada: Appl. Phys. Lett. **40**, 280 (1982)
2.93 R. O. Loutfy, A. M. Hor, A. Rucklidge: J. Imaging Technol. **31**, 31 (1987)
2.94 A. Kakuta, Y. Mori: "Near Infrared Sensitive Organic Photoreceptors," at the SPSE Annual Meeting, San Francisco, June 1983

2.95 S. Grammatica, J. Mort: Appl. Phys. Lett. **38**, 445 (1981)

2.96 F. Nakagawa, S. Itoh, M. Otsuki, K. Itoh, K. Tsuji: In *Third International Congress on Advances in Non-Impact Printing Technologies, Advance Printing of Paper Summaries* (SPSE, Springfield, VA 1986) p. 19

2.97 L. B. Schein: In *Electrophotography Second International Conference*, ed. by D. R. White (SPSE, Washington, DC 1974) p. 65

2.98 G. Starkweather: In *Laser Applications*, Vol. 4, ed. by J. Goodman, M. Ross (Academic, New York 1980) p. 125;
R. A. Sprague, J. C. Urback, T. S. Fisli: Laser Focus/Electro-Optics, **19**(10) 101 (1986);
D. McMurtry, M. Tinghitella, R. Svendsen: IBM J. Res. Dev. **28**, 257 (1984)

2.99 D. E. Grant: Symp. on Laser Recording and Information Handling, SPIE Proc. **200**, 195 (1979)

2.100 G. Paul: SPIE Proc. **396**, 204 (1983)

2.101 J. C. Urback, T. S. Fisli, G. K. Starkweather: Proc. IEEE **70**, 597 (1982)

2.102 Y. Hoshino, K. Tateishi: In *Advances in Non-Impact Printing Technologies for Computer and Office Applications*, ed. by J. Gaynor, (Van Nostrand Reinhold, New York 1981) p. 390

2.103 Many papers on LED arrays were presented at the *Second International Congress on Advances in Non-Impact Printing Technologies* (SPSE, Springfield VA 1984), see Advance Printing of Paper Summaries p. 168-175

2.104 K. Tateishi, Y. Ikeda, S. Kotani, S. Nakaya: SID Proc. **23**, 2, 1982

2.105 T. Nakamura, H. Morita, M. Maeda: J. Imaging Technol. **12**, 300 (1986)

2.106 K. Tateishi, Y. Hoshino: IEEE Trans. **IA-19** 169 (1983)

2.107 P. F. Heidrich, R. A. Laff, T. B. Light: U.S. Patent 4447126 (1984)

2.108 S. W. Depp, J. M. Eldridge, A. K. Juliana, M. H. Lee: IBM Tech. Discl. Bull. **25-12**, 6325 (1983)

2.109 H. Nakamura, K. Aoki, M. Yonekubo: In *Second International Congress on Advances in Non-Impact Printing Technologies, Advance Printing of Paper Summaries* (SPSE, Springfield, VA 1984) p. 213

2.110 B. Hill: Philips J. Res. **38**, 159 (1983)

2.111 B. Hill, K. P. Schmidt, L. Borgmann, H. Meyer, G. Much: J. Imaging Technol. **13**, 15 (1987)

2.112 B. Hill, K. Schmidt: Philips J. Res. **33**, 211 (1978)

2.113 J. Revelli, W. Hoston, D. Kinzer, R. V. Johnson: In *Second International Congress on Advances in Nonimpact Printing Technologies, Advance Printing of Paper Summaries* (SPSE, Springfield, VA 1984) p. 185;

J. T. Cutchen, J. O. Harris, Jr., G. R. Lagvna: Appl. Opt. **14**, 1986 (1975)
2.114 Z. Kun, D. Leksell, P. Malmberg, J. Asars, G. Brandt: SID International Symposium, Digest of Technical Papers, Vol. XVII (Palisades, New York 1986) p. 270
2.115 T. Nylund, C. Cowan, J. Spence, L. Steele: In *Third International Congress on Advances in Non-Impact Printing Technology Advance Printing of Paper Summaries* (SPSE, Springfield, VA 1986) p. 74
2.116 M. R. Spect, L. Contois, D. Santilli: In *Third International Congress on Advances in Non-Impact Printing Technology Advance Printing of Paper Summaries* (SPSE, Springfield, VA 1986) p.76
2.117 R. W. Gundlach: U.S. Patent 3084043 (1963);
N. Lindblad, R. Tift, P. K. Watson: P. K. Watson, N. Lindblad: In *Third International Congress on Advances in Non-Impact Printing Technologies*, ed. by J. Gaynor (SPSE, Springfield, VA 1986) pp. 113, 120
2.118 W. Gesierich, E. Weyde, H. Haydn: U.S. Patent 3285741 (1966)
2.119 Y. Moradzadeh: U.S. Patent 4272599 (1981)
2.120 A. J. Butler, J. F. Holburg, Z. J. Cendes: IEEE-IAS Annu. Conf. Proc. 1626 (1987)
2.121 B. Cherbuy: J. Imaging Technol. **13**, 215 (1986)
2.122 J. C. Azar, A. W. Henry, R. W. Ferguson: U.S. Patent 4372246 (1983)
2.123 A. W. Henry, J. C. Azar, J. Sagal: U.S. Patent 4372239 (1983)
2.124 P. D. Jachimiak: IEEE-IAS Annu. Conf. Proc. 295 (1977)
2.125 J. C. Minor: U.S. Patent 4357388 (1982)
2.126 N. L. Giorgini: U.S. Patent 4363862 (1982)
2.127 R. D. Archibald: Hewlett-Packard J. **33-6**, 24 (1982)
2.128 G. L. Holland: Hewlett-Packard J. **33-7**, 13 (1982)
2.129 H. Mugraner: U.S. Patent 4311723 (1982)
2.130 G. Hausmann: German Patent DE2838864C3 (1982)
2.131 H. S. Kocher: IEEE-IAS Annu. Conf. Proc. 34 (1979)
2.132 G. W. Baumann: IBM J. Res. Dev. **23**, 292 (1984)
2.133 T. Narusawa, N. Sawatari, H. Okuyama: J. Imaging Technol. **2**, 284 (1985)
2.134 C.C. Wilson: J. Appl. Photogr. Eng. **6**, 148 (1979)
2.135 J. C. Minor: Seminar on Trends in Office Automation, New York, (May 1983)
2.136 J. Newkirk: U.S. Patent 4375505 (1983)

Chapter 3

3.1 H. E. J. Neugebauer: In *Xerography and Related Processes*, ed. by J. Dessauer, H. Clark (Focal, New York 1965) Chap. 8
3.2 E. M. Williams: *The Physics and Technology of Xerographic Processes* (Wiley, New York 1984) p. 102

3.3 M. Scharfe: *Electrophotography, Principles and Optimization* (Research Studies Press, Letchworth, England 1984)
3.4 R. M. Schaffert: *Electrophotography* (Focal, New York 1980)
3.5 J. Dessauer, H. Clark (eds.): *Xerography and Related Processes* (Focal, New York 1965)
3.6 E. N. Wise: U.S. Patent 2618552 (1952);
L. E. Walkup: U.S. Patent 2618551 (1952);
L. E. Walkup and E. N. Wise: U.S. Patent 2638416 (1953)
3.7 D. Winkelmann: J. Electrost. **4**, 193 (1977)
3.8 C. J. Young: U.S. Patent 2786439 (1957);
E. Giaimo: U.S. Patent 2786440; (1957);
C. J. Young; U.S. Patent 2786441 (1957)
3.9 G. Kasper and J. May: U.S. Patent 4076857 (1978)
3.10 A. R. Kotz: U.S. Patent 3909258 (1975)
3.11 D. R. Field: IEEE Trans. **IA-19**, 759 (1983)
3.12 T. Takahashi, N. Hosono, J. Kanbe, T. Toyono: Photogr. Sci. Eng. **26**, 254 (1982)
3.13 F. Takeda, K. Sakamoto, K. Kobayashi: IEEE-IAS Annu. Conf. Proc. 1491 (1985)
3.14 M. Hosoya, S. Tomura, T. Uehara: IEEE-IAS Annu. Conf. Proc. 1485 (1985)
3.15 C. J. Claus, E. F. Mayer: In *Xerography and Related Processes*, ed. by J. Dessauer, H. Clark (Focal, New York 1965) p. 342
3.16 B. Landa: In *Third International Congress on Advances in Non-Impact Printing Technologies, Advance Printing of Paper Summaries* (SPSE, Springfield, VA 1986) p. 307
3.17 T. Nylund, C. Cowan, J. Spence, L. Steele: In *Third International Congress on Advances in Non-Impact Printing Technologies, Advance Printing of Paper Summaries* (SPSE, Springfield, VA 1988) p. 74
3.18 M. Specht, L. Contois, D. Santilli: In *Third International Congress on Advances in Non-Impact Printing of Paper Technologies, Advance Printing of Paper Summaries* (SPSE, Springfield, VA 1988) p. 76

Chapter 4

4.1 J. Lowell, A. C. Rose-Innes: Adv. Phys. **29**, 1947 (1980)
4.2 L. B. Loeb: *Static Electrification* (Springer, Berlin 1958)
4.3 W. R. Harper: *Contact and Frictional Electrification* (Oxford University Press, Oxford 1967)
4.4 C. F. Gallo, S. J. Ahuja: IEEE Trans. **IA-13**, 348 (1977)
4.5 D. J. Montgomery: Solid State Phys. **9**, 139 (1959)
4.6 W. T. Morris: Plast. Polym. **38**, 41 (1970)
4.7 D. A. Seanor: "Triboelectrification of Polymers – A Chemist's Viewpoint," in *Physicochemical Aspects of Polymer Surfaces*, Vol. 1, ed. by K. L. Mittal (Plenum, New York 1983) p. 477

4.8 Conferences on Static Electrification, Inst. Phys. Conf. Ser., No. 11 (1971); No. 27 (1975); No. 48 (1979)
4.9 H. R. Harper: Proc. R. Soc. London, Ser. A **205**, 83 (1951)
4.10 J. Lowell: J. Phys. D **8**, 53 (1975)
4.11 A. Wahlin, G. Bäckström: J. Appl. Phys. **45**, 2058 (1974)
4.12 D. K. Davies: J. Phys. D **2**, 1533 (1969)
4.13 A. R. Akande, J. Lowell: J. Electrost. **16**, 147 (1985)
4.14 R. Elsdon, F. R. G. Mitchell: J. Phys. D **9**, 1445 (1976)
4.15 I. I. Inculet, E. P. Wituschek: Static Electrification, Inst. Phys. Conf. Ser. **4**, 37 (1967)
4.16 W. D. Greason, I. I. Inculet: IEEE-IAS Annu. Conf. Proc. (1975) p. 428
4.17 F. Nordhage, G. Bäckström: J. Electrost. **3**, 371 (1971)
4.18 H. T. M. Haenen: J. Electrost. **2**, 151 (1976)
4.19 T. J. Fabish, C. B. Duke: J. Appl. Phys. **48**, 4256 (1977);
T. J. Fabish, H. M. Saltsburg, M. L. Hair: J. Appl. Phys. **47**, 940 (1976)
4.20 G. A. Cottrell, J. Lowell, A. C. Rose-Innes: J. Appl. Phys. **50**, 1374 (1979)
4.21 M. W. Williams: IEEE-IAS Annu. Conf. Proc. (1984) p. 131
4.22 H. W. Gibson: J. Am. Chem. Soc. **97**, 3832 (1975)
4.23 H. W. Gibson, F. C. Bailey: Chem. Phys. Lett. **51**, 352 (1977)
4.24 I. Shinohara, F. Yamamoto, H. Anzai, S. Endo: J. Electrost. **2**, 99 (1976)
4.25 P. J. Cressman, G. C. Hartmann, J. E. Kuder, F. D. Saeva, D. Wychick: J. Chem. Phys. **61**, 1740 (1974)
4.26 D. A. Hays: J. Chem. Phys. **61**, 1455 (1974)
4.27 H. Bauser: DECHEMA-Monogr. **72**, 11 (1974)
4.28 S. Kittaka, Y. Murata: Jpn. J. Appl. Phys. **18**, 515 (1979)
4.29 H. R. Harper: Proc. R. Soc. London **218**, 111 (1953)
4.30 G. A. Cottrell, C. Reed, A. C. Rose-Innes: Static Electrification, Inst. Phys. Conf. Ser. **48**, 249 (1979);
G. A. Cottrell, C. E. Hatto, C. Reed, A. C. Rose-Innes: J. Phys. D **17**, 989 (1984)
4.31 P. E. Shaw, C. S. Jex: Proc. R. Soc. London **118**, 108 (1928); P. E. Shaw: Proc. R. Soc. London **94**, 16 (1917);
H. Freundlich: *Colloid and Capillary Chemistry*, 3rd ed. (Methuen, London 1926) p. 284
4.32 V. J. Weber: J. Appl. Polym. Sci. **7**, 1317 (1963)
4.33 P. J. Sereda, R. F. Feldman: J. Text. Inst. **55**, T288 (1964)
4.34 J. A. Medley: Nature **171**, 1077 (1953)
4.35 W. A. Rudge: Philos. Mag. **25**, 481 (1913)
4.36 O. Knoblauch: Z. Phys. Chem. **39**, 225 (1902)
4.37 S. P. Rowland (ed.): *Water in Polymers* (American Chemical Society, Washington, DC 1980);

J. A. Barrie: In *Diffusion in Polymers* ed. by J. Crank, G. S. Park (Academic, New York 1968) Chap. 8
4.38 F. P. Bowden, W. R. Throssel: Nature **167**, 601 (1951)
4.39 M. I. Kornfield: J. Phys. D **9**, 1183 (1976)
4.40 P. S. H. Henry: Br. J. Appl. Phys. **4**, Suppl. 2, S6 (1957)
4.41 E. S. Robins, J. Lowell, A. C. Rose-Innes: J. Electrost. **8**, 153 (1980)
4.42 K. P. Homewood: J. Electrost. **10**, 229 (1981)
4.43 W. R. Salanek, A. Paton, D. T. Clark: J. Appl. Phys. **47**, 144 (1976)
4.44 M. W. Williams: J. Macromol. Sci., Rev. Macromol. Chem. **C14**, 251 (1976)
4.45 C. B. Duke, T. J. Fabish: Phys. Rev. Lett. **37**, 1075 (1976)
4.46 A. M. Cowley, S. M. Sze: J. Appl. Phys. **36**, 3212 (1965)
4.47 H. Krupp: Static Electrification, Inst. Phys. Conf. Ser. **11**, 1 (1971)
4.48 H. Bauser, W. Klopffer, H. Rabenhorst: Proc. 1st Int. Conf. on Static Electricity, Vienna, Austria, 4–6 May 1970, In *Adv. Stat. Electrification* **1**, 2 (1971)
4.49 F. R. Ruckdeschel, L. P. Hunter: J. Appl. Phys. **48**, 4898 (1977)
4.50 J. Henniker: Nature **196**, 474 (1962)
4.51 S. P. Hersh, D. J. Montgomery: Text. Res. J. **25**, 279 (1955)
4.52 G. S. Rose, S. G. Ward: Br. J. Appl. Phys. **8**, 121 (1957)
4.53 W. Schumann: Plaste Kautsch **10**, 526, 590, 654 (1963)
4.54 A. Coehn: Ann. Phys. (Leipzig) **64**, 217 (1898)
4.55 F. P. Bowden, D. Tabor: *The Friction and Lubrication of Solids*, Part 2 (Clarendon, Oxford 1964)
4.56 W. A. Zisman: Adv. Chem. **43**, 1 (1964)
4.57 E. Fukada, J. F. Fowler: Nature **181**, 693 (1958)
4.58 D. K. Davies: Proc. 1st Int. Conf. on Static Electricity, Vienna, Austria, 4-5 May 1970, in *Adv. Stat. Electrification* **1**, 10 (1971)
4.59 C. B. Duke, T. J. Fabish: J. Appl. Phys. **49**, 315 (1978)
4.60 K. T. Whitby, B. T. H. Liu: In *Aerosol Science*, ed. by C. N. Davies, (Academic, London 1966) Chap. 3
4.61 L. Cheng, S. L. Soo: J. Appl. Phys. **41**, 585, (1970)
4.62 A. Y. H. Cho: J. Appl. Phys. **35**, 2561 (1964)
4.63 W. B. Kunkel: J. Appl. Phys. **21**, 833 (1950)
4.64 G. Röbin, Porstendörfer: J. Colloid Interface Sci. **69**, 183 (1979)
4.65 C. Hendricks: "Charging Macroscopic Particles", in *Electrostatics and Its Applications*, ed. by A. D. Moore (Wiley, New York 1973) p. 57
4.66 L. B. Schein: Photogr. Sci. Eng. **19**, 255 (1975)
4.67 E. H. Lehmann, G.R. Mott: In *Xerography and Related Processes*, ed. by J. Dessauer, H. Clark (Focal, New York 1965) Chap. 10
4.68 L. B. Schein, J. Cranch: J. Appl. Phys. **46**, 5140 (1975)

4.69 P. M. Cassiers, J. van Engeland: Photogr. Sci. Eng. **9**, 273 (1965)
4.70 L. B. Schein: In *Electrophotography, Second International Conference*, ed. by D. R. White (SPSE, Washington, DC 1974) p. 65
4.71 D. A. Hays: J. Appl. Phys. **48**, 4430 (1977)
4.72 D. Winkelmann: J. Electrost. **4**, 193 (1977)
4.73 J. McCabe: U.S. Patent 3795617 (1974)
4.74 Lieng-Huang Lee: Photogr. Sci. Eng. **22**, 228 (1978)
4.75 C. R. Raschke: In *Electrophotography, Second International Conference*, ed. by D. R. White (SPSE, Washington,DC 1974) p. 104
4.76 H. Fielder, H. Stottmeister: "Zu einigen Beziehungen zwischen Tonerladung, Tonerkorngrösse und Haftkraft von Kaskadenentwicklern," Signal, AM **4** 317 (1976) (taken from [4.72])
4.77 J. H. Daly, D. Hayward, R. A. Pethrick: J. Phys. D **19**, 885 (1986)
4.78 G. T. Brewington: In *Colloids and Surfaces in Reprographic Technology*, ed. by M. Hair, M. D. Croucher (ACS Symp. Ser. 200, Washington, DC 1982) p. 183
4.79 T. J. Fabish, M. L. Hair: J. Colloid Interface Sci. **62**, 16 (1977)
4.80 P. C. Julien: In *Carbon Black-Polymer Composites*, ed. by E. Sichel (Marcel Dekker, New York 1982) p. 189
4.81 W. M. Prest, R. A. Mosher: In *Colloids and Surfaces in Reprographic Technology*, ed. by M. Hair, M. D. Croucher (ACS Symp. Ser. 200, Washington, DC 1982) p. 225
4.82 R. J. Gruber: *SID International Symposium Digest of Technical Papers* (Palisades, New York 1987) p. 272
4.83 K. L. Birkett, K. L. Gregory: Dyes Pigm. **7**, 341 (1986)
4.84 G. Harpavat, R. Orr: IEEE-IAS Annu. Conf. Proc. (1975) p. 158
4.85 L. F. Collins: J. Appl. Phys. **48**, 4569 (1977)
4.86 E. M. Williams: *The Physics and Technology of Xerographic Processes* (Wiley, New York 1984) p. 134
4.87 R. W. Stover, P. C. Schoonover: SPSE Annu. Conf. Proc. (1969) p. 156
4.88 R. Hölz: Data given in [4.72]
4.89 R. B. Lewis, E. W. Connors, R. F. Koehler: Jpn. J. Electrophotography **22**, 85 (1983) and U.S. Patent 4375673
4.90 B. D. Terris, A. B. Jaffe: Inst. Phys. Conf. Ser. No. 85: Section 1, paper presented at Electrostatics 1987, Oxford
4.91 H. Demizu, T. Saito, K. Aoki: In *Third International Congress on Advances in Non-Impact Printing Technologies*, ed. by J. Gaynor (SPSE, Springfield, VA 1987) p. 84
4.92 Y. Takahashi, H. Horiguchi, T. Sakata: In *Third International Congress on Advances in Non-Impact Printing Technologies*, ed. by J. Gaynor (SPSE, Springfield, VA 1987) p. 49

4.93 N. Kutsuwada, H. Kashimada, M. Fukuda, T. Suzuki, K. Ohkawa: J. Imaging Technol. **12**, 220 (1986)
4.94 N. Kutsuwada, Y. Nakamura: IEEE-IAS Annu. Conf. Proc. (1987) p. 1597
4.95 M. K. Mazumder, R. E. Ware, T. Yokoyama, B. Rubin, D. Kamp: IEEE-IAS Annu. Conf. Proc. (1987) p. 1606

Chapter 5

5.1 L. Walkup: U.S. Patent 2618551 (1952);
E. Wise: U.S. Patent 2618552 (1952)
5.2 T. T. Thourson: IEEE Trans. ED-**19**, 495 (1972)
5.3 J. Bickmore, K. W. Gunther, J. F. Knapp, W. A. Sullivan: Photogr. Sci. Eng. **14**, 42 (1970)
5.4 M. Levy, L. Walkup; R. Gundlack: In *Xerography and Related Processes*, ed. by J. Dessauer, H. Clark (Focal, New York 1965) Chap. 2, Chap. 9
5.5 E. Lehmann, G. Mott: In *Xerography and Related Processes*, ed. by J. Dessauer, H. Clark (Focal, New York 1965) Chap. 10
5.6 W. A. Sullivan, T. L. Thourson: Photogr. Sci. Eng. **11**, 115 (1967)
5.7 D. K. Donald, P. K. Watson: Photogr. Sci. Eng. **14**, 36 (1970)
5.8 D. K. Donald, P. K. Watson: IEEE Trans. ED-**19**, 458 (1972)
5.9 S. C. Maitra, H. Scher, J. Knapp: IEEE-IAS Annu. Conf. Proc. (1974) p. 31
5.10 N. Herbert, D. K. Donald, L. Collins: IEEE Trans. IA-**13**, 183 (1977)
5.11 J. D. Jackson: *Classical Electrodynamics* (Wiley, New York 1965) p. 112
5.12 R. W. Stover: IEEE-IAS Annu. Conf. Proc. (1974) p. 43
5.13 H. E. J. Neugebauer: In *Xerography and Related Processes*, ed. by J. Dessauer, H. Clark (Focal, New York 1965) Chap. 8
5.14 O. G. Hauser, R. S. Menchel: SPSE Annu. Conf. Proc. (1968) p. 36
5.15 P. M. Cassiers, J. van Engeland: Photogr. Sci. Eng. **9**, 273 (1965)
5.16 L. B. Schein: In *Electrophotography, Second International Conference,* ed. by D. R. White, (SPSE, Washington DC, 1974) p. 65
5.17 R. W. Stover, P. C. Schoonover: SPSE Annu. Conf. Proc. (1969) p. 156
5.18 M. Mukherjee, P. Mukherjee, A. Ghosh: IEEE Trans. IA-**21** 535 (1985)

Chapter 6

6.1 C. Young: U.S. Patents 2786439 (1957); 2786441 (1957)
6.2 E. Giaimo: U.S. Patent 2786440 (1957)
6.3 T. B. Jones, G. L. Whittaker, T. J. Sulenski: Powder Technol. **49**, 149 (1987);

G. Harpavat: IEEE Trans. MAG-**10**, 919 (1974)
6.4 R. W. Gundlach: In *Xerography and Related Processes*, ed. by J. Dessauer, H. Clark (Focal, New York 1965) Chap. 9
6.5 C. Young, H. Greig: RCA Rev. **15**, 471 (1954)
6.6 J. A. Amick: RCA Rev. **20**, 753 (1959)
6.7 T. Kimura, M. Yukozawa: Denshi Shashin (Electrophotogr.) **5**, 33 (1963)
6.8 Y. Moradzadeh, D. Woodwood: Photogr. Sci. Eng. **10**, 96 (1966)
6.9 H. Hasegawa, S. Sugihara, S. Nishikawa: Denshi Shashin (Electrophotogr.) **6**, 65 (1966)
6.10 T. L. Thourson: IEEE Trans. ED-**19**, 495 (1972)
6.11 L. B. Schein: In *Electrophotography, Second International Conference*, ed. by D. R. White (SPSE, Washington, D.C. 1974) p. 65
6.12 L. B. Schein: Photogr. Sci. Eng. **19**, 3 (1975)
6.13 L. B. Schein: Photogr. Sci. Eng. **19**, 255 (1975)
6.14 L. B. Schein, K. J. Fowler: J. Imaging Technol. **11**, 295 (1985)
6.15 D. Burland, L. B. Schein: Phys. Today **39**, 46 (May 1986)
6.16 G. Harpavat: IEEE-IAS Annu. Conf. Proc. 128 (1975)
6.17 E. Williams: IEEE-IAS Annu. Conf. Proc. 215 (1978)
6.18 E. M. Williams: *The Physics and Technology of Xerographic Processes* (Wiley, New York 1984)
6.19 A. Kondo, M. Kamiya: Tappi **59**, 94 (1976)
6.20 W. Verlinden, J. Van Engeland, J. Van Biessen: In *Electrophotography, Third International Conference, Advance Printing of Summaries* (SPSE, Springfield, VA 1977) p. 49
6.21 J. Van Engeland: Photogr. Sci. Eng. **23**, 86 (1979)
6.22 M. Scharfe: *Electrophotography, Principles and Optimization* (Research Studies Press, Letchworth, England 1984)
6.23 K. B. Paxton: Photogr. Sci. Eng. **22**, 159 (1978)
6.24 E. R. Hill, J. J. Griesmer: Photogr. Sci. Eng. **17**, 47 (1972)
6.25 T. Takahashi, T. Sakata: In *Electrophotography, Second International Conference,* ed. by D. R. White (SPSE, Washington, D.C. 1974) p. 100
6.26 J. Nakajima, M. Kimura, H. Takahashi: Fujitsu Sci. Tech. J. 115 (Sept. 1979)
6.27 O. G. Hauser, R. S. Menchel: "Deposition and Scavenging during Electroded Cascade Development," presented at the Annu. Symp. SPSE, Washington, D.C., October 31, 1968
6.28 P. M. Cassiers, J. Van Engeland: Photogr. Sci. Eng. **9**, 273 (1965)
6.29 J. C. Maxwell: *Electricity and Magnetism* (Clarendon, Oxford 1873) p. 365
6.30 J. C. M. Garnett: Philos. Trans. R. Soc. Lond. **205**, 237 (1906)
6.31 Lord Rayleigh: Philos. Mag. **34**, 481 (1892)
6.32 AIP Conference Proceedings, *Electrical Transport and Optical Prop-*

erties of Inhomogeneous Media, ed. by J. C. Garland, D. B. Tanner (AIP, New York 1978)
6.33 R. C. McPhedran, D. R. McKenzie: [Ref. 6.32, p. 294]
6.34 D. A. Hays: Photogr. Sci. Eng. **22**, 232 (1978)
6.35 M. H. Lee, G. Beardsley: In *Third International Congress on Advances in Non-Impact Printing Technologies*, ed. by J. Gaynor (SPSE, Springfield, VA 1987) p. 75
6.36 W. A. Sullivan, T. L. Thourson: Photogr. Sci. Eng. **11**, 115 (1967)
6.37 J. J. Folkins: IEEE-IAS Annu. Conf. Proc., 15 (1985)
6.38 U. Vahtra: Photogr. Sci. Eng. **26**, 292 (1982)
6.39 J. A. Benda, W. J. Wnek: IEEE Trans. IA-**17**, 610 (1981)
6.40 J. Nakajima, T. Matsuda: IEEE-IAS Annu. Conf. Proc. (1978) p. 225
6.41 S. Jen, A. R. Lubinsky: In *Electrophotography, Fourth International Conference,* ed. by S. Ing, M. Tabak, W. Haas (SPSE, Springfield, VA 1981) p. 239
6.42 J. De Lorenzo, P. A. Garsin: IEEE-IAS Annu. Conf. Proc. (1981) p. 980
6.43 G. Goldmann: In *Advances in Non-Impact Printing Technologies for Computer and Office Applications*, ed. by J. Gaynor, (Van Nostrand Reinhold, New York 1981) p. 148
6.44 T. Teshigawara, H. Tachibana, K. Terao: IEEE-IAS Annu. Conf. Proc. (1985) p. 151
6.45 E. T. Miskinis, T. A. Jadwin: U.S. Patent 4546060 (1985)
6.46 D. A. Hays: U.S. Patent 4370056 (1983)
6.47 A. R. Lubinsky, G. A. Denton, P. D. Keller, J. E. Williams: U.S. Patent 4537494 (1985)

Chapter 7

7.1 G. P. Kasper, J. W. May: U.S. Patent 4076847 (1978)
7.2 W. S. Jewett: IEEE-IAS Annu. Conf. Proc. (1977) p. 557
7.3 J. A. Benda, W. J. Wnek: IEEE Trans. IA-**17**, 610 (1981)
7.4 J. Nakajima, T. Matsuda: IEEE-IAS Annu. Conf. Proc. (1978) p. 225
7.5 J. J. Folkins: IEEE-IAS Annu. Conf. Proc. (1985) p. 1510, IEEE Trans. IA-**24**, 250 (1988)
7.6 M. Scharfe: *Electrophotography, Principles and Optimization* (Research Studies Press, Letchworth, England 1984)
7.7 D. A. Hays: IEEE-IAS Annu. Conf. Proc. (1985) p. 1515
7.8 Y. Hoshino: Jpn. J. Appl. Phy. **19**, 2413 (1980)
7.9 M. H. Lee, G. Beardsley: In *Third International Congress on Advances in Non-Impact Printing Technologies*, ed. by J. Gaynor (SPSE, Springfield, VA 1987) p. 75
7.10 L. B. Schein, K. J. Fowler, G. Marshall, V. Ting: J. Imaging Technol. **13**, 60 (1987)

7.11 L. B. Schein, K. J. Fowler: J. Imaging Technol. **11**, 295 (1985)
7.12 E. M. Williams: *The Physics and Technology of Xerographic Processes* (Wiley, New York 1984)
7.13 M. H. Lee, J. Ayala: J. Imaging Technol. **11**, 279 (1985)

Chapter 8

8.1 A. Y. H. Choi: J. Appl. Phys. **35**, 2561 (1964)
8.2 A. R. Kotz: U.S. Patent 3909258 (1975)
8.3 D. R. Field: IEEE Trans. IA-**19**, 759 (1983)
8.4 A. Shimada, M. Anzai, K. Noguchi: J. Imaging Sci. **29**, 209 (1985); A. Shimada, M. Anzai, A. Kakuta, T. Kawanishi: IEEE Trans. IA-**23**, 804 (1987)
8.5 R. J. Faust: In *Advances in Non-Impact Printing Technologies for Computer and Office Applications*, ed. by J. Gaynor, (Van Nostrand Reinhold, New York 1982) p. 162
8.6 W. L. Buehner, J. D. Hill, T. H. Williams, J. W. Woods: IBM J. Res. Dev. **21**, 2 (1977)
8.7 K. Nelson: U.S. Patent 4121931 (1978)
8.8 J. Nakajima, A. Teshima, M. Horie: Trans. Inst. Electron. Commun. Eng. Jpn. **E63**, 240 (1980)
8.9 M. H. Lee, W. Imaino, D. Brandt: Photogr. Sci. Eng. **28**, 24 (1984)
8.10 M. H. Lee, W. Imaino: Photogr. Sci. Eng. **28**, 19 (1984)
8.11 W. Imaino, K. Loeffler, R. Balanson: In *Colloids and Surfaces in Reprographics Technology*, ed. by M. Hair, M. Croucher (ACS, Washington, DC 1982) p. 249
8.12 W. Imaino, A. C. Tang: Appl. Opt. **22**, 1875 (1983)
8.13 J. Alward, W. Imaino: IEEE Trans. MAG-**22**, 128 (1986)
8.14 T. Takashi, N. Hosono, J. Kanbe, T. Toyona: Photogr. Sci. Eng. **26**, 254 (1982)
8.15 H. Demizu, T. Saito, K. Aoki: In *Third International Congress on Advances in Non-Impact Printing Technologies*, ed. by J. Gaynor (SPSE, Springfield, VA 1987) p. 84
8.16 K. Sakamoto, F. Takeda, K. Kobayashi: IEEE-IAS Annu. Conf. Proc. (1985) p. 1502
8.17 F. Takeda, K. Sakamoto, K. Kobayashi: IEEE-IAS Annu. Conf. Proc. (1985) p. 1491
8.18 M. Hosoya, S. Tomura, T. Vehara: IEEE-IAS Annu. Conf. Proc. (1985) p. 1495
8.19 R. W. Gundlach: U.S. Patent 4556013 (1985)
8.20 M. Yoshikawa: U.S. Patent 4606990 (1986)
8.21 K. Terao, S. Inaba, K. Ito, IEEE-IAS Annu. Conf. Proc. (1987) p. 1615
8.22 T. Flint: U.S. Patent 3552355 (1971)
8.23 T. S. Chang, C. V. Wilbur: In *Electrophotography, Second Interna-*

tional Conference, ed. by D. R. White (SPSE, Washington, DC 1974) p.74
8.24 C. Hendricks: In *Electrostatics and its Applications* ed. by A. D. Moore (Wiley, New York 1973) p. 57
8.25 J. F. Hughes: *Electrostatic Powder Coating* (Research Studies Press, Wiley, New York 1984)
8.26 S. Masuda, A. Mizuno, S. Tanaka: IEE-IAS Annu. Conf. Proc. (1983) p. 1020
8.27 F. W. Schmidlin: U.S. Patent 4647179 (1987)
8.28 J. R. Melcher, E. P.Warren, R. H. Kotwal: IEEE-IAS Annu. Conf. Proc. (1987) pp. 1591, 1595
8.29 S. Masuda, K. Fajibayashi, K. Ishida: Electr. Eng. in Jpn. **92**, 43 (1972)
8.30 S. Masuda, T. Kamimura: J. Electrost. **1**, 351 (1975)
8.31 S. Gan-mor, S. Law: IEEE-IAS Annu. Conf. Proc. (1987) p. 1578

Chapter 9

9.1 W.E. Bixby, P. G. Andrus, L. E. Walkup: Photogr. Eng. **5**, 195 (1954)
9.2 R. E. Rayford, W. E. Bixby: Photogr. Eng. **6**, 173 (1955)
9.3 J. H. Dessauer, G. R. Mott, H. Bogdonoff: Photogr. Eng. **6**, 250 (1955)
9.4 J. T. Bickmore, M. Levy, J. Hall: Photogr. Sci. Eng. **4**, 37 (1960)
9.5 R. B. Lewis, H. M. Stark: In *Current Problems in Electrophotography*, ed. by W. F. Berg, K. Hauffe (de Gruyter, Berlin 1972)
9.6 J. T. Bickmore; J. T. Bickmore, C. R. Mayo, G. R. Mott, R. G. Vyverberg: In *Xerography and Related Processes*, ed. by J. H. Dessauer, H. E. Clark (Focal, New York 1965) pp. 310, 467
9.7 M. Scharfe: *Electrophotography Principles and Optimization* (Research Studies Press, Letchworth, England 1984)
9.8 W. A. Sullivan, T. L. Thourson: Photogr. Sci. Eng. **11**, 115 (1967)
9.9 R. G. Andrus, J. M. Hardenbrook, O. A. Ullrich: *Electrophotography, Second International Conference,* ed. by D. R. White (SPSE, Washington, DC 1974) p. 62
9.10 H. G. Greig: U.S. Patent 2811465 (1957)
9.11 J. C. Wilson: U.S. Patent 2846333 (1958)
9.12 C. R. Mayo: U.S. Patent 2895847 (1959)
9.13 R. W. Gundlach: U.S. Patent 3166432 (1965)
9.14 R. W. Willmott: U.S. Patent 3232190 (1966)
9.15 R. Lowrie: U.S. Patent 2803177 (1957)
9.16 L. S. Chang, C. V. Wilbur: *Electrophotography, Second International Conference,* ed. by D. R. White (SPSE, Washington, DC 1974) p. 74

9.17 R. M. Schaffert: *Electrophotography* (Focal, London 1980)
9.18 E. M. Williams: *The Physics and Technology of Xerographic Processes* (Wiley, New York 1984) Chap. 9
9.19 T. Takahashi, N. Hosono, J. Kanbe, T. Toyona: Photogr. Sci. Eng. **26**, 254 (1982)
9.20 M. Hosoya, S. Tomura, T.Vehara: IEEE-IAS Annu. Conf. Proc., (1985) p. 1485
9.21 K. Sakamoto, F. Takeda, K. Kobayashi: IEEE-IAS Annu. Conf. Proc., (1985) p. 1502
9.22 A. R. Kotz: U.S. Patent 3909258 (1975)
9.23 K. Nelson: U.S. Patent 4121931 (1978)
9.24 A. Shimada, M. Anzai, K. Noguchi: J. Imaging Sci. **29**, 209 (1985)
9.25 A. Shimada, M. Anzai, A. Kakuta, T. Kawanishi: IEEE Trans. IA-**23**, 804 (1987)
9.26 D. R. Field: IEEE Trans. IA-**19**, 759 (1983)
9.27 R. J. Faust: In *Advances in Non-Impact Printing Technologies for Computer and Office Applications,* ed. by J. Gaynor (Van Nostrand Reinhold, New York 1982) p. 162
9.28 F. Takeda, K. Sakamoto, K. Kobayashi: IEEE-IAS Annu. Conf. Proc. (1985) p. 1491
9.29 H. Demizu, T. Saito, K. Aoki: In *Third International Congress on Advances in Non-Impact Printing Technologies* ed. by J. Gaynor (SPSE, Springfield, VA 1987) p. 84
9.30 M. Kohyama, T. Kasai, M. Yamashita: J. Imaging Technol. **12**, 47 (1986)
9.31 G. S. P. Castle, A. Dean, L. B. Schein: to be published

Chapter 10

10.1 K. A. Metcalfe: J. Sci. Instrum. **32**, 74 (1955)
10.2 K. A. Metcalfe, R. J. Wright: J. Oil Colour Chem. Assoc. **39**, 845 (1956)
10.3 V. E. Straughan, E. F. Mayer: Proc. Nat. Electron. Conf. **13**, 959 (1957)
10.4 C. J. Claus, E. F. Mayer: In *Xerography and Related Processes*, ed. by J. Dessauer, H. Clark (Focal, New York 1965) Chap. 12
10.5 G. Jacobson, J. Hillkirk: *Xerox, an American Samurai* (Macmillan, New York 1986) p. 122
10.6 T. Nylund, C. Cowan, J. Spence, L. Steele: In *Third International Congress on Advances in Non-Impact Printing Technologies, Advance Printing of Paper Summaries*, (SPSE, Springfield, VA 1986) p. 74
10.7 M. R. Specht, L. Contois, D. Santelli: In *Third International Congress on Advances in Non-Impact Printing Technology, Advance Printing of Paper Summaries* (SPSE, Springfield, VA 1986) p. 76

10.8 S. Glasstone, D. Lewis: *Elements of Physical Chemistry* (van Nostrand, Princeton 1966) p. 580
10.9 S. Stotz: In *Current Problems in Electrophotography*, ed. by W. F. Berg, K. Hauffe (de Gruyter, Berlin 1972) p. 336
10.10 J. Halfdanarson, K. Hauffe: Photogr. Sci. Eng. **23**, 27 (1979)
10.11 V. Novotny: Colloids Surf. **2**, 373 (1981)
10.12 H. M. Stark, R. S. Menchel: J. Appl. Phys. **41**, 2905 (1970)
10.13 M Schleusener: J. Signalaufzeichnungsmaterialien **5**(1), 39 (1977); ibid. **5** (2), 93 (1977)
10.14 I. Brodie, J. A. Dahlquist, A. Sher: J. Appl. Phys. **39**, 1618 (1968)
10.15 T. Kimura, M. Yukozawa: Denshi Shashin (Electrophotogr.) **5**, 33 (1956)
10.16 Y. Moradzadeh, D. Woodwood: Photogr. Sci. Eng. **10**, 96 (1966)
10.17 H. Hasegawa, S. Sugihara, S. Nishikawa: Denshi Shashin (Electrophotogr.) **6**, 65 (1966)
10.18 E. C. Hutter: Photogr. Sci. Eng. **15**, 251 (1971)
10.19 T. Kurita: Denshi Shashin, (Electrophotogr.) **3**, 26 (1961)
10.20 E. Mohn: Photogr. Sci. Eng. **15**, 451 (1971)
10.21 R. Stechemesser: Photogr. Sci. Eng. **26**, 27 (1982)
10.22 J. M. Schneider, P. K. Watson: Phys. Fluids **13**, 1948, 1955 (1970)
10.23 N. Felici: Rev. Gen. Electr. **78**, 717 (1969)
10.24 L. D. Reed, F. A. Morrison, Jr.: J. Colloid Interface Sci. **54**, 117 (1976)
10.25 V. Novotny, M. Hair: J. Colloid Interface Sci. **71**, 273 (1979); V. Novotny: Colloids Surf. **2**, 373 (1981)
10.26 V. Novotny: J. Electrochem. Soc. **133**, 1629 (1986)
10.27 H. G. Junginger, R. F. Schmidt, R. Strunk: Photogr. Sci. Eng. **22**, 213 (1978)
10.28 H. G. Junginger, R. Strunk: J. Appl. Phys. **47**, 3021 (1976)
10.29 H. G. Junginger, R. Strunk: J. Photogr. Sci. **25**, 109 (1977)
10.30 R. B. Crofoot, Y. C. Cheng: J. Appl. Phys. **50**, 6583 (1979)
10.31 J. C. Lacroix, P. Atten, E. J. Hopfinger: J. Fluid Mech. **69**, 529 (1975)
10.32 P. Atten, J. C. Lacroix: J. Mec. **18** 469 (1979)
10.33 S. Stotz: J. Colloid Interface Sci. **65**, 118 (1978)
10.34 R. Kohler, D. Giglberger, E. Bestenreiner: Photogr. Sci. Eng. **22**, 218 (1968)
10.35 P. H. Wiersema, A. L. Loeb, J. Th. G. Overbeek: J. Colloid Interface Sci. **22**, 78 (1966)
10.36 E. Huckel: Phys. Z. **24**, 204 (1924)
10.37 J. A. Dahlquist, I. Brodie: J. Appl. Phys. **40**, 3020 (1969)
10.38 L. B. Harris: Rev. Sci. Instrum. **40**, 905 (1969)
10.39 A. Kondo, J. Yamada: Proc. TAPPI Repr. Conf. (Tappi, Boston 1973) p. 39

10.40 J. van Engeland, W. Verlinden, J. Marien, W. Palmans: Proc. Fourth Int. Congr. Repr. Inf. (Hannover, 1975) Spec. Rep. p. 117
10.41 V. Novotny: Colloids Surf. **21**, 219 (1986)
10.42 H. Murray: IEEE Trans. IA-**23**, 831 (1987)
10.43 B. Landa, E. P. Charlap: U.S. Patent 4378422 (1983)
10.44 B. Landa: U.S. Patents 4413048 (1983); 4582774 (1984)
10.45 B. Landa: U.S. Patents 4454215 (1984); 4460667 (1984)
10.46 T. Kuroturi, M. Mochizuki, S. Tatsumi: U.S. Patent 441533 (1983)
10.47 B. Landa, O. Sagiv: U.S. Patent 4538899 (1985)
10.48 B. Landa: In *Third International Congress on Advances in Non-Impact Printing Technologies, Advance Printing of Paper Summaries* (SPSE, Springfield, VA 1986) p. 307
10.49 V. Levy, R. Nethaniel, Y. Niv, Y. Krumberg: In *Third International Congress on Advances in Non-Impact Printing Technologies, Advance Printing of Paper Summaries* (SPSE, Springfield, VA 1986) p. 51
10.50 Y. Niv, Y. Adam, Y. Krumberg: In *Third International Congress on Advances in Non-Impact Printing Technologies, Advance Printing of Paper Summaries* (SPSE, Springfield, VA 1986) p. 57

Chapter 11

11.1 O. Sahni: In *Proceedings of the SPIE/IS&T Symposium on Electronic Imaging: Science and Technology*, Vol. 1458 (IS&T, Springfield, VA 1991) p. 4
11.2 P. G. Roetling: In *Proceedings of the SPIE/IS&T Symposium on Electronic Imaging: Science and Technology*, Vol. 1458 (IS&T, Springfield, VA 1991) p. 17
11.3 J. Kasson, W. Plouffe: In *Proceedings of the SPIE/IS&T Symposium on Electronic Imaging: Science and Technology*, Vol. 1460 (IS&T, Springfield, VA 1991) p. 11
11.4 R. C. Durbeck, S. Sherr: *Output Hardcopy Devices* (Academic, New York 1988)
11.5 J. A. C. Yule: *Principles of Color Reproduction* (Wiley, New York 1967)
11.6 R. W. G. Hunt: *The Reproduction of Colour*, 4th ed. (Fountain Press, Tolworth, England 1987) Chap. 6
11.7 R. W. G. Hunt: Color Res. Appl. **4**, 39 (1979)
11.8 D. A. Hays: J. Imaging Technol. **17**, 252 (1991)
11.9 R. M. Shaffert: Reprographics, pp. 16–19 (Sept. 1966)
11.10 J. T. Bickmore, C. R. Mayo, G. R. Mott, R. G. Vyverberg: In *Xerography and Related Processes*, ed. by J. Dessauer, H. Clark (Focal, New York 1965) p. 491

11.11 J. S. Rydz, S. W. Johnson: RCA Rev. **19**, 465 (1958)
11.12 D. J. Parker: RCA Eng. **6**, 7 (1960)
11.13 S. W. Johnson: U.S. Patent 3150976 (1964)
11.14 C. J. Claus, E. F. Mayer: In *Xerography and Related Processes* ed. by J. Dessauer, H. Clark (Focal, New York 1965) p. 370
11.15 M. Kuehnle: In *Principles of Color Proofing*, ed. by M. H. Bruno (Gama Communications, Salem, NH 1986) p. 181
11.16 T. Nylund, C. Cowan, J. Spence, L. Steele: In *Advance Printing of Paper Summaries, Third International Congress on Advances in Non-Impact Printing Technologies* (SPSE, Springfield, VA 1986) p. 74
11.17 M. R. Specht, L. Centois, D. Santelli: In *Advance Printing of Paper Summaries, Third International Congress on Advances in Non-Impact Printing Technologies* (SPSE, Springfield, VA 1986) p. 76
11.18 R. Goren: J. Appl. Photogr. Eng. **8**, 233 (1982)
11.19 N. Burnigham, W. B. Sherwood: In *Sixth International Congress on Advances in Non-Impact Printing Technologies* (IS&T, Springfield, VA 1990) p. 437
11.20 Robert Ullichney: *Digital Halftoning* (MIT Press, Cambridge, MA 1987)
11.21 W. Lama, S. Feth, R. Loce: J. Imaging Technol. **15**, 130 (1989)
11.22 H. Yamamoto, Y. Takashima, H. Terada, H. Kunishige, T. Saitoh: J. Imaging Technol. **16**, 228 (1990); J. Jpn. Electrophotogr. Soc. **28**, 284 (1989); ibid. **29**, 9, 14 (1990)
11.23 Y. Takashima, H. Yamamoto: U.S. Patent 4778740 (1988)
11.24 I. Chen: J. Imaging Sci. **35**, 365 (1991)
11.25 E. H. Yamamoto: In *Sixth International Congress on Advances in Non-Impact Printing Technologies* (IS&T, Springfield, VA 1990) p. 34
11.26 M. Kohyama, T. Kasai, M. Yamashita: J. Imaging Technol. **12**, 47 (1986)
11.27 E. T. Miskinis, T. A. Jadwin: U.S. Patent 4546060 (1985)
11.28 E. T. Miskinis: In *Sixth International Congress on Advances in Non-Impact Printing Technologies* (IS&T, Springfield, VA 1990) p. 101
11.29 R. J. Gruber, E. N. Dalal: In *Seventh Toner and Developer Industry Conference*, Imaging Material Seminar Series, Sept. 16–19, Santa Barbara (Diamond Research Corp., Santa Barbara, CA 1990)
11.30 S. Chiba, S. Inoue: In *Advance Printing of Paper Summaries, Fourth International Congress on Advances in Non-Impact Printing Technologies* (SPSE, Springfield, VA 1988) p. 105
11.31 R. N. Goren: J. Appl. Photogr. Eng. **2**, 17 (1976)
11.32 K. Hirakura, Y. Nagahara, M. Hasebe: In *Seventh International Congress on Advances in Non-Impact Printing Technologies* (IS&T, Springfield, VA 1991) p. 49

11.33 M. Hasebe, T. Miura, N. Sawayma, K. Aoki, T. Maruta: In *Seventh International Congress on Advances in Non-Impact Printing Technologies* (IS&T, Springfield, VA 1991) p. 56
11.34 M. Itaya, S. Haneda: Proc. SPIE **1253**, 360 (1990)
11.35 S. Karasawa, F. Matsumoto, T. Ishida, S. Sakuma, K. Sakai: In *Electrophotography Fourth International Conference*, ed. by S. W. Ing, M. D. Tabok, W. E. Haas (SPSE, Springfield, VA 1981) p. 423
11.36 Y. Tamura, Y. Nakao, Y. Chujo, M. Kashimagi: In *Fourth International Congress on Advances in Non-Impact Printing Technologies* (SPSE, Springfield, VA 1988) p. 184
11.37 D. G. Parker, J. E. May: In *Seventh International Congress on Advances in Non-Impact Printing Technologies*, Vol. I (IS&T, Springfield, VA 1991) p. 63
11.38 N. Hoshi, I. Komatsu, M. Anzai, K. Tanno, H. morishita, S. Saito, S. Katagiri: In *Advances in Non-Impact Printing Technologies for Computer and Office Equipment*, ed. by J. Gaynor (Van Nostrand, New York 1981) p. 518
11.39 D. G. Parker, W. M. Allen, Jr., R. P. Germain: U.S. Patent 4771314 (1988)

Chapter 12

12.1 N. Tetsutani, Y. Hoshino, M. Ozawa: J. Imaging Technol. **16**, 1 (1990)
12.2 M. Kimura, S. Sasaki, J. Nakajima: In *Fourth International Congress on Advances in Non-Impact Printing Technologies* (SPSE, Springfield, VA 1988) Meeting, p. 190
12.3 F. W. Schmidlin: In *Sixth International Congress on Advances in Non-Impact Printing Technologies* (SPSE, Springfield, VA 1990) p. 170
12.4 T. Nishigaito, K. Okuna, S. Ohara, G. Sato: In *Seventh International Congress on Advances in Non-Impact Printing Technologies*, Vol. II (IS&T, Springfield, VA 1991) p. 462
12.5 M. Hosoya, M. Saito, T. Uehara: In *Seventh International Congress on Advances in Non-Impact Printing Technologies*, Vol. I (IS&T, Springfield, VA 1991) p. 113
12.6 T. Kawanishi, M. Igarashi, M. Obu, T. Wada: In *Seventh International Congress on Advances in Non-Impact Printing Technologies*, Vol. II (IS&T, Springfield, VA 1991) p. 511
12.7 K. Ito, M. Watanabe, Y. Watanabe, Y. Suwabe, K. Terao: In *Seventh International Congress on Advances in Non-Impact Printing Technologies*, Vol. II (IS&T, Springfield, VA 1991) p. 519
12.8 D. E. Bugner: J. Imaging Technol. **35**, 377 (1991)

12.9 M. C. Tam, A. L. Pundsack, R. W. Gundlach, P. S. Vincett, G. J. Kovacs, C. A. Jennings, R. O. Loutfy: J. Imaging Sci. **32**, 247 (1988)

12.10 M. C. Tam, R. O. Loutfy, G. J. Kovacs, A. L. Pundsack, J. Meester, H. Aboushaka: J. Imaging Sci. Technol. **36**, 81 (1992)

12.11 J. J. Eltgen, J. Magnenet, J. P. Bresson: J. Imaging Sci. **15**, 224 (1989). This was published previously as *Third International Congress on Advances in Non-Impact Printing Technologies*, ed. by J. Gaynor (SPSE, Springfield, VA 1987) p. 547

12.12 J. J. Eltgen, J. Magnenet, C. Guerin, J. P. Bresson, J. Estavoyer: In *Fifth International Congress on Advances in Non-Impact Printing Technologies* (SPSE, Springfield, VA 1989) p. 710

12.13 B. Cherbury, P. Raulin, R. Mercier: In *Fifth International Congress on Advances in Non-Impact Printing Technologies* (SPSE, Springfield, VA 1989) p. 674

12.14 G. E. Keefe, E. J. Yarmchuk, D. B. Dove: J. Imaging Sci. **33**, 69 (1989)

12.15 W. J. Caley, W. R. Buchan, T. W. Pape: In *Fifth International Congress on Advances in Non-Impact Printing Technologies* (SPSE, Springfield, VA 1989) p. 690

12.16 M. Omodani, M. Ohta, T. Taneka, Y. Hoshino: In *Fifth International Congress on Advances in Non-Impact Printing Technologies* (SPSE, Springfield, VA 1989) p. 670

12.17 M. Takeuchi, A. Onose: In *Seventh International Congress on Advances in Non-Impact Printing Technologies*, Vol. I (IS&T, Springfield, VA 1991) p. 200

12.18 F. W. Schmidlin: In *Seventh International Congress on Advances in Non-Impact Printing Technologies*, Vol. I (IS&T, Springfield, VA 1991) p. 136

12.19 D. A. Hays, W. H. Wayman: J. Imaging Sci. **33**, 160 (1989)

12.20 D. S. Rimai, L. P. DeMejo, R. C. Bowen: J. Appl. Phys. **68**, 6234 (1990)

12.21 D. S. Rimai, L. P. DeMejo, R. C. Bowen: J. Appl. Phys. **66**, 3574 (1989)

12.22 D. S. Rimai, L. P. DeMejo, R. C. Bowen: J. Appl. Phys. **65**, 755 (1989)

12.23 S. Nakamura, H. Hirabayashi, J. Araya, N. Koitabashi: U.S. Patent 4851960 (1989)

12.24 F. N. Rothacher, D. P. Bujese: U.S. Patent 3778690 (1973)

12.25 T. Kimura: U.S. Patent 3626260 (1969)

12.26 M. Zaretsky: In *Seventh International Congress on Advances in Non-Impact Printing Technologies*, Vol. I (IS&T, Springfield, VA 1991) p. 73

12.27 M. Kimura, W. Wanou, T. Nakashima, T. Inagak: In *Seventh International Congress on Advances in Non-Impact Printing Technologies*, Vol. II (IS&T, Springfield, VA 1991) p. 503
12.28 G. Fletcher: In *Sixth International Congress on Advances in Non-Impact Printing Technologies, Advance Printing of Paper Summaries* (SPSE, Springfield, VA 1990) p. 18
12.29 M. H. Lee, G. Beardsley: In *Fourth International Congress on Advances in Non-Impact Printing Technologies* (SPSE, Springfield, VA 1988) p. 168
12.30 L. B. Schein, M. LaHa, D. Novotny: Phys. Lett. A **167**, 79 (1992)
12.31 J. Lowell, A. C. Rose-Innes: Adv. Phys. **29**, 1947 (1980)
12.32 A. Kondo: Denshi Syasin Gakkai **43**, 26 (1979)
12.33 R. Stover: private communication
12.34 T. Kurita: private communication
12.35 L.-H. Lee: Photogr. Sci. Eng. **22**, 228 (1978)
12.36 J. H. Anderson: J. Imaging Sci. **33**, 200 (1989)
12.37 J. H. Anderson, D. E. Bugner: In *Fourth International Congress on Advances in Non-Impact Printing Technologies* (SPSE, Springfield, VA 1988) p. 79
12.38 R. J. Nash, J. T. Bickmore: In *Fourth International Congress on Advances in Non-Impact Printing Technologies* (SPSE, Springfield, VA 1988) p. 113
12.39 J. H. Anderson: J. Imaging Technol. **16**, 204 (1990)
12.40 W. R. Hutcheson, A. Sukovich: In *Fourth International Congress on Advances in Non-Impact Printing Technologies* (SPSE, Springfield, VA 1988) p. 127
12.41 M. Suzuki, T. Isoda: In *Seventh International Congress on Advances in Non-Impact Printing Technologies*, Vol. I (IS&T, Springfield, VA 1991) p. 226
12.42 L. B. Schein, M. LaHa, G. Marshall: J. Appl. Phys. **69**, 6817 (1991)
12.43 H.-T. Macholdt, A. Sieber: J. Imaging Technol. **14**, 89 (1988)
12.44 H. Yamazaki, H. Seki, J. Takahashi: In *Sixth International Congress on Advances in Non-Impact Printing Technologies* (SPSE, Springfield, VA 1990) p. 132
12.45 N. Hayashi: In *Sixth International Congress on Advances in Non-Impact Printing Technologies* (SPSE, Springfield, VA 1990) p. 144
12.46 A. Diaz, D. Fenzel-Alexander, D. C. Miller, D. Wollman, A. Eisenberg: J. Poly. Sci. Part C: Poly. Lett. **28**, 75 (1990)
12.47 A. Diaz, D. Fenzel-Alexander, D. Wollman: J. Poly. Sci. Part B: Poly. Phys. **29**, 1559 (1991)
12.48 A. F. Diaz, D. Wollman, D. Deblow: Chem. Mater. **3**, 997 (1991)
12.49 H. A. Mizes, E. Conwell, D. P. Salamida: Appl. Phys. Lett. **56**, 1597 (1990)

12.50 N. Matsui, K. Oka, Y. Inaba: In *Sixth International Congress on Advances in Non-Impact Printing Technologies* (SPSE, Springfield, VA 1990) p. 123
12.51 L. M. Folan, S. Arnold, T. R. O'Keeffe, D. E. Spock, L. B. Schein, A. F. Diaz: J. Electrostatistics **25**, 155 (1990)
12.52 N. Kutsuwada, Y. Nakamura: J. Imaging Technol. **15**, 1 (1989)
12.53 R. B. Lewis, E. W. Connors, R. F. Koehler: Jpn. J. Electrophotogr. **22**, 85 (1983)
12.54 B. D. Terris, A. B. Jaffe: Inst. Phys. Conf. Ser., No. 85, p. 63 (1987)
12.55 B. D. Terris, K. J. Fowler: J. Imaging Technol. **17**, 215 (1991)
12.56 P. C. Julien: In *Sixth International Congress on Advances in Non-Impact Printing Technologies, Advance Printing of Paper Summaries* (SPSE, Springfield, VA, 1990) p. 185
12.57 Y. Takahashi, J. Nakabayashi: In *Fifth International Congress on Advances in Non-Impact Printing Technologies* (SPSE, Springfield, VA 1989) p. 206
12.58 Y. Takahashi, H. Horiguchi, T. Sakata: J. Imaging Technol. **15**, 235 (1989)
12.59 M. Kimura, M. Wanou, S. Sasaki, T. Inagaki: In *Fifth International Congress on Advances in Non-Impact Printing Technologies* (SPSE, Springfield, VA 1989) p. 202
12.60 Y. Tabata: In *Sixth International Congress on Advances in Non-Impact Printing Technologies* (SPSE, Springfield, VA 1990) p. 233
12.61 M. K. Mazumder, R. E. Ware, T. Yokoyama, B. Rubin, D. Kamp: IEEE-IAS Annu. Conf. Proc. (1987) p. 1606
12.62 R. H. Epping: In *Fourth International Congress on Advances in Non-Impact Printing Technologies* (SPSE, Springfield, VA 1988) p. 102;
R. H. Epping, R. N. Hess: In *Sixth International Congress on Advances in Non-Impact Printing Technologies* (SPSE, Springfield, VA 1990) p. 225
12.63 D. Schulze-Hagenest: In *Sixth International Congress on Advances in Non-Impact Printing Technologies* (SPSE, Springfield, VA 1990) p. 217
12.64 N. Kutsuwada, Y. Nakamura: J. Imaging Technol. **16**, 48 (1990)
12.65 S. C. Hart, J. J. Folkin, C. G. Edmunds: In *Sixth International Congress on Advances in Non-Impact Printing Technologies* (SPSE, Springfield, VA 1990)
12.66 M. H. Lee, G. Beardsley: In *Third International Congress on Advances in Non-Impact Printing Technologies*, ed. by J. Gaynor (SPSE, Springfield, VA 1987) p. 75
12.67 D. A. Hays, J. Imaging Technol. **15**, 29 (1989)

12.68 L. B. Schein, G. Beardsley, M. Moore: J. Imaging Technol. **16**, 129 (1990)
12.69 T. Yamazaki, S. Takahashi, N. Kobayashi, T. Nakanishi: J. Imaging Sci. **35**, 334 (1991)
12.70 D. A. Hays: Photogr. Sci. Eng. **22**, 232 (1978)
12.71 T. Kurita: In *Seventh International Congress on Advances in Non-Impact Printing Technologies*, Vol. I (IS&T, Springfield, VA 1991) p. 147
12.72 D. A. Hays: In *Seventh International Congress on Advances in Non-Impact Printing Technologies*, Vol. I (IS&T, Springfield, VA 1991) p. 93
12.73 F. W. Schmidlin: In *Seventh International Congress on Advances in Non-Impact Printing Technologies*, Vol. I (IS&T, Springfield, VA 1991) p. 136
12.74 L. B. Schein: Photogr. Sci. Eng. **19**, 255 (1975)
12.75 D. A. Hays: J. Imaging Technol. **16**, 209 (1990)
12.76 L. B. Schein: J. Imaging Technol. **16**, 217 (1990)
12.77 L. B. Schein, G. Beardsley, C. Eklund: J. Imaging Technol. **17**, 84 (1991)
12.78 F. W. Schmidlin: In *Photoconductivity and Related Phenomena*, ed. by J. Mort, D. M. Pai (Elsevier, New York, 1976) Chap. 11
12.79 R. Nash: In *Fifth International Congress on Advances in Non-Impact Printing Technologies* (SPSE, Springfield, VA 1989) p. 158
12.80 J. J. Folkins: In *Sixth International Congress on Advances in Non-Impact Printing Technologies* (IS&T, Springfield, VA 1991) p. 15
12.81 J. Thompson: In *Sixth International Congress on Advances in Non-Impact Printing Technologies* (SPSE, Springfield, VA 1990) p. 347
12.82 A. Shinozaki, F. Takeda, T. Motohashi: In *Sixth International Congress on Advances in Non-Impact Printing Technologies* (SPSE, Springfield, VA 1990) p. 55
12.83 K. Hirose, H. Kamaji, M. Ikeda: In *Seventh International Congress on Advances in Non-Impact Printing Technologies*, Vol. I (IS&T, Springfield, VA 1991) p. 123
12.84 H. Kamaji, K. Hirose, Y. Nishio, M. Kimura: In *Seventh International Congress on Advances in Non-Impact Printing Technologies*, Vol. I (IS&T, Springfield, VA 1991) p. 129
12.85 K. Yanagida, K. Oka, T. Imai: J. Imaging Technol. **15**, 178 (1989)
12.86 J. Agui, P. Sanda, D. Dove: In *SPIE Proceedings Feb. 1990*, Vol. 1253 (SPIE, Washington, DC 1990) p. 340
12.87 J. Agui, P. Sanda, D. Dove: In *Fifth International Congress on Advances in Non-Impact Printing Technologies* (SPSE, Springfield, VA 1989) p. 129
12.88 N. Goal, P. Spencer: Polym. Sci. Technol. **9B**, 76 (1975)

12.89 M. H. Lee, T. C. Reiley, C. I. Dodds: In *Sixth International Congress on Advances in Non-Impact Printing Technologies* (SPSE, Springfield, VA 1990) p. 196
12.90 H. Tachibana, N. Hyakutake, K. Terao: In *Conf. Rec. of IEEE Industry Applications Society Annual Meeting* (San Diego, Oct. 1989) p. 2260
12.91 G. S. P. Castle, M. A. Dean, L. B. Schein: In *Sixth International Congress on Advances in Non-Impact Printing Technologies* (SPSE, Springfield, VA 1990) p. 62
12.92 L. B. Schein, G. S. P. Castle, A. Dean: J. Imaging Technol., **15**, 9 (1989)
12.93 M. Huijben: In *Seventh International Congress on Advances in Non-Impact Printing Technologies*, Vol. II (IS&T, Springfield, VA 1991) p. 453
12.94 H. Demizu, T. Saito, K. Aoki: J. Imaging Technol. **15**, 227 (1989)
12.95 H. Yamamoto, Y. Takashima, H. Terada, H. Kunishige, T. Saitoh: J. Imaging Technol. **16**, 228 (1990)
12.96 J. R. Larson: In *Fourth International Congress on Advances in Non-Impact Printing Technologies* (SPSE, Springfield, VA 1988) p. 142
12.97 F. M. Fowkes, H. Jinnai, M. A. Mostafa, F. W. Anderson, R. J. Moore: ACS Symp. Ser. 200 (ACS, Washington, DC 1984) p. 282
12.98 M. D. Croucher, S. Drappel, J. Duff, K. Lok, R. W. Wong: Colloid and Surfaces **11**, 303 (1984)
12.99 K. Pearlstine, L. Page, L. El-Sayed: J. Imaging Sci. **35**, 55 (1991)
12.100 T. G. Davis, G. A. Gibson, R. H. Luebbe, K. Yu: In *Fifth International Congress on Advances in Non-Impact Printing Technologies* (SPSE, Springfield, VA 1989) p. 417
12.101 J. R. Larson, G. A. Lane, J. R. Swanson, T. J. Trout, L. El-Sayed: In *Fifth International Congress on Advances in Non-Impact Printing Technologies* (SPSE, Springfield, VA 1989) p. 430
12.102 G. A. Gibson, R. H. Luebbe: J. Imaging Technol. **17**, 207 (1991)
12.103 V. Novotny, M. Hair: J. Colloid Interface Sci. **71**, 273 (1979)
12.104 V. Novotny: Colloids Surf. **2**, 373 (1981)
12.105 V. Novotny: J. Electrochem. Soc. **133**, 1629 (1986)
12.106 G. P. Buxton: In *Fourth International Congress on Advances in Non-Impact Printing Technologies* (SPSE, Springfield, VA 1988) p. 133
12.107 Y. Almag, E. Eilon, M. Stern: In *Fifth International Congress on Advances in Non-Impact Printing Technologies* (SPSE, Springfield, VA 1989) p. 401
12.108 B. Landa, Y. Niv, Y. Almag, P. B. Avraham: In *Fifth International Congress on Advances in Non-Impact Printing Technologies* (SPSE, Springfield, VA 1989) p. 392

12.109 B. Landa: In *Third International Congress on Advances in Non-Impact Printing Technologies, Advance Printing of Paper Summaries* (SPSE, Springfield, VA 1986) p. 307
12.110 B. Nathaniel, R., Y. Almag, P. B. Avraham: In *Fifth International Congress on Advances in Non-Impact Printing Technologies* (SPSE, Springfield, VA 1989) p. 401
12.111 V. Levy, R. Nethaniel, Y. Niv. Y. Krumberg: In *Third International Congress on Advances in Non-Impact Printing Technologies, Advance Printing of Paper Summaries* (SPSE, Springfield, VA 1986) p. 51

Subject Index

Springer Series in Electronics and Photonics

Editors: I.P. Kaminov W. Engl T. Sugano

Managing Editor: H.K.V. Lotsch

Volume 22 **High-Speed Electronics**
Basic Physical Phenomena and Device Principles
Editors: B. Källbäck and H. Beneking

Volume 23 **Guided-Wave Acoustooptics**
Interactions, Devices, and Applications
Editor: C.S. Tsai

Volume 24 **Picosecond Electronics and Optoelectronics II**
Editors: F.J. Leonberger, C.H. Lee, F. Capasso, and H. Morkoc

Volume 25 **Photonic Switching**
Editors: T.K. Gustafson and P.W. Smith

Volume 26 **Guided-Wave Optoelectronics**
Editor: T. Tamir 2nd Edition

Volume 27 **Ultra-Fast Silicon Bipolar Technology**
Editors: L. Treitinger and M. Miura-Mattausch

Volume 28 **Physics of Quantum Electron Devices**
Editor: F. Capasso

Volume 29 **Photonic Switching II**
Editor: K. Tada

Volume 30 **Nonlinear Photonics**
Editors: H.M. Gibbs, G. Khitrova, and N. Peyghambarian

Volume 31 **Single-Electron Tunneling and Mesoscopic Devices**
Editors: H. Koch and H. Lübbig

This series was originally published under the title
Springer Series in Electrophysics
and has been renamed starting with Volume 22.